First published 1977 by CRC Press
Taylor & Francis Group
6000 Broken Sound Parkway NW, Suite 300
Boca Raton, FL 33487-2742

Reissued 2018 by CRC Press

Library of Congress Cataloging in Publication Data

Main entry under title:

Chemoprophylaxis and virus infections of the respiratory
tract.

Bibliography: p.
Includes index.
1. Respiratory organs – Diseases – Prevention.
2. Antiviral agents. I. Oxford, John Sidney. [DNLM:
1. Influenza vaccine. 2. Virus diseases -- Prevention and
control. 3. Respiratory tract infections – Prevention
and control. WC505 C517]
RC732.C44 616.2 76-45674
ISBN 0-8493-5136-7

A Library of Congress record exists under LC control number: 76045674

ISBN 13: 978-1-315-89151-4 (hbk)
ISBN 13: 978-1-351-07061-4 (ebk)

Visit the Taylor & Francis Web site at http://www.taylorandfrancis.com and the
CRC Press Web site at http://www.crcpress.com

PREFACE

Some of the problems associated with any future chemoprophylactic control of the most serious respiratory virus infection of man (namely influenza) and also relevant in some respects to the control of parainfluenza and picornavirus infections have been highlighted recently with the isolation of A/New Jersey/8/76 ($H_{swine\,1}N_1$) virus from a military recruit who died of influenzal pneumonia this winter at Fort Dix, New Jersey. The total. number of cases of infection by the swine-like virus at Fort Dix was serologically estimated to be 500 in a camp of 12,000 persons. The virus A/NJ/8/76 is antigenically related to the swine influenza virus of Shope (A/swine/30 $H_{swi\,1}N_1$), implicated as the causative agent of the influenza pandemic of 1918 which resulted in 20 million deaths (548,000 of these in the United States) and perhaps 200 million clinical cases of influenza world-wide.[1] Will the A/NJ/76 virus spread and, if so, are we able to prevent a new pandemic by vaccination alone? If we had a very effective inhibitor of the virus, how could we use it in practice? Amantadine has antiviral activity against influenza A viruses, but is it sufficiently effective to play a role in a preventative campaign? Should we look more carefully at the possibility of combined chemo- and immunoprophylaxis? It must be realized that the chances of eliminating the virus from the pig population will be very low;[1] thus, the specter of this influenza A virus crossing the interspecies barrier to man will be with us for a long time to come, especially as the potentially susceptible antibody-free group in the population increases each year while older, presumably immune people die.

To date, influenza has remained completely uncontrolled as an epidemic disease. The policy in most countries has been to offer inactivated vaccine to groups at special risk, such as chronic bronchitics, asthmatics, and persons with heart disease of any etiology. More widespread vaccination with live, attenuated virus has been carried out in the Soviet Union and China but with rather variable results. For the first time, therefore, a country (the U.S.) is attempting to prevent a possible influenza outbreak by a mass immunization campaign. Canada will also produce enough vaccine to immunize half of its population. This brings us to a consideration of one of the main obstacles which to date has not been overcome with influenza vaccines. Influenza viruses are antigenically variable: faced with immunological pressure from an immune or partly immune population, antigenic mutants with a survival value are selected. In the face of a mass immunization campaign, the A/NJ/8/76 virus may simply "drift" antigenically and thus still be able to infect and spread in the community. A mutant thus selected may have an even greater capacity to spread than the original parent.

Large-scale trials conducted by the Medical Research Council in the U.K. over the last 20 years have shown extremely variable protection resulting from vaccination against influenza subtypes, varying from negative results to a maximum of 70% protection.[2] More recent trials in the public service in England[3] have highlighted another difficulty of large-scale influenza immunization – low initial acceptance of the vaccine (around 42%) and even lower acceptance in succeeding years (14% in one factory). In these industrial trials, it was concluded that the use of inactivated influenza vaccine in healthy adults was unlikely to produce any great benefit in winters where the incidence of influenza infection is low. Two contributing factors comprised the incomplete protective effect of the vaccine, the low attack rate and the poor acceptance of the vaccine. The saving in sickness absence was approximately 7 days per 100 employees in each winter influenza period, during which about 270 days were lost by each 100 employees. A similar situation is likely to occur in the U.S., especially if there is not an outbreak of A/NJ/76 in the winter of 1977. Will the whole population be vaccinated in the following and succeeding years as well? An alternative plan would be to stockpile vaccine and rely on close surveillance to indicate any spread of virus, whereupon groups of the population could be vaccinated. Given an effective anti-influenza drug, this could be used during the first 10 to 15 days following immunization to give protection during a period when vaccine-induced antibody levels would be low; it could also be used at times of peak exposure to virus to augment any vaccine-induced immunity. Certainly the economic, social, and health gains which would accrue from preventing influenza outbreaks would be considerable, as will be seen in different chapters of this book. For example, in England working days lost through influenza sickness outnumber days lost through industrial strikes. In the U.S. when the Asian (H2N2) influenza arrived in 1957 to 1958, nearly 70,000 deaths occurred, and in 1968 to 1969 the Hong Kong (H3N2) variant caused 33,000 deaths and was costed at 3.8 billion dollars.[4]

Would this supplementation be the main role of influenza chemoprophylaxis? Amantadine has been shown to prevent infection in humans with influenza A subtypes and to resolve clinical symptoms in persons already infected and showing signs of illness. The main problem is the practical question of when to administer the drug. Obviously, a whole population cannot take amantadine daily. The question is partially answered in this book with the description of trials of the compound in general practice in the U.K. Most persons are made aware by the media of the advent of influenza in the community, either at home or at work. Under these conditions, a general practitioner could prescribe a course of compound for 10 or 15 days during a high-risk period. The ideal situation would arise with an antiviral compound which was not effective in completely suppressing viral infection, but rather reduced replication and abrogated clinical illness while allowing subsequent development of protective antibody. As mentioned above, further consideration should be given to the possibility of combined chemo- and immunoprophylaxis. Even with a highly mutable and variable biological entity like influenza, the chances of simultaneous development of drug resistance and antigenic drift would be low.

Even if we find a really effective drug against upper respiratory tract infections, there are strategic problems of administration, particularly if the compound is considered a prophylactic. The tremendous practical advance of recent years resulting from the detailed studies of amantadine both in clinical practice and in the laboratory indicates that the compound has therapeutic activity. Of course the details of the immunopathology of influenza are not clear at present, particularly in molecular terms. Why should a virus infection, predominantly of the upper respiratory tract in man, produce such a manifest clinical response? Respiratory infections have always been difficult to treat. Even a physician presented with a sore throat caused by a β-hemolytic streptococcus will not observe a dramatic effect through treatment with penicillin to which the organism is highly sensitive. Thus, the prevention and, in particular, cure of infections of the upper respiratory tract are fraught with difficulties. However, the whole concept merits more intensive efforts. Respiratory infections are global in impact and, with the exception of influenza, occur throughout the year in all races and classes of people. Although the clinical manifestations are not as dramatic as smallpox, for example, this family of respiratory viruses causes untold misery, morbidity, and mortality. They are worthy of considerable respect and research.

At present, there are few compounds that are known to be effective against influenza and common cold viruses; therefore, the main function of this book is to lay the groundwork for future work and to encourage more active studies in this area. The book has been organized so as to interrelate all aspects of the problem – clinical, epidemiological, pharmacological, molecular biological, etc. Some chapters offer practical advice and techniques, while others are purely theoretical and speculative. Some chapters have been written by specialists in fields other than virology. Thus, fresh ideas can be introduced into an area which to date has been characterized by hard work and no dramatic breakthroughs. Therefore, a wide range of authors have contributed, many of whom are not convinced that these respiratory infections can be controlled by interfering with virus replication through the use of specific inhibitors. However, we are all convinced of the importance of bringing this large group of viruses under control.

There is also controversy over the strategy to be used in the search for new virus-inhibitory molecules. Should we rely on random screening which has been so successful in finding antibacteria, antiprotozoa and antihelminth compounds or should we try to use more contrived and specific approaches? I feel that active searchers have not examined closely enough the explosive increase in our knowledge of the biochemistry or molecular biology of the viruses in question. However, alternative considerations, are given in other chapters of this book particularly in the context of the immense cost of producing a new drug.

Originally, it was our intention to produce a single-volume book covering all aspects and approaches to the problem of specific inhibitors of respiratory viruses. However, as the work progressed, it became obvious that certain chapters, because of the research interests of the authors, concentrated particularly on influenza viruses. It seemed logical, therefore, to divide the book into two volumes, the first emphasizing influenza and the second concentrating on other viruses as well as discussing important general aspects of drug screening and clinical testing, although the second volume does have some chapters which deal mainly with influenza. The two volumes obviously interrelate very closely. For example, Professor Sir Charles Stuart-Harris' chapter on clinical aspects of respiratory infections (Chapter 1, Volume I) is immediately relevant to both volumes, as is Professor Albert's contribution on selective toxicity (Chapter 2, Volume I). Clinical testing of compounds, either in volunteers or in general practice, should be viewed in light of applicability to either influenza or other respiratory virus infections. The two volumes together, then, form the multidisciplinary approach to the problem which we desired.

I hope readers will find the book useful, controversial, and stimulating. It is not meant to be an exhaustive summary of all past work on inhibitors of these virus groups; rather, it is intended to help us define where we are and to offer some suggestions for the future. There is certainly plenty of room for future antiviral compounds outside the respiratory virus group. Work has already been started to search for inhibitors of the cancer-producing oncornaviruses, and a few clinical trials are under way with antiviral compounds and hepatitis B infection, for example.

Obviously, past experience has shown that some viral and bacterial infections, such as smallpox, can be controlled or even eliminated by effective immunization programs. Since this latter approach takes advantage of a natural protective mechanism rather than introducing foreign low molecular weight compounds with all the associated difficulties of toxicity, it would seem to be preferable. Thus, research into more effective influenza vaccines is being intensified, and some work with rhinovirus vaccines is being continued. Inactivated influenza virus vaccines are now composed of relatively pure preparations of whole or partially disrupted virions prepared by continuous-flow rate zonal centrifugation. However, we still are unclear as to whether they induce a high-quality antibody of the correct specificity and why the immunity is incomplete and short-lived. There is an increasing number of experiments with attenuated mutant, recombinants, and low-temperature adapted viruses as candidate live vaccine strains, but the virus is elusive and success seems a long way off. The fascination of the possible intervention with Ehrlich's "magic bullet" continues, and a glance at the bacterial and protozoal world indicates the tremendous changes in the pattern of infective diseases which have resulted from the introduction of chemotherapy. Tuberculosis control, for example, takes advantage of both immunization and chemotherapy. Perhaps as virologists we can learn from this broad approach. If this book encourages a dual approach to the problem of prophylaxis for respiratory virus infections, then it will have succeeded in its objective.

J. Oxford
London
July 1976

REFERENCES

1. Editorial, Swine influenza campaign. Should we vaccinate the pigs?, *Science,* 192, 870, 1976.
2. Belyavin G., Experience in the control of influenza by vaccination – a brief review, in *Chemotherapy and Control of Influenza,* Oxford, J. S. and Williams, J. D., Eds., Academic Press, London, 1976.
3. **Smith, J. W. G., Fletcher, W. B., and Wherry, P. J.,** Vaccination in the control of influenza, *Postgrad. Med. J.,* Vol. 52, 1976.
4. Editorial, *JAMA,* 235, 2753, 1976.
5. **Stuart-Harris, C.,** Swine influenza virus in man. Zoonosis or human pandemic, *Lancet,* 2, 31, 1976.
6. Pandemic of Influenza, 1918–1919, Ministry of Health Report No. 4, HMSO, London, 1920.

THE EDITOR

J. S. Oxford, Ph.D., is Member of the Scientific Staff, National Institute for Biological Standards and Control, Division of Virology, London, England.

Dr. Oxford obtained his B.Sc. degree in 1963 from the University of Reading and received his Ph.D. degree in 1966 from the University of Sheffield.

Prior to his present position, Dr. Oxford was Member of the Scientific Staff, Medical Research Council, National Institute for Medical Research, Mill Hill. Dr. Oxford has also served as Research Fellow, Australian National University and Lecturer in Medical Microbiology, The Medical School, University of Sheffield.

Dr. Oxford is a member of the Society for General Microbiology and the British Society for Antimicrobial Chemotherapy.

Dr. Oxford is currently a member of the Editorial Board of the *Journal of Antimicrobial Chemotherapy* and *Experimental Cell Biology*. He has published more than 70 scientific papers, mainly on the immunology and replication of influenza virus. Dr. Oxford's current research interest is in aspects of the control of influenza by immunoprophylaxis and chemoprophylaxis.

CONTRIBUTORS

Adrien Albert, Ph.D., D.Sc.
Research Professor
Department of Pharmacological Sciences
State University of New York at Stony Brook
Long Island, New York

Richard D. Barry, M.A., B.V.Sc., Ph.D. Sc.D.
Huddersfield Lecturer in Special Pathology (Virology)
University of Cambridge
Cambridge, England

Sylvia E. Glass, Ph.D.
Medical Research Council Scholar
Virus Department, Laboratories Block
New Addenbrook's Hospital
Cambridge, England

Peter Palese, Ph.D., Mag. Pharm.
Associate Professor, Department of Microbiology
Mount Sinai School of Medicine
City University of New York
New York, New York

G. C. Schild, Ph.D.
Head
Division of Virology
National Institute of Biological Standards and Control
London, England

Jerome L. Schulman, M.D.
Professor
Department of Microbiology
Mount Sinai School of Medicine
City University of New York
New York, New York

Sir Charles Stuart-Harris, C.B.E., M.D., D.Sc., F.R.C.P.
Postgraduate Dean
Faculty of Medicine
University of Sheffield Medical School
Sheffield, England

ACKNOWLEDGMENTS

I would like to thank Gillian Oxford for assistance with many aspects of the book and Alan Hodge for the line drawings. The photographic departments of the Medical Research Council Laboratories at Mill Hill and Hampstead were of considerable assistance. The staff of CRC Press has provided assistance at every stage in the preparation of the book.

TABLE OF CONTENTS

Chapter 1

CLINICAL ASPECTS OF THE RESPIRATORY TRACT

Sir Charles Stuart-Harris

TABLE OF CONTENTS

I. INTRODUCTION

The chemotherapeutic era has already witnessed great changes in the pattern of infective disorders. Bacterial infections of the lung and of the air passages have shared in this conquest, as shown, for instance, by the decline in tuberculosis in the past 30 years. Acute bacterial disorders, such as pneumonia, streptococcal tonsillitis, suppurative otitis media, have also changed greatly. Thirty years ago, pneumonia killed many in infancy or early adult life, but now only the elderly or chronically ill fail to respond to appropriate antibiotic therapy. Yet the common acute disorders of the respiratory tract due to viruses have altered little, and their clinical patterns and incidence remain unchanged. Even though relatively benign, acute respiratory virus infections cause loss of time from schooling and from work unsurpassed by any other group of infections.

Among the respiratory illnesses, those due to the influenza viruses stand apart from the others, for they cause significant mortality and recurrent epidemics on a world scale, which render them a formidable challenge. In spite of many attempts to prevent the spread of infection by immunoprophylaxis, the influenza viruses remain unchecked, except to a limited extent in cases where immunization has been practiced. The first faltering steps in chemoprophylaxis have now been taken against influenza. Though the scope of this advance is yet inappreciable, the successful demonstration that antiviral activity against influenza can be produced by drugs has been both timely and heartening.

Apart from influenza, however, the respiratory viruses remain largely unaffected by any form of prophylaxis. Their very number and variety has thwarted attempts to develop immunoprophylaxis, as will be described in Chapter 3. It is the purpose of this book to review the status and potentiality of chemoprophylaxis against these viruses. In this chapter, an outline of present knowledge concerning the viruses will be followed by a brief account of the epidemiology of their infections, and then the relationship between viruses and clinical events will be discussed. The viruses themselves are an array of unrelated agents, some containing DNA, most containing RNA, and varying greatly in size and structural complexity. They form eight groups, sometimes of a small number, but often of a large number of antigenic serotypes, and some have homologs in the animal kingdom. They are fully described in Andrewes and Pereira's compendium.[1] In this book only an

TABLE 1

The Viruses of Acute Respiratory Disease

Group	Nucleic acid	Symmetry	Serotypes	Principal human disease
Influenza viruses				
Influenza type A	RNA	Helical	Multiple, with variations	Epidemic influenza
Influenza type B	RNA	Helical	Multiple, with variations	Epidemic influenza
Influenza type C	RNA	Helical	More than 1	Sporadic respiratory disease
Paramyxoviruses				
Parainfluenza	RNA	Helical	4	Childhood lower-respiratory-tract disease; croup
Respiratory syncytial	RNA	Helical	More than 1	Childhood bronchiolitis
Picornaviruses				
Enteroviruses				
(Coxsackie, Echovirus)	RNA	Icosahedral	Multiple – at least 70	Febrile illnesses; aseptic meningitis
Rhinoviruses	RNA	Icosahedral	At least 100	Common cold
Coronaviruses	RNA	Helical	More than 1	Common cold
Adenoviruses	DNA	Icosahedral	At least 33	Febrile pharyngitis; tonsillitis
Herpesviruses				
Herpes simplex	DNA	Icosahedral	2	Primary stomatitis; recurrent skin rash

outline of the viruses will be attempted, and Table 1 lists their principal characters in summary. Knowledge concerning some of the viruses is relatively recent, but it is already clear that they behave independently of each other from an epidemiological and immunological standpoint. The character they possess in common – namely, that they multiply in the superficial epithelium of the human respiratory tract – does not distinguish them from other viruses (such as those of poliomyelitis, measles, and rubella) that gain access to the blood from a portal of entry in the nasopharynx. The ability of the respiratory viruses to induce pathological lesions in the respiratory tract is, however, the reason for the symptoms and signs that they induce. The fact that, in general, they limit their invasion to this system is an important negative character.

II. RNA VIRUSES

A. The Influenza Viruses (*Myxovirus influenzae*)

1. Classification

Three types of influenza viruses are distinguished by the antigenic characters of their nucleocapsid, determined by complement fixation or immunodiffusion tests.[2] Only types A and B are regularly associated with epidemics of clinical influenza; type C should properly be regarded as a cause of mild human respiratory illness. Types B and C viruses appear to occur only in man, but type A viruses belong to a large group of myxoviruses that cause respiratory disease in pigs and horses (swine and equine influenza) and in many avian species (fowl plague and other avian influenzas – more fully described in Chapter 3).

The influenza viruses are subclassified according to their surface antigens, and these are of two types of glycoproteins – hemagglutinins (H) and neuraminidases (N). The particles of the viruses are roughly spherical in outline and measure 80 to 120 nm in diameter. They are pleomorphic, some strains containing numerous long filaments (Figure 1). Each nucleocapsid is helical in form and consists of ribonucleic acid combined with protein subunits. Almeida and Brand[3] have recently shown the helix to be double in outline in electronmicrographs. The helix exists in the form of a skein, which breaks into segments during intracellular replication. It can be released from the particle by treatment with ether or with certain detergents. Surrounding the nucleocapsid is a second protein (M or membrane protein), which reacts serologically in a manner similar to the RNA

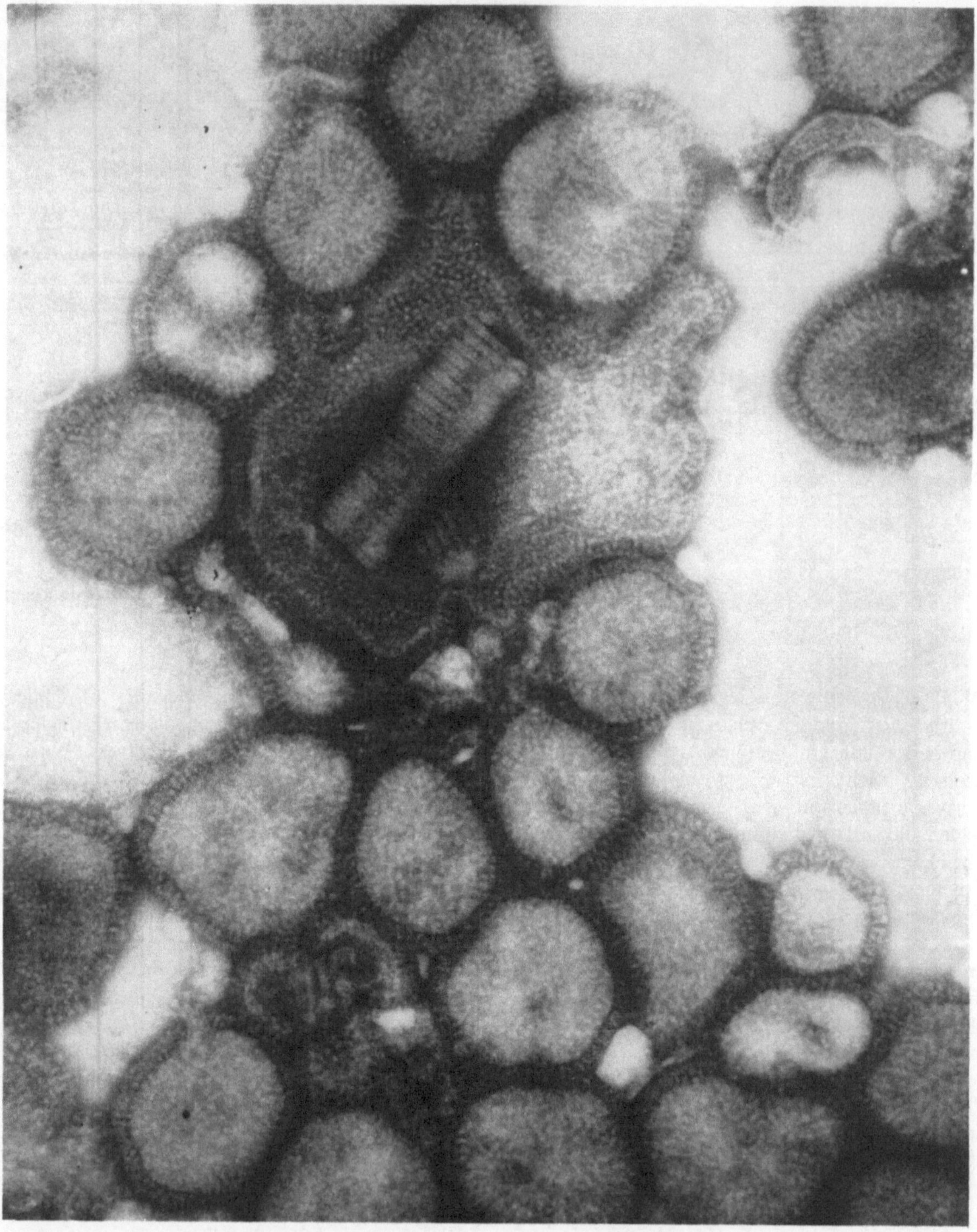

FIGURE 1. Influenza A virus, A/HK/1/68 strain, showing surface antigens projecting from the envelope and internal helix of RNA. (Photo courtesy of J. Heather, Division of Virology, National Institute for Biological Standards and Control, London.)

protein. The whole particle is enclosed in a lipoprotein envelope, partly derived from host cellular material as the particle emerges from an infected cell. Through this envelope project the surface structural proteins, H and N, already mentioned. The latter give the particle its subtype character, determined in the case of the H antigen by the hemagglutination inhibition test or by the neutralization of infectivity. The hemagglutinins are concerned with the adsorption of the particles to mucoprotein receptor areas on red cells or on other susceptible cell surfaces (see Chapter 5).

The N antigens are enzymes able to split mucoproteins with the release of N-acetyl neuraminic acid. Their function, though still obscure, is probably concerned with the release of nonaggregated particles from the host cell; it does not appear to be concerned with viral entry into the cell, even though, in vitro, the neuraminidase can elute virus adsorbed to red cells. Both H and N antigens stimulate the formation of local and serum antibodies during infection or after immunization of the intact host. The antibodies to the hemagglutinins can prevent infection and are obviously concerned with host immunity. Those inhibitory to the neuraminidase reduce the quantity of virus released from infected cells and play a subsidiary role in immunity by reducing the shedding of virus.

The antigenic formula of human influenza A viruses is described according to the subtype of the H and N antigens as H0N1, H1N1, H2N2, or H3N2.[4,5] Table 2 gives the chronological sequence of the subtypes, the years in which they appeared and later disappeared as they were replaced by their successors, and the reference strains. New major antigenic subtypes have appeared at a particular time and place at roughly 10-year intervals, and it is a remarkable property of the influenza A viruses thus to undergo major antigenic variation. The origin of the variants of a new subtype is unknown. Some investigators favor their emergence by a form of hybridization (genetic recombination) of human and animal strains, whereby an unfamiliar animal H or N antigen is transferred to a human virus. There exists other evidence of serological character that human viruses of a particular antigenic character may reappear as infecting agents after more than 50 years of absence from the human scene. Until this matter is resolved, it will not be known whether the number of variant subtypes is limited or unrestricted. In any case, it is clear that major antigenic variation is a principal factor in the periodic appearance of influenza A world epidemics (pandemics) and in the failure of human vaccines made from previous strains.

Minor variations of both H and N antigens occur year by year with both influenza A and influenza B viruses. They are believed to result from the selective effect of multiplication in a population possessing partial serological immunity. Known as antigenic drift, this effect involves both antigens, though not necessarily at one and the same time. It is clearly concerned with the recurrence of epidemics of both influenza A and B, and it also has to be taken into account in the formulation of immunizing materials for human

TABLE 2

Surface Antigens of Human Influenza A Viruses

Strain	First isolation	Last isolation	Hemagglutinin	Neuraminidase
WS	1932			
PR8	1934		H0	N1
Others	–	1946		
FM1	1946		H1	N1
Others	–	1957		
Asian	1957		H2	N2
Others	–	1968		
Hong Kong	1968	Still current		
ENG/42	1972	(1976)	H3	N2
Others	–			

use. Chakraverty,[6] Schild et al.,[7] and Curry et al.[8] have described the antigenic modifications of the B viruses.

The H and N antigens of the animal influenza A viruses are mostly distinct from those of the human A viruses. There are, however, some interesting examples of shared antigens, such as the A/Hong Kong (H3N2) and the A/Equine/2/63 and A/Duck/Ukraine/63 viruses, which share certain polypeptide links in the light chains of their hemagglutinins.[9] There are other examples of N antigens shared between human viruses and animal strains, such as swine influenza virus (N1) and A/Turkey/65 avian virus (N2).[10]

Information concerning the influenza C virus antigens is limited, but it is known that variation of the hemagglutinins occurs.

2. Cultural and Biological Properties

The viruses can be cultivated from human specimens (throat swabs or garglings) collected from persons with influenza and inoculated into the amniotic cavity of 10- to 12-day-old fertile hens' eggs. Passage from amnion to allantoic cavity is later successful. Some human strains, chiefly of influenza B, grow better in primary monkey kidney tissue cultures cultivated at lower than normal temperatures (33 to 35°C) than in hens' eggs.

In tissue cultures such as those of monkey kidney, the influenza viruses produce little obvious alteration in appearance of the cells. By adding red cells to infected cultures after incubation and then washing off the surplus cells, virus-infected cells can be outlined by the adsorbed red cells. This hemadsorption technique of Vogel and Shelokov[11] provides a useful method for the titration of virus in tissue cultures and for measurement of the neutralization of infectivity.

The replication of virus within susceptible cells is fully described in Chapter 5. The virus RNA polymerase is a structural component distinct from the polymerases of the host cell. It has been identified by sucrose-gradient centrifugation[12] and assayed biochemically.[13] The enzyme is concerned with the transcription of virus RNA. Studies with the electron microscope indicate that virus particles become attached to susceptible cells and penetrate by viropexis prior to uncoating and release of RNA into the cytoplasm.[14]

Although virus-specific antigens are identifiable within the host cell nucleus and cytoplasm a few hours after infection, particles can only be visualized at the cell membrane immediately prior to their actual release. The various biochemical processes concerned with replication of the genome and with formation of structural proteins occur, therefore, in the absence of identifiable morphological elements.

The pathological effects of influenza virus in man remain uncertain, because only those in cases of fatal influenzal pneumonia have been well-studied. In these, epithelial lesions are found in the air passages, in the larger bronchi, and in the trachea. Figures 2 and 3 show reduction of the stratified columnar epithelium in width consequent on necrosis of the superficial cells. In the rare case of influenza virus pneumonia without

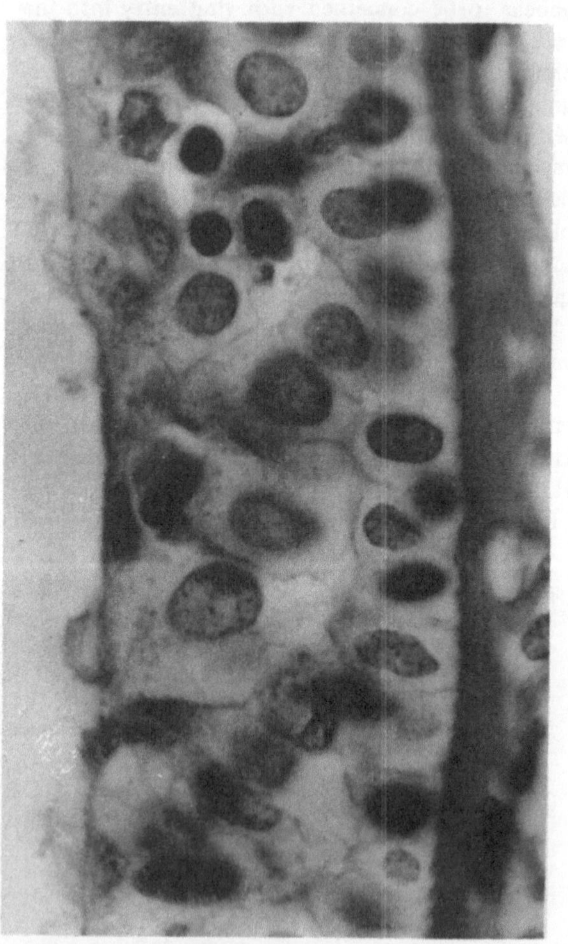

FIGURE 2. Bronchial lesions in human influenza, A/Asian (H2N2) strain. Early epithelial lesions. Magnification × 2,400. (Courtesy of the late Professor J. Mulder and Dr. J. F. Ph. Hers, Department of Medicine, University of Leyden, The Netherlands.)

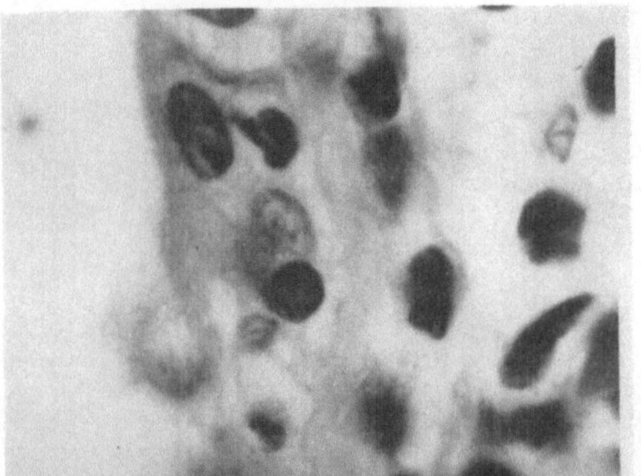

FIGURE 3. Bronchial lesions in human influenza, A/Asian (H2N2) strain. Advanced epithelial lesions. Ciliated cells and deeper cells have undergone necrosis and detachment. There is subepithelial infiltration. Magnification × 2,400. (Courtesy of the late Professor J. Mulder and Dr. J. F. Ph. Hers, Department of Medicine, University of Leyden, The Netherlands.)

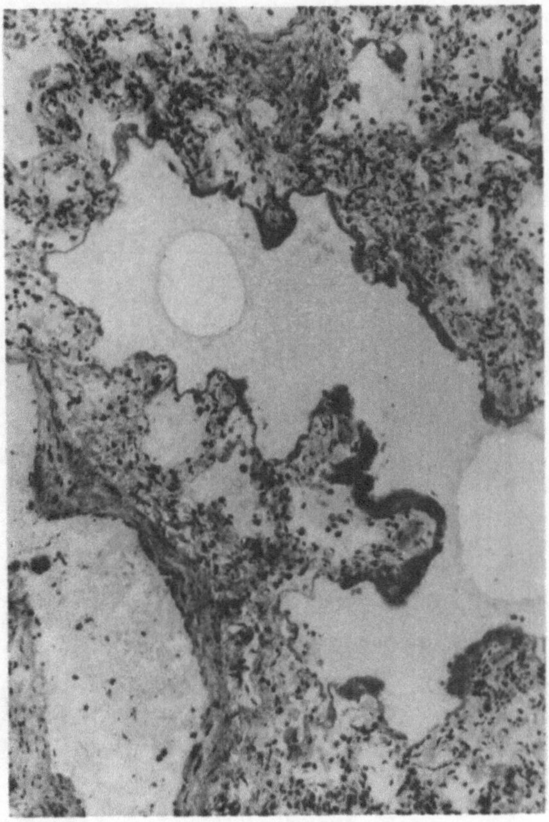

FIGURE 4. Influenza virus pneumonia in man, A/Asian (H2N2) strain. The alveolar ducts and terminal airways are lined by a hyaline membrane. Note the surrounding infiltration. Magnification × 480. (Courtesy of the late Professor J. Mulder and Dr. J. F. Ph. Hers, Department of Medicine, University of Leyden, The Netherlands.)

secondary bacterial infection, the lung shows thrombo-necrosis of the alveolar wall, the alveolar ducts are lined by a hyaline fibrinous membrane, and infiltration of the alveoli with red cells, leukocytes, and edema fluid occurs (Figures 4 and 5).

Experimental study of the disease produced by inoculating ferrets or mice intranasally with egg-cultivated virus has shown similar histological changes in the air passages and alveoli (see Chapter 1, Volume II). Ferrets inoculated intranasally with human secretions from patients with influenza develop fever and epithelial-cell necrosis in the nasal turbinates' ciliated epithelium (see Chapter 1, Volume II). Recovery is associated with speedy repair of the cell damage and restoration of normal epithelium. It is possible that similar changes in the nose and upper airways occur in uncomplicated influenza in man. Serial passage in ferrets results in virus adaptation, and a more virulent infection with pneumonia occurs if the inoculation is carried out in anesthetized animals. Mice develop lung lesions after intranasal inoculation of egg-cultivated virus; after adaptation by passage, the virulence of the infection increases, and the mice die from consolidation of the lungs.

Thus, laboratory strains of the influenza virus exhibit widely different pathogenic properties. In spite of this, there is little evidence of enhanced virulence of influenza in man, such as might be anticipated during epidemics. Indeed, only the 1918–1919 pandemic of influenza exhibited ex-

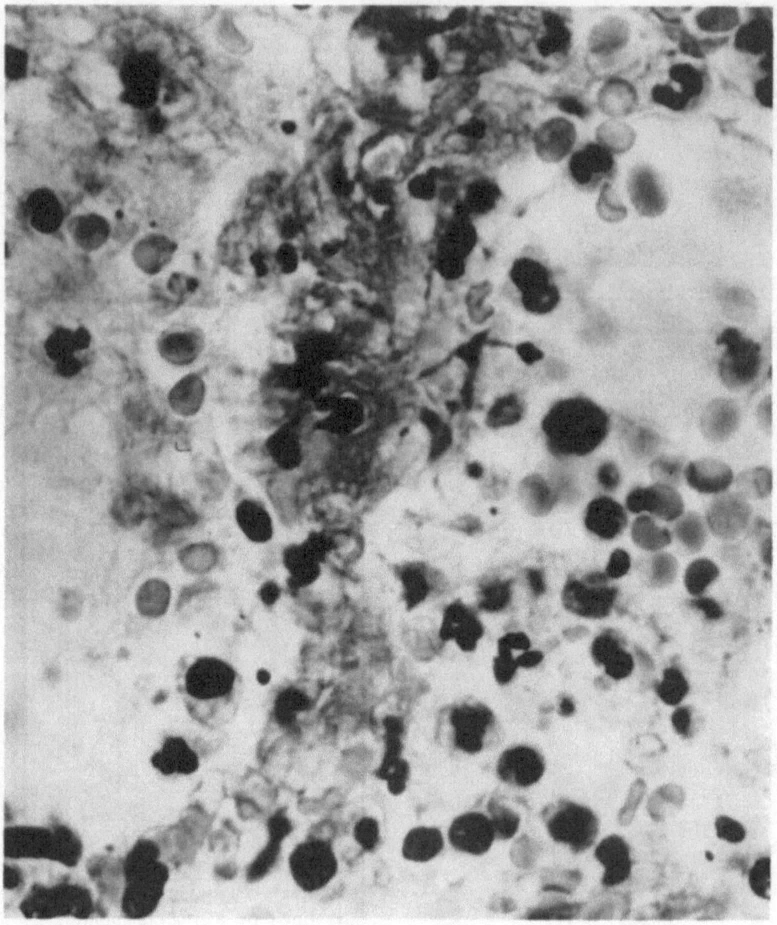

FIGURE 5. Influenza virus pneumonia in man, A/Asian (H2N2) strain. The alveolar walls show necrosis and thrombosed capillaries. The alveoli contain leukocytes, red cells, and desquamated epithelium. Magnification × 860. (Courtesy of the late Professor J. Mulder and Dr. J. F. Ph. Hers, Department of Medicine, University of Leyden, The Netherlands.)

ceptional mortality, particularly affecting adults aged 20 to 40. Though many virologists believe that the virus causing this epidemic had enhanced pneumotropic action, this view is contested. It is certainly true that most of the differences in the prevalence and severity of epidemics since 1930 can be explained by the ages of those chiefly attacked, by the presence or absence of persons in the community with chronic chest disease, and by lack of previous exposure to the antigens of the virus strains concerned. In spite of this, the emergence of a new influenza virus A as a major antigenic variant is a cause of anxiety, as was the case when the Asian (H2N2) and Hong Kong (H3N2) viruses first appeared on the human scene. Though epidemics of influenza A occur in most countries annually or biennially, and those of influenza B less frequently, the whereabouts of viruses of either type is obscure in interepidemic periods. Serological evidence of sporadic infection in the community exists between epidemics, but the viruses are rarely isolated. An alternative view is that the viruses pass from the northern to the southern hemisphere with the seasons in an unbroken chain of infections.

Changes in the virulence of the influenza virus for man are readily produced in the laboratory, so that attenuated strains can be developed. Strains resistant to normal serum inhibitors, as well as cold-adapted or temperature-sensitive (ts) strains that fail to grow at higher than normal body temperatures, are often attenuated for man. Viruses have been selected by cultivation with or without previous treatment by mutagenic agents such as fluorouracil.[15-17] Genetic recombination has also been much used to confer attenuation on a virus whose antigens are suitable for human vaccines. Genetic interchange during simultaneous infection with two viruses is readily achieved in the laboratory, and this has demonstrated that the genes conveying the information for the antigens are separate from those concerned with virulence. Plaque-picking is necessary in relation to the purification of viruses used as seeds for vaccine production, but particular cultural conditions are required to obtain plaques in infected tissue cultures.

B. Paramyxoviruses

According to Andrewes and Pereira,[1] the paramyxoviruses include two members of the respiratory virus groups – parainfluenza and respiratory syncytial (RS) viruses – as well as measles, mumps, and certain animal viruses. Only the respiratory viruses are considered here.

1. Parainfluenza Viruses

a. Classification

These viruses resemble morphologically the influenza viruses, but their particles are larger (100 to 200 nm in diameter). They possess an enveloped helical nucleocapsid in the form of a double helix, and through the envelope there project surface proteins with hemagglutinin and neuraminidase activity (Figure 6). The reaction between viruses and red cells ultimately results in hemolysis. Little is yet known about the neuraminidase. The viruses are classified into four different antigenic serotypes, I to IV, recognized by hemagglutination inhibition, neutralization, and complement fixation tests.[18,19] There is no common antigen based on the nucleocapsid, as in the case of the influenza viruses. Cross reactions between the types occur to some extent with convalescent sera, but more importantly with mumps virus. Thus, human infections with mumps cause antibody responses to type I parainfluenza virus. Some animal viruses appear to be closely related to the human strains. Parainfluenza III virus is identical to the virus of bovine shipping fever, type II is similar to a simian virus (SV5) commonly found in normal monkey kidney cultures, and the mouse Sendai virus recovered from Japan is antigenically similar to type I virus, although it is mouse-virulent.

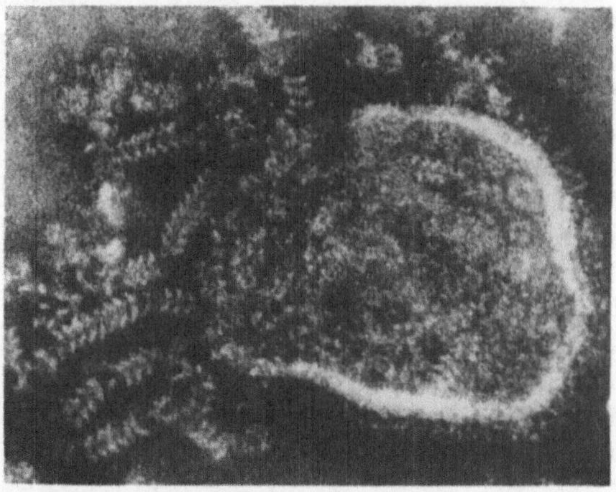

FIGURE 6. Single-particle parainfluenza virus, ruptured. Note the surface antigens projecting from the envelope and the internal RNA helix emerging from the disrupted virion. Magnification × 250,000. (Courtesy of Dr. June Almeida, Wellcome Research Foundation, Beckenham, Kent.)

b. Cultural and Biological Properties

The parainfluenza viruses are most readily cultivated from human throat swabs by inoculation into primary monkey kidney tissue cultures. There are no apparent cellular changes, but hemadsorption can be demonstrated. Passage from culture into the amniotic cavity of fertile hens' eggs is more likely to yield virus than direct inoculation of eggs with human material. Once established in tissue cultures, the viruses induce syncytial-cell formation. They have also been cultivated in organ cultures of human trachea.[20] Guinea pigs and hamsters can be infected intranasally, but the infection is inapparent.

The parainfluenza viruses are common endemic infectious agents mainly affecting children. Winter outbreaks occur chiefly with croup and other lower-respiratory-tract infections. Antibodies develop in an increasing percentage of children during infancy, and in most childrens' sera they are active against types I to III by the age of 6. The rate of the rise of antibodies in the community is most rapid against type I parainfluenza virus.[21,22] Recognition of the viruses in throat secretions and in tissue cultures is assisted by immunofluorescence techniques.[23]

2. Respiratory Syncytial Virus

a. Structure

This myxovirus is both more fragile and more labile than the parainfluenza viruses. Strains recovered from man are relatively homogeneous antigenically, though antigenic differences were recognized by Coates et al.[24] by virus neutralization tests. It is difficult to study the structure of RS virus by electron microscope. The particles are pleomorphic; spherical particles are 80 to 500 nm in diameter, and filamentous forms also occur.[25] The nucleocapsid is helical. Surface projections are visible on the virus envelope, but no hemagglutinins have ever been demonstrated. Complement-fixing antigens can be separated from the specific RS antigen demonstrated by neutralization or gel diffusion techniques.

b. Cultural and Biological Properties

RS virus can be recovered from throat swabs inoculated into serial cell cultures, such as Hep-2 or HeLa cells. Specific immunofluorescence can be used to demonstrate virus antigens in nasopharyngeal secretions or in infected tissue cultures.[23] The name of this virus is derived from the fact that tissue cultures develop large syncytial cells, and these provide a method of assay of virus or of neutralizing antibody.[24]

RS virus will infect a number of animals, including ferrets and hamsters by intranasal inoculation, and mice intracerebrally. Chimpanzees in captivity may acquire RS infection naturally. Most animal infections, including those of the ferret and hamster, are subclinical. The bovine strain of RS virus recovered from calves with respiratory-tract symptoms is antigenically similar to the human virus.[26]

Human infection occurs chiefly in infants less than 6 months of age, often during epidemics; 80 to 90% of children possess serum neutralizing antibodies by the age of 4. Disease in children appears to be determined in part by immunopathological processes, as will be described below in considering the major clinical illness caused by RS virus – acute bronchiolitis.

C. Picornaviruses

This large and heterogeneous collection of viruses includes relatively few groups that attack the human respiratory tract. However, the enormous number of rhinoviruses derived from the large number of serotypes makes up for the relative lack of numbers of respiratory pathogens among the other enteroviruses. The classification of a virus as "picorna" is based upon the small size, icosahedral symmetry, and possession of RNA (see Chapter 4, Volume II). Picornaviruses do not possess an envelope – a fact that explains their resistance to disintegration by ether. The enteroviruses, which include Poliovirus, Coxsackievirus, and Echovirus, are also acid-resistant, whereas rhinoviruses are inactivated at pH 3.0 to pH 5.0. The basis of this inactivation of infectivity appears to be disruption of the capsid. The capsomeres of Poliovirus have been estimated by various authors to number 32 or 64.

1. Coxsackieviruses and Echoviruses

Unlike the rhinoviruses, Coxsackieviruses and Echoviruses appear to be relatively infrequently concerned in respiratory illnesses. Their potential for causing respiratory-tract symptoms nevertheless exists, and a description of their general character is necessary.

a. Classification

The Coxsackieviruses were discovered during tests on human feces from persons involved in outbreaks resembling poliomyelitis. Their classification into Coxsackieviruses A and B was made on the basis of the different effects observed in suckling mice inoculated with the two groups. Coxsackievirus A causes necrosis of striated muscle as part of a generalized infection, whereas Coxsackievirus B induces meningoencephalitis in similar baby mice. Neither group of viruses can be readily cultivated in tissue cultures during primary isolation from man, though there are important exceptions, such as Coxsackievirus A21, which multiplies in HeLa cells. Antigenic analysis by neutralization tests in suckling mice has led to the recognition of 24 types of A and 6 types of B viruses. Minor cross reactions exist between A and B viruses.

The Echoviruses were also recovered from human fecal extracts, but by inoculation into tissue cultures, and for some time it proved difficult to establish their relation to human illnesses. The word "orphan" was attached to the group for this reason, to emphasize the ready recovery of the viruses from children in a normal state of health. It has been estimated that approximately half the members of families from whom Coxsackieviruses or Echoviruses are recovered have no observable illness at the time when the viruses are being actively excreted.[27] Antigenic analysis of the Echoviruses has yielded 34 serotypes by neutralization tests. Complement fixation tests can also be used for type analysis. Some Echovirus types agglutinate human Group O red blood cells, thus permitting hemagglutination inhibition, which gives less specific results. Only three of the Coxsackieviruses react similarly with human red cells. A slight antigenic relationship exists between certain types of Echovirus and Coxsackievirus.

b. Characters

The particles of both groups of viruses are similar in size and morphology to other members of the picornaviruses. The tiny icosahedra, 20 to 29 nm in diameter, readily pack together to form intracellular crystals, and electron micrographs are relatively easily obtained. Monkey kidney tissue cultures inoculated with Echovirus develop focal cytopathic lesions that are much less distinctive than those of Poliovirus. Whereas the Coxsackieviruses readily cause lesions in baby mice, the Echoviruses are generally noninfective for experimental animals other than monkeys. Some types – for instance, Echovirus type 9 – are mouse-pathogenic, and this particular variety was at first classified as a Coxsackievirus because of this. Monkeys inoculated intraspinally and intracerebrally with several serotypes of Echovirus develop CNS lesions, though symptoms are usually absent. One type of Coxsackievirus (A7) is sometimes regarded as a fourth type of Poliovirus because of its pathogenic effects in man as well as in monkeys.

In man, both Coxsackieviruses and Echoviruses appear to be frequent inhabitants of the alimentary tract, and during illnesses the viruses are also found in the nasopharynx. Recovery of Coxsackieviruses by the Public Health Laboratory Service[28] and by other laboratories in the United Kingdom between 1958 and 1962 showed that the majority of illnesses temporally related to virus recovery were associated with meningitis or other neurological symptoms. The condition known as pleurodynia or Bornholm disease is related only to Coxsackieviruses B, as are also certain cardiac effects. Rashes occur more often with A than with B Coxsackieviruses. Respiratory illnesses occurred in about 10% of infections, more often involving the upper than the lower tract. Coxsackievirus A21 behaves wholly as a respiratory pathogen.[29,30]

Echovirus infections appear in the community as relatively localized, occasionally widespread outbreaks, often during the summer months in temperate climates. They spread through the population in a relatively narrow stream, being chiefly manifest by benign acute meningitis of the lymphocytic or aseptic variety. Occasionally, more serious neurological effects occur. Rashes occur with infections by certain serotypes, such as Echo type 9 and type 16. Respiratory illnesses appear to occur to a minor extent, but types 11, 19, 20, and 25 have been described as respiratory viruses by individual workers. When adult volunteers are inoculated with secretions containing type 20 virus, symptoms are elicited resembling those of the common cold.[31] Similar studies with type 11 virus by Saliba et al.[32] also suggest that the virus causes cold-like illnesses. Over the period from 1959 to 1963, 89 of 1,062 illnesses related to Echoviruses in England and Wales were of respiratory character.[33] This must be compared with the 187 respiratory illnesses from 1,631 Coxsackie infections described by Vernon[28] in a similar period of time. An epidemic form of conjunctivitis of the eye has recently been shown to be due to an Echovirus[34,35] designated as Enterovirus 70.[170]

2. Rhinoviruses

a. Classification and Cultivation

These viruses resemble other picornaviruses in structure in that their particles are uniform icosahedra, 27 to 28 nm in diameter. Thin sections of infected cells show particles packed together into semicrystalline arrays. The virions are relatively fragile, in spite of being ether-resistant. By treating virus preparations with immune serum prior to preparation for electron microscopy, Kapikian et al.[36] obtained excellent photographs of Rhinovirus particles in three forms: complete particles, "empty" capsids of ring-like appearance, and mixtures of both types (Figure 7). Study of the many serotypes of Rhinovirus, on which their classification is based, has been hampered by the need to perform tissue culture neutralization tests, even though immunodiffusion and complement fixation techniques are also possible. The first attempt at a numbering system was published by a group of collaborators in 1967,[37] when 55 serotypes were recognized. Among these is type 1A, which is a reclassified virus originally described as Echovirus type 28.[38] Type 1B is a rhinovirus recovered in England and named B632. Later, additional serotypes were added, bringing the number to 89,[39] and a further 20 types are under consideration.[40] Although some serological cross

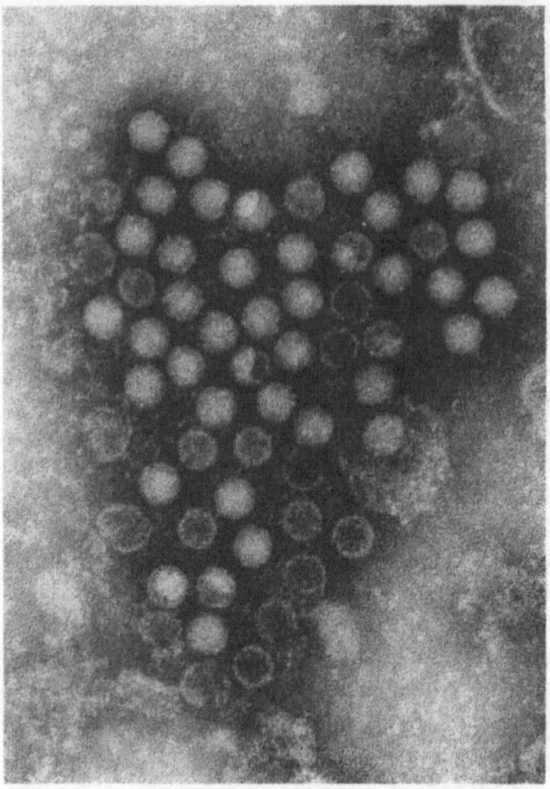

FIGURE 7. Assay of Rhinovirus particles. Both "full" and "empty" particles are present. Magnification × 250,000. (Courtesy of Dr. June Almeida, Wellcome Research Foundation, Beckenham, Kent.)

relationships have been observed, the striking character of the rhinoviruses is the relative distinctness of the serotypes and the lack of cross immunity between them.[41]

Rhinoviruses have for some time also been classified by their ability to multiply in either primary monkey kidney cultures or human cells (M) as compared to exclusively requiring human cells (H), such as cells of human embryo kidney (HEK) or the diploid cell cultures of human embryonic fibroblasts (W.I.38).[42] In fact, most strains grow in HEK cells, and only a small proportion replicate as M viruses. In addition to those rhinoviruses growing in ordinary tissue cultures, strains exist that can only be propagated in organ cultures, such as of human embryo trachea or nasal epithelium, in which they cause a reduction of ciliary activity.[43,44]

b. Biological Properties

The study of the properties of rhinoviruses has proved as difficult as their cultivation. Ether resistance and acid lability have been mentioned above. Growth of the viruses is improved by culturing in rolled tissue cultures at a lower temperature than normal (33°C), and by using medium at a pH between 6.8 and 7.3. Cytopathic effects are found with many strains. Under suitable cultural conditions, focal microplaques can be obtained.[45] Experimental infection is possible in chimpanzees and gibbons, but not in smaller animals. Some bovine strains of rhinoviruses have been recovered from cattle, and other strains have been identified in horses.

Studies of the replication of rhinoviruses in tissue culture indicate that RNA can be detected 4 to 5 hr after infection, with maximum synthesis occurring 2 hr later. The RNA is single-stranded, and its synthesis is dependent upon an RNA polymerase[46] (see Chapter 4, Volume 2). Its synthesis is not inhibited by actinomycin D, and therefore is not dependent on DNA.[47] RNA has been extracted from virions in a still infectious form, using cold phenol.[48] A number of structural proteins have been identified in rhinoviruses, and there appear to be three major and at least one minor component.[49] The attachment of rhinoviruses to cells is said to be by adsorption to specific receptors that are sensitive to trypsin. It is enhanced by high concentrations of magnesium ions.

Studies of the epidemiology of rhinovirus infections in man have proved difficult. In particular, serological antibody measurements are handicapped by the multiplicity of serotypes. Gwaltney[40] quotes Hamparian et al.,[50] whose findings with 56 serotypes show that the percentage of viruses neutralized by sera from children and adults rises steadily from 5% in infants less than 1 year old to 50% at ages 31 to 40. The proportion of sera with antibodies to a particular serotype varies greatly at all ages, however; types 1A and 2 are neutralized by 50% or more of sera, whereas others, such as types 3 and 7, are neutralized by only 12% and 10% respectively. This indicates merely that some rhinoviruses are more prevalent and possibly more infectious than others, as has indeed been suggested by such isolations of viruses as those at Tecumseh by Monto and Cavallaro.[51] Community studies also indicate that, taking all serotypes together, the rhinoviruses are endemic and at the lowest level of infection in summer (June and July) in temperate climates. A sharp rise in illnesses due to the viruses occurs in September, and this is succeeded by a plateau with minor oscillations, amounting to a total of 6 to 8 illnesses per 1,000 persons per day.[52] The family studies in Seattle conducted by Fox et al.[53] showed first the great importance of rhinoviruses in childrens' illnesses, with children under 1 year of age having the highest rate of infection and often the most severe illness. Secondly, infections within the community are spread by children more frequently than by adults, but the most effective introducers of infection into the family are the fathers. The overall importance of rhinoviruses in acute upper respiratory illness in man is apparent when the frequency of the common cold, largely the result of infection by this group, is compared with that of other clinical illnesses.

D. Coronaviruses

This is the name given to a group of human and animal viruses including those producing infectious avian bronchitis, mouse hepatitis, and transmissible gastroenteritis of pigs. Human coronaviruses were first identified in Britian by Tyrrell and Bynoe,[43] and in the U.S.A. by Hamre and Procknow.[54] Details are presented in Chapter 5, Volume II.

1. General Characters of Human Coronaviruses

The particles of Coronavirus resemble myxoviruses in being enveloped with projecting club-shaped spikes[55] and having an overall spherical shape measuring 80 to 160 nm in diameter (Figure 8). They are ether-sensitive and contain RNA. The first British strains could only be cultivated by using organ cultures of human embryo trachea, but a series of other strains, including the prototype 229E of Hamre and Procknow,[54] will multiply in human embryo tissue cultures. Hemagglutination can be obtained at 4°C by using human or vervet monkey red cells. Some viruses have been adapted to suckling mice by intracerebral inoculation.

Studies by McIntosh et al.[56] of relatively few strains revealed antigenic diversity. A cluster of viruses resembles prototype 229E, but the organ culture strains differ antigenically from these by neutralization or complement fixation tests. Some strains show a strong antigenic relationship to mouse hepatitis virus. The fact of antigenic diversity makes studies of the human epidemiology of the viruses difficult, but it has already been shown that antibody rises in human sera occur during outbreaks of upper respiratory illness, including colds.[57] Both in Britain[58] and in the U.S.A.[59] the antigenically diverse strains 229E and OC43 appeared to be prevalent in alternate years. Yet the estimate of the contribution of these viruses to the total of respiratory illnesses over an 8-year period by Kaye and Dowdle[59] was only 7%. The clinical spectrum of illness will be discussed later in this chapter.

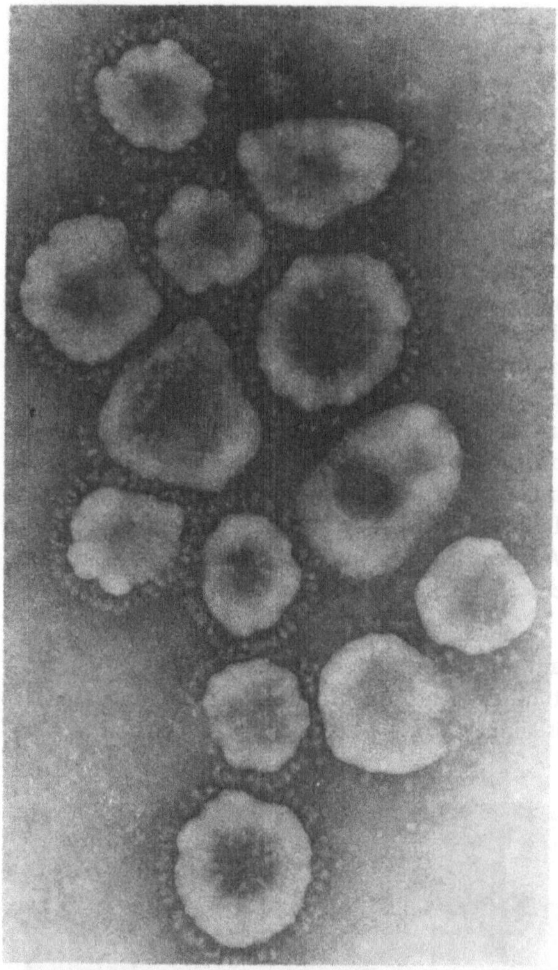

FIGURE 8. Coronavirus particles. Note the characteristic halo of club-shaped projections. Magnification × 165,000. (Courtesy of Dr. June Almeida, Wellcome Research Foundation, Beckenham, Kent.)

III. DNA VIRUSES

A. Adenoviruses

1. Structure and Classification

Although adenovirus particles resemble picornaviruses in possessing icosahedral symmetry, they

virus, and the Epstein-Barr virus. The monkey virus (B virus) can cause fatal illnesses in man. The virions of the herpesviruses are icosahedral in shape and measure 100 nm in diameter. They are surrounded by an envelope, and therefore exhibit ether sensitivity. The capsomeres number 162 and are in the shape of hollow prisms (Figure 10). The DNA forms an internal double-stranded core; a large number of proteins have been identified in infected cells, of which at least 25 are structural proteins, including 12 glycoproteins.

The antigenic structure of herpes simplex virus is complex. Up to 18 antigens have been identified. The antigens are of two main types, so that types 1 and 2 have been separated, and the serotypes differ in frequenting different parts of the body. An antigen can be separated from the virus particles, and this has been found to exhibit cross reactions between *Herpesvirus hominis*, EB virus, and Marek's disease of chickens. Other cross reactions have been reported. Complement fixation, neutralization, and immunodiffusion tests can all be used.

2. Cultural Characters

Herpesvirus hominis is recovered frcm man by cultivation in tissue cultures of HeLa cells, by inoculation onto the chorioallantoic membrane of fertile hens' eggs, or by inoculating suckling mice. Virus can be assayed by counting pocks on the chorioallantoic membrane or plaques in tissue culture. Infected cells develop cytopathic effects and may show eosinophilic nuclear inclusions. Guinea pigs and rabbits are susceptible, and strains adapted to guinea pig skin or rabbit testis have been developed. Keratoconjunctivitis is produced in the rabbit after corneal inoculation; this has been useful in experimental chemotherapy studies with compounds such as cytosine arabinoside and 5-iododeoxyuridine.

Entry of the virus into the cell is facilitated by fusion of the envelope with the cell membrane. Once within the cell, the virion undergoes uncoating, and DNA protein enters the nucleus. The morphology of this process was described by Miyamoto and Morgan.[79] For the description of the replication process the reader is referred to Andrewes and Pereira[1] and to the review by Nahmias and Roizman.[80]

3. Pathologic Effects in Man

Type 1 herpes simplex virus is found in the mouth and pharynx, and also by cultivating sensory-nerve ganglia after death.[81] Type 2 virus is generally regarded as a genital-tract inhabitant, though it is also associated with skin lesions on the

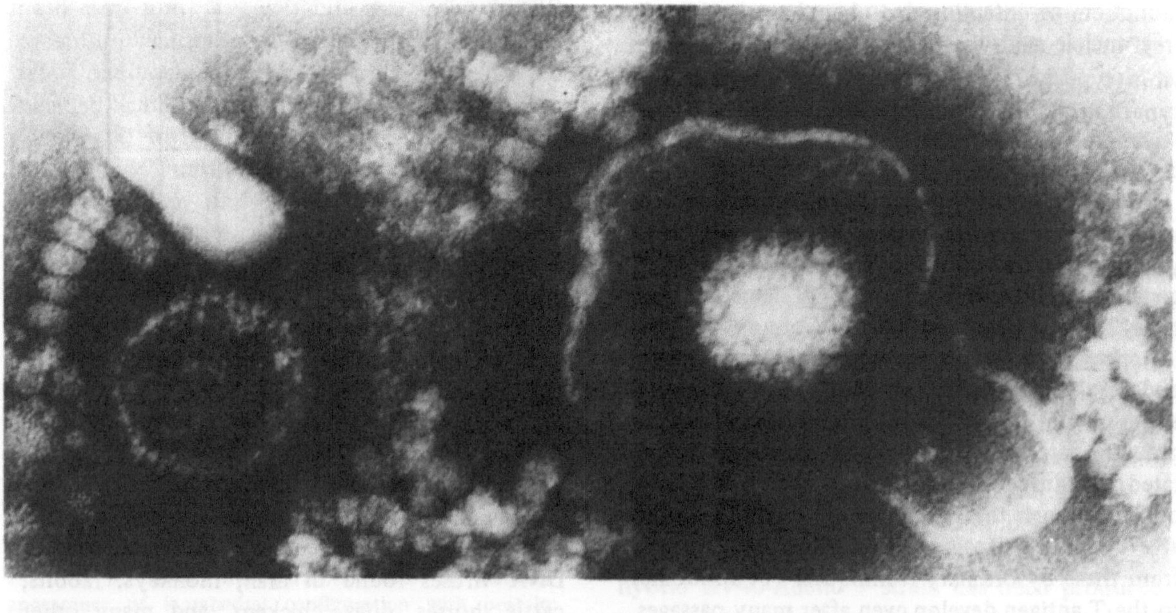

FIGURE 10. Two particles of herpesvirus. A free capsid is shown at the left side, and a complete virion is shown at the right side. Note the surrounding cell-derived membrane. Magnification × 250,000. (Courtesy of Dr. June Almeida, Wellcome Research Foundation, Beckenham, Kent.)

lower half of the body. It has been recovered from ganglia of sacral nerves.[82] There is general agreement that the type I virus enters the body through the oropharynx as a result of airborne infection. Primary infection in children usually causes acute gingivostomatitis, which may be ulcerative, but can also be silent. Thereafter, the virus probably spreads systemically, and occasionally disseminated disease may occur in babies, with lesions in the liver. The virus apparently persists during life in sensory-nerve ganglia, and particularly in the ganglion of the trigeminal nerve. Periodic reactivation of infection caused by illness of the host, especially fever, permits the virus to spread along nerve fibers to the skin, where it produces the familiar "cold sores" or fever blisters on the face around the mouth. Virus is shed from such lesions and will again infect susceptible persons. Those subject to recurrent herpes infection have high titers of neutralizing antibody in their sera. A primary type of herpes virus lesions of the fingers, resembling whitlows, has been described. The two most serious forms of herpes infections – encephalitis and generalized skin eruption in children with eczema – do not directly involve the respiratory tract.

IV. THE EPIDEMIOLOGY OF ACUTE RESPIRATORY DISEASE

In order to obtain statistical evidence of the incidence of acute respiratory illnesses, it has proved necessary to mount special studies of families and other groups. Those carried out in the U.S.A. and in Britain have usually concentrated upon relatively small numbers, largely because of the need to employ special methods of recording numbers and clinical categories of illness as well as the results of laboratory investigations. The first family studies, performed in Cleveland, Ohio, by Dingle and his co-workers,[78] suggested a total of 5.6 acute respiratory illnesses per person per year. In Britain, Hope-Simpson[83] evolved a diary system of records in Cirencester, which suggested that there were 7 illnesses per person per year. In 1967, Sutton et al.,[84] working in Sheffield, again with family diaries, found the figure to be 6.8. This vast toll of morbidity has since been studied in far greater detail and, in general, has been confirmed. The variations imposed by different environmental and personal factors are not fully understood, but will be briefly described.

A. The Effect of Season and Climate

Respiratory infections reach their peak during the winter and are at their lowest level in summer. Every year, the seasonal pattern in the northern hemisphere begins in September with a rapid rise in incidence of total acute respiratory infections; a plateau follows from October to March, consisting of a series of peaks and intervening troughs; and in April or May there is a rapid fall in incidence to the nadir in July and August. Figure 11, reproduced from Dingle's survey, *Illness in the Home*,[78] shows the pattern of acute respiratory disease in each of three separate years. Figure 12 shows the effect of season on illness in general practice in a report of the study developed by a Medical Research Council's Working Party on Acute Respiratory Disease.[85,86]

Chief among the sharp peaks in incidence are those due to influenza epidemics, usually occurring between December and February and superimposed on the curve representing undifferentiated illnesses. These are not shown in Dingle's Cleveland survey, because influenza was classified separately from common unclassified disease. Figure 13, obtained from the British general practice study of 1964 to 1966, shows the two epidemics of influenza recorded during the study. The annual curve of total respiratory infections is not smooth, but spiky, and the causes of sharp increases are now known to result from activity of the different respiratory viruses preponderating at different times. The influenza, parainfluenza, and respiratory syncytial viruses cause separate outbreaks, as was clearly shown by Hope-Simpson and Higgins[87] in their studies in Cirencester. Similarly, the study of viruses and illnesses among British children in hospitals, organized by the Medical Research Council's Working Party and the Public Health Laboratory Service, clearly showed periods of prevalence derived from individual virus outbreaks in the community (Figure 14).[88]

The manner in which season affects the prevalence of acute respiratory infections remains unexplained. Lower-respiratory-tract illnesses are more affected than upper-tract illnesses, as is clearly shown in Figure 12. This applies even more strikingly in regard to children whose illness requires their admission to a hospital. Some investigators believe that the chief factor by which climate affects the respiratory tract is the environmental temperature. Hope-Simpson[83] thus plotted the prevalence of colds in a group of 80 volunteers

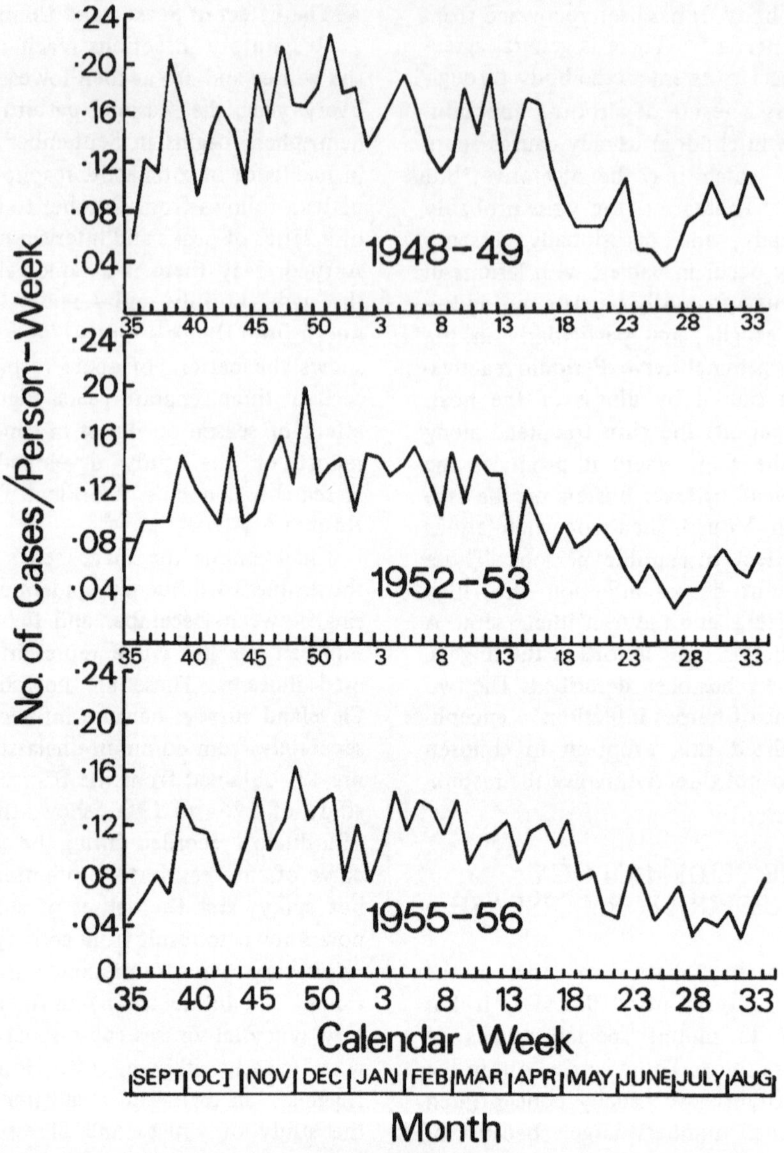

FIGURE 11. Effect of season on total acute respiratory disease. (From Dingle, J. H., Badger, G. F., and Jordan, W. S., Jr., *Illness in the Home*, Press of Case Western Reserve University, Cleveland, 1964. With permission.)

at Cirencester together with the daily external temperature, measured in the earth 1 ft below the surface. There was a remarkable inverse correlation, suggesting that the old adage that chills induce colds has truth. However, Andrewes[89] and Dowling et al.[90] were unable to find changes in the susceptibility of volunteers to an inoculum of infective cold secretions when the volunteers were subjected to experimental chilling.

A study of families in Tecumseh, Michigan, conducted by the Department of Epidemiology of the School of Public Health, Ann Arbor, showed a similar inverse correlation between external temperature and the prevalence of acute respiratory disease throughout the year.[91] A particularly strong negative correlation occurred during the two periods of early fall and spring. Rainfall was not so correlated, but in tropical countries, where the environmental temperature shows little variation, the two seasons of dry and wet weather are also correlated with changes in the prevalence of respiratory disease. Thus, Sutton,[92] working with

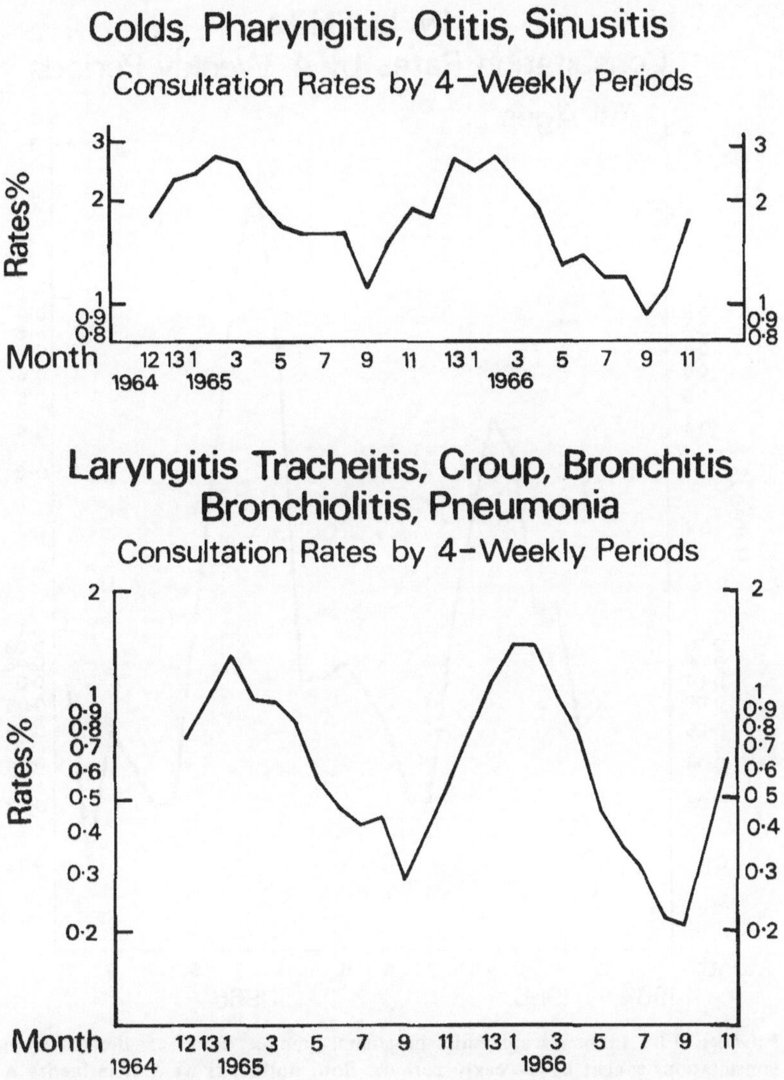

FIGURE 12. Effect of season on upper-respiratory-tract disease (top) and lower-respiratory-tract disease (bottom). Rates are derived from consultation sought in general practice by 4-weekly periods. (Unpublished data, courtesy of D. L. Miller, Collaborative Study, Medical Research Council/Public Health Laboratory Service, London.)

families in Trinidad, found that at the start of the rainy season, in June, respiratory infections increased sharply. Monto and Johnson[93] studied 92 families in Paraiso, Panama, over a 4-year period; respiratory infections increased *pari passu* with increased rainfall at both the beginning and the end of the rainy season; these were also periods of high humidity.

One suggestion made to account for the influence of climate has been that external cold assists the survival of viruses outside the body, and that the relative humidity has some influence on the stability of airborne droplets in which viruses can travel from one person to another. Yet the same environmental conditions exert different influences on different viruses. Hemmes et al.[94] and Buckland and Tyrrell[95] found that, whereas influenza virus remained infective longer in dry air with low relative humidity, enteroviruses, such as Poliovirus, preferred high humidity. Tyrrell[96] also found that adenoviruses and rhinoviruses were less stable in dry air than in moist air. The relative humidity measured out-of-doors cannot be regarded as a good index of the relative humidity inside houses. Since most respiratory viruses are acquired by contact indoors rather than out-of-

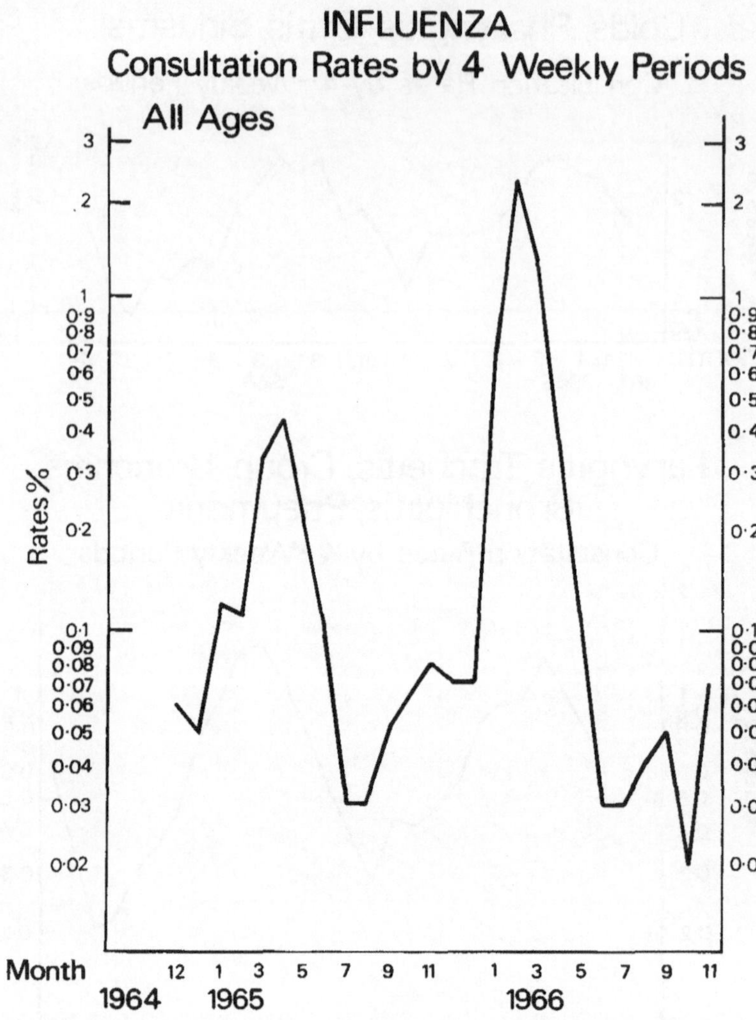

FIGURE 13. Influenza epidemics in general practice. Rates are derived from consultations sought by 4-weekly periods. Both outbreaks were of influenza A and B infections. (Unpublished data, courtesy of D. L. Miller, Collaborative Study, Medical Research Council/Public Health Laboratory Service, London.)

doors, no simple direct correlation between virus transmission and external humidity seems likely. A more significant effect may be that of the numbers of personal contacts made indoors. Thus, the crowding of persons together in boarding schools, barracks, or other residential institutions clearly enhances the attack rate of influenza virus during epidemics, compared to that occurring in ordinary homes.

Finally, the effect of weather on the mechanism of nonspecific defense of the respiratory tract must be considered. Cold reduces the rate of ciliary action, and thus the clearance of inhaled organisms from the lung is impaired by chilling. So little is as yet known concerning host factors of susceptibility and resistance of a nonimmunological character that it would be unwise to assume that a winter season enhances respiratory infections by an effect on the viruses rather than on their hosts. Both are probably concerned.

B. The Effect of Age on the Prevalence of Acute Respiratory Disease

The incidence of acute respiratory infections varies greatly at different ages. Every parent knows that children are infected more frequently than adults. In Tecumseh, Michigan, during the 6-year study of 4,905 residents, 14,600 cases of respiratory illnesses were recorded.[97] The mean number of infections per person per year was 6.1 in infants

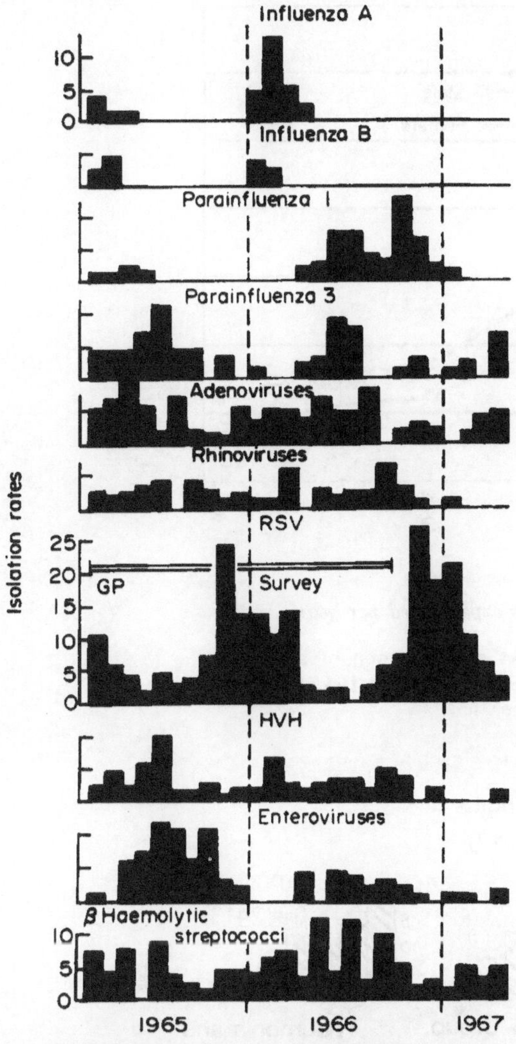

Isolation rates (%) by time. Hospital survey.

FIGURE 14. Viruses isolated from children in hospitals. (From Poole, P. M. and Tobin, J. O'H., *Postgrad. Med. J.*, 49, 778, 1973. With permission.)

less than 12 months of age, and the rate fell with each year of life, reaching 3.5 in the age group from 5 to 9 years. From age 10 to age 29, the rates of illness varied from 2.4 to 2.8, and a further fall occurred above the age of 30. Figure 15 is reproduced from Monto and Ullman's[97] paper. A somewhat similar curve for patients from less than 1 to 17 years of age was obtained by Dingle et al.[78] during the Cleveland study, but at all ages higher rates of illness were found than in Tecumseh; also, boys showed a higher incidence than girls. In Tecumseh, higher rates of illness were found in boy infants than in girl infants, but above the age of 3, females had higher rates than males throughout the age range.

In Britain, Miller[86] found that age had a significantly different influence on different categories of illness in general practice. Figure 16 shows that the common cold syndrome affected the preschool child under age 5 to a much greater extent than older children and adults, whereas pharyngitis achieved its highest incidence in 5-year-old children, the incidence being less in both younger and older children. These two diagnostic categories will be described below. They had the highest overall consultation rates in general practice. Lower-respiratory-tract infections in general practice affected preschool children to a greater extent than any other age group, but caused a relatively smaller total of consultations than did upper-respiratory-tract infections. A particularly striking age–incidence relationship occurred in the illnesses of children admitted to hospitals, as shown in Figure 17.[86] Bronchitis and bronchiolitis predominated in the admissions of babies under 1 year of age, and bronchiolitis actually caused more admissions of babies less than 6 months old than of those above this age. Speculation concerning the influence of age upon the incidence of respiratory disease has pointed to the lack of immunological experience in babies. Once maternal antibody has been lost, resistance to infection depends upon its acquisition during exposure to the community pool of respiratory viruses. There is no humoral umbrella to restrain the multiplication of virus in the respiratory tract.

Within the family setting, the introduction of viruses by members of the family depends on extrafamilial contacts, but adults, and particularly mothers, are relatively less effective introducers of infection than children. This probably reflects the acquired resistance of adults, who are poorer transmitters of the infecting agents that they may subsequently acquire. However, a contrary view has been expressed by Fox et al.[53] in discussing rhinoviruses, as already mentioned (see page 11). According to Dingle et al.,[78] school children under 6 years of age are most often to blame for introducing the common cold syndrome into the family, next are nursery school attenders, and finally school children over 6 years of age. Clearly, an infant under 3 years of age in a family containing older children is heavily exposed to the agents introduced into the family by the other members.

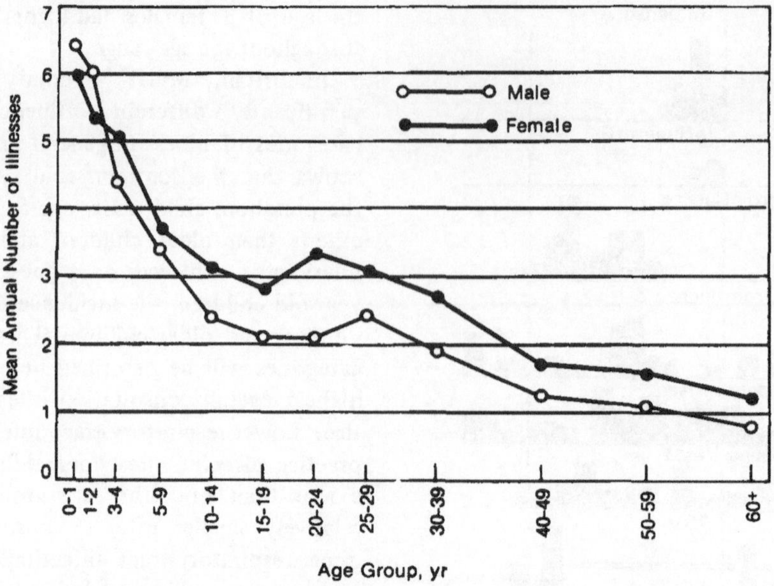

Mean number of respiratory illnesses experienced per year.

FIGURE 15. Incidence of respiratory illnesses as a function of age. (From Monto, A. S. and Ullman, B. M., *J. Am. Med. Assoc.*, 227, 164, 1974. With permission.)

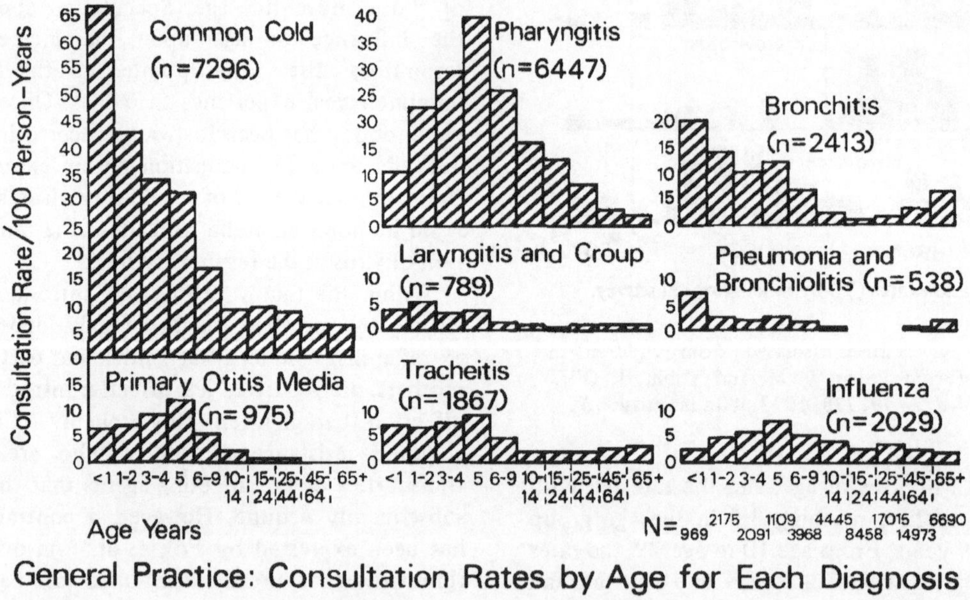

FIGURE 16. Acute respiratory disease syndromes according to age. (From Miller, D. L., *Postgrad. Med. J.*, 49, 749, 1973. With permission.)

When the age–incidence relationship of different virus infections is considered, specific experience also varies. The influenza viruses and the rhinoviruses cause infections at all ages. In the case of the influenza viruses, a greater dominance of overt infection occurs in childhood with influenza B than with influenza A. But even though some influenza A epidemics attack children and adolescents to a greater extent than adults, a different age distribution may occur in a subsequent epi-

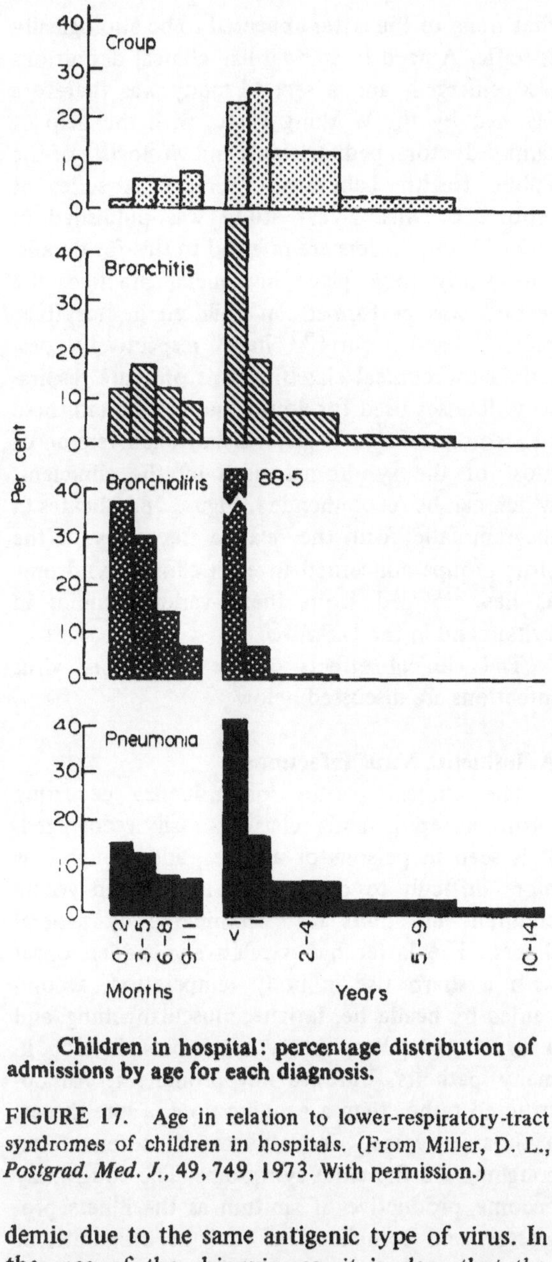

Children in hospital: percentage distribution of admissions by age for each diagnosis.

FIGURE 17. Age in relation to lower-respiratory-tract syndromes of children in hospitals. (From Miller, D. L., *Postgrad. Med. J.*, 49, 749, 1973. With permission.)

demic due to the same antigenic type of virus. In the case of the rhinoviruses, it is clear that the multiplicity of serotypes prevents any simple interpretation of the age–incidence relationship of the whole group. Nevertheless, rhinoviruses cause higher rates of infections in children than in adults, and in babies less than 1 year of age than in older infants.[40] Immunological factors seem likely to be the chief reason for variations of prevalence of infection with the age of the host.

C. Residential Factors

Just as the youngest child in the family is the one most subject to respiratory infections, so the new entrant to a residential school, a service establishment, or any other organization involving residence outside the home has the highest rate of respiratory infections. The resistance exhibited by older members of the community has long been known as "seasoning," a factor totally unconnected with climate. Studies on influenza and acute respiratory disease in military groups, carried out in Britain by the author[98] and in the U.S.A. by the Commission on Acute Respiratory Disease,[99] confirmed the importance of these infections to recruits. A period of 6 months of residence in a military group resulted in a sharp decrease in the rate of febrile respiratory infections, including pharyngitis. In due course,[100] much of the phenomenon in the U.S. military groups was shown to be due to adenovirus infections, the resident viruses in the community. These may have persisted in latent fashion without causing overt disease, but able to renew outbreaks of illness as each new batch of recruits arrived.

However, the phenomenon of the high incidence of disease in recruits is not confined to adenovirus infections. It existed in the epidemic of influenza in 1937 at Chatham,[101] and it is also characteristic of respiratory disease in military groups caused by Coxsackievirus A21,[30] an agent that persists longer in the pharynx than most other respiratory viruses. Experimental study involving change of the composition of a group by adding fresh susceptibles to the older residents was first explored by Greenwood et al.,[102] using *Salmonella typhimurium* in groups of mice. After the initial wave of mortality due to this infection, periodic additions of previously unexposed mice enabled the infection to persist within the herd and to cause recurrent outbreaks. The analogy to respiratory disease due to certain viruses seems close. Its explanation is believed to be immunological, at least in part.

Residence also has another important feature; it appears to enhance the incidence of acute respiratory infection compared with the rate experienced in a similar population living at home. Influenza, in fact, is the classic example of a disease that may cause explosive outbreaks with a high attack rate in residential schools at a time when only sporadic disease occurs in the community. This was striking in the first outbreaks in England of the then "new" antigenic variant of influenza A1 (H1N1) in 1947. Outbreaks in military camps of 20% and more were observed in the north of England, though the general population experienced little evidence of an epidemic.[98] In the winter of 1968–1969, the first attack by the Hong Kong variant of influenza A (H3N2) caused epidemics with rates of illness ranging from 20 to 80% in residential schools,[103] whereas the community as a whole had a minor, though protracted, outbreak. The fact is that the combination of bringing people together with frequent contacts during waking hours and a dormitory existence at night has an enhancing effect on influenza and other respiratory viruses, which is separate from immunological factors. It seems probable that this enhancement is in some way connected with airborne transmission, yet there is no simple explanation for the explosive effect that is witnessed.

V. CLINICAL AND VIROLOGICAL CORRELATIONS

When one views the respiratory infections in the natural setting of the home, place of work, or hospital, one encounters difficulties in relating viruses to human events. On the one hand, there is a multiplicity of viruses, some of which are technically difficult to study; on the other hand, clinical events in the community fall into a relatively small number of diagnostic categories. The lack of readily recognizable diagnostic criteria and the essential need to classify illnesses by syndromes rather than by anatomically affected areas confuse the clinician and have hampered correlations with causative viruses. Therefore, the need for organized special collaborative studies to compare laboratory, clinical, and epidemiological findings has become apparent.

The pioneer family studies by Dingle et al.[78] occurred before the full range of respiratory viruses was known, but they may have inspired the later studies of Fox and his co-workers in New York[104] and Seattle[105] families. In Tecumseh, Michigan, the late Thomas Francis inspired family studies by the School of Public Health, Ann Arbor, with a similar purpose.[91] In Britain, the first collaborative plan of the Medical Research Council's Working Party on Acute Respiratory Virus Infections was a 3-year study in general practice from 1961 to 1964.[106] This established the fact that the various virus groups were all concerned with each of the clinical syndromes, so that none of the latter appeared to be etiologically specific. A need to standardize clinical definitions also emerged, and a second study was therefore planned by the Working Party, with the help of family doctors, pediatricians, and virologists of the Public Health Laboratory Service. A series of reports of this 2-year study was published in 1973,[85] and readers are referred to this for details. One study took place in general practice, the second was performed on children in hospitals. Miller[86] and Court[107] have, respectively, described the clinical classifications of acute respiratory illnesses used for adults and children in these two studies. Table 3 gives a shortened version of most of the syndromes used by the clinicians, which can be recommended. Figure 18 indicates in diagrammatic form the relative frequency of the virus groups concerned in each clinical syndrome as has emerged from these various studies in Britain and in the U.S.A.

The clinical effects of the respiratory virus infections are discussed below.

A. Influenza Virus Infections

The clinical picture of influenza occurring during an epidemic is relatively easily recognized. It is seen in persons of all ages, and though it is more difficult to diagnose in infants and young children, in adults it is dominated by general illness. The latter has a relatively sudden onset with a sharp rise in body temperature, accompanied by headache, fatigue, muscular aching, and a dry cough. Respiratory symptoms develop in many patients, but are not prominent. An obstructed rather than a running nose, a sore throat, a hoarse voice, and substernal soreness on coughing are the chief symptoms. The cough may become productive of sputum as the illness progresses, and complications such as bronchitis or pneumonia occur subsequently. They are more common in persons with previous chest disease than in healthy people. The aged, who may succumb to influenza, often do so without extensive involvement of the respiratory tract. Bacterial complications causing pneumonia, however, are the principal overall cause of death, occurring predominately in persons over 55 years of age. Influenza in children is usually febrile, but often with few symptoms beyond a sore throat and a cough. Objective signs of abnormality in both children and adults are scant in uncomplicated

TABLE 3

Classification of Acute Respiratory Disease by Syndromes

I. Upper-respiratory-tract infections

1. Influenza
   Sudden onset, fever, malaise, headache, and cough
2. Common cold
   Nasal discharge, sneezing, and nasal blockage. Most common colds are apyrexial or with a temperature of less than 99.5°F. Febrile common cold is a variant.
3. Sore throat
   Pharyngitis is indicated by generalized erythema of pharynx and posterior nasal airway. Tonsillitis may be localized or share with general pharyngitis. Exudate may or may not be present.
4. Laryngitis and croup
   Hoarseness or loss of voice occur in both. Croup in infants and children is a form of obstructed breathing, causing a noise (stridor), a crowing cough, and hoarse voice. General illness is likely in croup.
5. Acute otitis media
   Pain in the ear, deafness, bulging drum, perhaps perforation with aural discharge. Usually secondary to other respiratory syndromes, occasionally primary.

II. Lower-respiratory-tract infections

1. Bronchitis
   Cough, wheezing, breathlessness, and fever. Groaning ronchi on auscultation. Sputum expectoration (mucoid or mucopurulent).
2. Bronchiolitis
   Cough, breathlessness, and evidence of distended chest. In children, excessive air in the lungs appears as translucency in chest radiographs. In adults as well as in children, cyanosis is due to lack of oxygen (anoxemia).
3. Pneumonia
   Fever, pleural pain, breathlessness, cough, and expectoration. Evidence is required of consolidation of the lungs clinically and/or of opacification of the lung fields in chest radiographs. Lobar, lobular (bronchial), and atypical varieties.

(Syndrome of "tracheitis" is not recommended for use)

illnesses. These signs last acutely from 2 to 5 days, and convalescence is usually complete in 7 to 10 days.

There are no reliable clinical criteria for distinguishing influenza A from influenza B. School outbreaks of either type of influenza are indistinguishable, and individual illnesses are of equal severity. In the general population, however, there is less prevalence of influenza B in adults, and this has given rise to the impression that influenza B is a less severe disease. Certainly its mortality is less than that of influenza A, doubtless because fewer persons of middle age or older are affected. Influenza B may also present as a common-cold-like illness with profuse nasal discharge, but usually fever in addition. The proportion of cases of "typical" influenza and common cold varies greatly. Influenza C causes an undistinctive mild respiratory illness resembling influenza.

Deliberate infection of volunteers with influenza A viruses cultivated in the laboratory has clearly shown that laboratory cultivation can result in attenuation of the virulence and infectivity of the virus so that it no longer produces fever and such general systemic effects as those obtained with freshly-isolated virus. The mild effects of attenuated viruses resemble the minor symptoms experienced by many adults during an epidemic, but in the latter case the benign character of the infection is due to partial immunity of the host. Serological studies of community groups before and after an outbreak of either influenza A or B show that infection occurs silently and without symptoms in persons of all ages. The exact ratio of subclinical to clinical infections undoubtedly varies; estimates have ranged from 1:3 to 1:1. In both influenza A and influenza B epidemics, the highest rate of infections, as measured by serological methods, occurs in children and adolescents.[108] It has usually been found that, when a strain of virus A with a "new" H or N antigen appears, the rate of infections is greatest in subjects 5 to 15 years of age, though all ages are ultimately affected.

The clinical syndromes shown in Figure 18 show that the influenza viruses are associated with

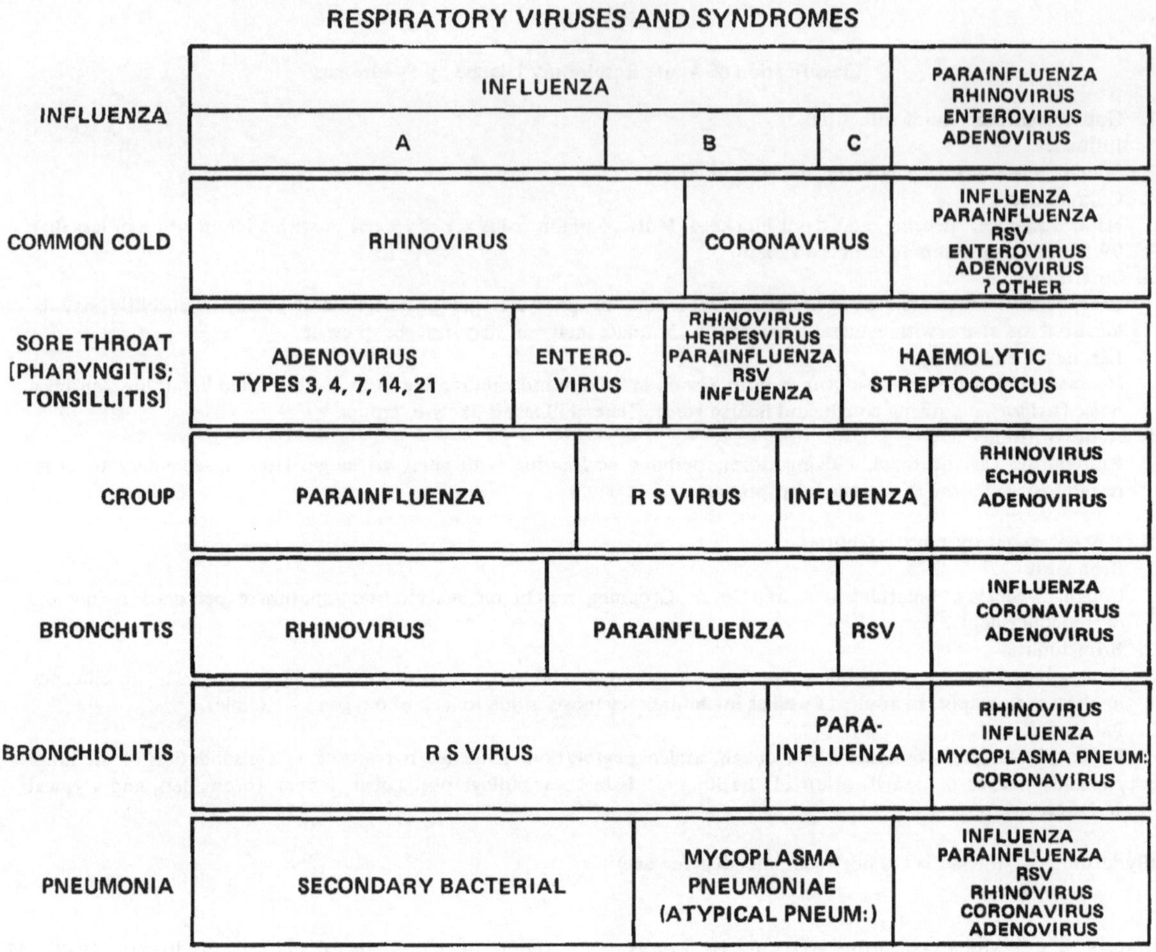

FIGURE 18. Approximate proportional representation of respiratory viruses and syndromes, shown in diagrammatic form.

each of the clinical syndromes, but occur in the greatest proportion in clinical influenza. Lower-respiratory-tract syndromes are very important, because they result in a need for adequate medical care. Complications involving the chest are relatively the most numerous; fortunately, those of the cardiac or nervous systems are infrequent. Tables 4 and 5, as well as a paper by Grist,[109] show the percentage of influenza A and influenza B viruses recovered from specimens in relation to syndromes during the British M.R.C./P.H.L.S. study in children and adults.

B. Parainfluenza Virus Infections

Both parainfluenza and RS viruses chiefly cause acute clinical illnesses in young children. The larynx, the bronchi, and the bronchioles are most involved, but the upper respiratory tract is also affected. The relationship of parainfluenza viruses to croup and laryngitis is well established, type 1 serotype being the culprit more often than the other serotypes. In both conditions the voice is hoarse or lost. Croup in infants causes noisy breathing (stridor), due to inflammatory swelling of the vocal cords, which narrows the air passage. This leads to obstruction, increased inspiratory efforts, and recession of the chest. Because of this, oxygenation is reduced, leading to cyanosis, and there is restlessness.

When the bronchi are involved by parainfluenza virus infection, there is fever, cough, increased rate of breathing, and audible respiration. Characteristic groaning or bubbling signs are heard in the chest. This is the picture of acute bronchitis. In adults, parainfluenza viruses are not commonly the cause of bronchitis, except in acute exacerbations of symptoms in patients with chronic bronchitis or emphysema.[110,111] Increased sputum and dyspnea result. In children, the alveoli may also be affected, with resultant pneumonia.

Parainfluenza viruses have been found throughout the world, and the World Health Organization studies[112,113] indicate that they are frequent causes of acute respiratory illnesses in children. Clarke[114] recorded the parainfluenza viruses recovered from patients with various illnesses during the British M.R.C./P.H.L.S. studies (Tables 4 and 5) and found that upper-respiratory-tract syndromes occurred as frequently as those involving the chest.

Volunteers inoculated with cultured virus[115,116] develop colds. Resistance to infection depends on the presence of virus-neutralizing antibodies in nasal secretion.[117] Serum antibodies alone are unable to prevent infection or reinfection. The evidence does not suggest that parainfluenza viruses, under normal circumstances, commonly cause colds in adults.[118,119] Community studies show that discrete periods of prevalence of the different serotypes occur during the winter season.[120]

C. Respiratory Syncytial Virus Infections

A characteristic illness known as acute bronchiolitis of infants, seen during winter outbreaks in many countries, is due to RS virus infection.[121] In babies less than 1 year of age, often only 3 to 6 months old, there is a sudden onset of rapid, distressed breathing. General illness, often without fever, as well as cyanosis, reduced oxygen uptake, and a clear or excessively translucent chest radiograph are found. The chest is distended, the rib spaces recede on inspiration because the fine air passages are blocked by the infection, and great respiratory effort is required. Complete recovery is usual, but the possible effects of bronchiolitis on the subsequent health of children are not fully known. RS virus has been identified as a cause in some cases of the sudden-infant-death syndrome ("cot death") and of pneumonia in children.[122,123]

In spite of the emphasis on the lower respiratory tract, Gardner[124] found that only 95 of the 256 viruses isolated during the British M.R.C./P.H.L.S. studies were linked with cases of bronchiolitis, and 61 were from cases described as pneumonia. The remainder included 60 from cases of bronchitis, 34 from colds, and 6 from cases of croup (Table 5). Nearly all the viruses came from infants and children up to 14 years of age; there were few recoveries from adult respiratory illnesses.

Deliberate infection of adult volunteers by Kravetz et al.[125] resulted in minor colds. Hamre and Procknow[126] also recovered RS viruses from young adults with common colds, but only in 4% of instances, and Stenhouse[127] obtained serological evidence of RS infection in acute illnesses in patients with chronic bronchitis. Resistance to infection in adults was correlated by Mills et al.[128] with high levels of neutralizing antibody in nasal secretions. Serum antibodies alone did not prevent infection of volunteers, as in the similar case of parainfluenza virus infection. In Seattle, Cooney et al.[129] found evidence of silent RS infection in older children and adults.

Studies of the pathogenesis of RS infection were intensified by the disastrous experience of children given inactivated RS virus vaccine in the U.S.A. Kim et al.[130] found that subsequent exposure to natural infection suggested that immunization had resulted in an apparent sensitizing effect. Vaccine-induced serum antibody was actually blamed for leading to bronchiolitis, perhaps by an interaction with virus antigen produced in the lung.[131] Chanock et al.[132] suggest that maternal antibody, passively acquired by the baby, is an immunological mediator of disease, but Gardner et al.[133] prefer to believe in a type 1 reagin-induced allergic response. Whatever may prove to be the true explanation of bronchiolitis, its restricted age range and the preventive effect of respiratory secretory antibody suggest that immunization will prove difficult to effect. Effective chemotherapy for the treatment of this condition would thus be welcomed.

D. Enterovirus Infection

There is little to add to the foregoing comments concerning enteroviruses (see page 10). Poole and Tobin[88] emphasized that, though members of this group of viruses are often recovered from healthy children, three times as many recoveries were made during the M.R.C./P.H.L.S. studies from children with respiratory symptoms than from controls. The most common clinical syndrome was one of acute sore throat, classified as pharyngitis or tonsillitis. Because adenoviruses are more often the cause of this syndrome than enteroviruses, its clinical description is given below. In the

TABLE 4

Relationship of Viruses to Clinical Syndrome[a]

Clinical syndromes	Influenza		Common cold		Pharyngitis and tonsilitis		Laryngitis and croup		Bronchitis		Bronchiolitis and pneumonia	
Number of specimens	(478)		(985)		(1377)		(191)		(302)		(163)	
Virus group	Number of viruses recovered	% of viruses recovered	Number of viruses recovered	% of viruses recovered	Number of viruses recovered	% of viruses recovered	Number of viruses recovered	% of viruses recovered	Number of viruses recovered	% of viruses recovered	Number of viruses recovered	% of viruses recovered
Influenza	94	19.7	19	1.9	11	0.8	4	2.1	8	2.6	5	3.1
Parainfluenza	22	4.6	57	5.8	38	2.7	23	12.0	17	5.6	10	6.1
RS	5	1.0	4	0.4	2	0.15	3	1.5	3	0.9	17	10.4
Rhinoviruses	15	3.0	62	6.3	38	2.7	32	16.7	15	5.0	9	5.6
Adenoviruses	5	1.0	27	2.7	39	2.8	7	3.7	9	3.0	6	3.7
Herpesviruses	6	1.3	9	0.9	35	2.5	6	3.2	2	0.6	4	2.4

[a] After the general practice M.R.C./P.H.L.S. study of children and adults of all ages.[85]

TABLE 5

Relationship of Viruses to Clinical Syndromes[a]

Clinical syndromes	Influenza		Common cold		Pharyngitis and tonsillitis		Croup		Bronchitis		Bronchiolitis		Pneumonia	
Number of specimens	(14)		(232)		(501)		(230)		(453)		(303)		(543)	
Virus group	Number of viruses recovered	% of viruses recovered	Number of viruses recovered	% of viruses recovered	Number of viruses recovered	% of viruses recovered	Number of viruses recovered	% of viruses recovered	Number of viruses recovered	% of viruses recovered	Number of viruses recovered	% of viruses recovered	Number of viruses recovered	% of viruses recovered
Influenza	6	42.7	6	2.6	11	2.2	3	1.3	7	1.5	4	1.3	9	1.7
Parainfluenza	1	7.0	13	5.6	30	6.0	50	21.7	15	3.3	9	3.0	16	2.9
RS	0	0	15	6.5	13	26.0	6	2.6	60	13.2	95	31.3	61	11.2
Rhinoviruses	0	0	7	3.0	6	1.2	5	2.2	11	2.4	4	1.3	11	2.0
Adenoviruses	0	0	7	3.0	42	8.4	3	1.3	7	1.5	0	0	19	3.5
Herpesvirus	1	7.0	5	2.2	26	5.2	1	0.4	9	2.0	3	1.0	13	2.4

[a]After the M.R.C./P.H.L.S. study of children up to 14 years of age in hospitals.[85]

M.R.C./P.H.L.S. study it was found that enteroviruses were isolated more commonly from infants 6 months to 2 years of age than from older children.

In the illnesses associated with Coxsackie A21 virus, fever, sneezing, nasal discharge, cough, pharyngitis, and myalgia were present both in British patients[134] and in American patients.[135] Volunteers develop respiratory symptoms, and in some cases abdominal symptoms, after intranasal infection.[136,137] Comments have already been made above on the symptoms in volunteers given respiratory echoviruses.

E. Rhinovirus Infections

The common cold syndrome, with which rhinoviruses are strongly associated, is as familiar to lay persons as to medical men. Sneezing, watery nasal discharge, a cough, and dry throat begin either suddenly or gradually. A mild degree of discomfort or slight elevation of body temperature develop, and the nasal discharge becomes abundant, thick, yellowish, and even frankly purulent. The cough may continue or increase in patients with previous chest disease; such patients also suffer increased sputum production. The nasal sinuses fill with secretion and do not empty properly. Children may suffer an earache, or even mild otitis media. There may be hoarseness of the voice. All degrees from mild to severe symptoms occur, including a variant with fever in excess of 38°C (febrile common cold). In 7 to 10 days the symptoms clear, leaving few after-effects. Thus, the condition is annoying and uncomfortable, but not life-threatening. Its repetitions throughout life make it one of the illnesses that, at present, have to be borne rather than prevented, but its consequences for those with chronic chest or heart disease can be serious. Even ordinary bacterial (pneumococcal) pneumonia may follow in its train.

Rhinoviruses are rarely isolated from healthy persons, except in cases of incubating infection. Inapparent infections are estimated to be one third as frequent as those that are clinically obvious.[40] Higgins[138] analyzed the relationship of virus isolations to clinical syndromes in the British M.R.C./P.H.L.S. studies. Tables 4 and 5 show that in hospitalized children as well as general-practice patients rhinoviruses were recovered from those whose illnesses were classified as pharyngitis, laryngitis and croup, bronchitis, bronchiolitis and pneumonia, or common cold. In general practice some illnesses were even specified as influenza. The overall frequency of isolation was greater in general-practice patients (4.4%) than in hospitalized children (1.8%). This overall rate of recovery is certainly an underestimation of the extent of rhinovirus infection, probably because of the technical difficulties of virus isolation. Fox et al.[53] estimate that among families in Seattle rhinoviruses caused 16.2% of respiratory illnesses in children under 5 years of age, and 16.5% of mothers' and fathers' illnesses. Lower rates were noted in school children and in adolescents (12.3%).

Volunteers inoculated with rhinovirus cultures usually develop mild common colds.[96] The inoculation of volunteers with a rhinovirus culture by means of a small-particle aerosol, which was inspired deeply,[139] produced tracheobronchitis. The responsibility of rhinoviruses for lower-respiratory-tract illnesses in ordinary life is difficult to estimate, though it is greater in children than in adults. In children with asthma, rhinoviruses may precipitate an asthma attack,[140,141] and in adults with chronic bronchitis, exacerbations of chest symptoms follow similar infection often with little evidence of upper-respiratory-tract involvement.[127]

F. Coronavirus Infections

Relatively few studies have been made from which it is possible to deduce the breadth of clinical response to coronavirus infection. Bradburne et al.,[142] working with volunteers, compared the clinical effects of the 229E virus with those of the British organ-cultured B814 strain, and contrasted both with those of two types of rhinoviruses. All viruses reproduced the common-cold syndrome in a proportion of the volunteers, and there were no essential differences between the two coronaviruses. The principal difference between coronavirus and rhinovirus infections was that a greater proportion of general symptoms and more abundant nasal discharge occurred in the former. In a group of children resident for several years in Atlanta, Kaye et al.[143] identified serological responses to the 229E virus in 93 instances. Of these, 49 occurred without reported illness, and the remaining 44 were associated with sore throat, cough, nasal discharge, and fever. Children under 10 years of age and those from 10 to 14 years of age contributed about the same proportion of serological responses. Some of the latter occurred despite preexisting antibody.

In a group of 34 younger children, less than 18 months of age and admitted to hospital because of illness, McIntosh et al.[144] found serological evidence of infection in 20 patients with pneumonia and in 10 patients with bronchiolitis, with the remaining 4 cases exhibiting tracheobronchitis and bronchitis. These authors considered that coronaviruses were important causes of lower-respiratory-tract illness in childhood. Monto and Lim[145] also found evidence of winter outbreaks of coronavirus infection among Tecumseh families, and these occurred in alternate years. The authors noted that reinfections were common both in children and in adults, and that exacerbations of chronic bronchitis could occur in the latter. Serological methods alone may be inadequate for a full study of the effects of the coronaviruses, because of the antigenic diversity exhibited by the latter. At least two antigens must be used.[58]

G. Adenovirus Infections

The patterns of infection caused by the adenoviruses were first described in military groups (see page 13). The account of epidemic acute respiratory disease in wartime by the Commission on Acute Respiratory Disease[146] fits that of more recent experiences in servicemen in the U.S.A. and elsewhere. Trainee (recruit) epidemics described in the U.S.A. from 1966 to 1971 were principally caused by types 4 and 7, but types 3, 14, and 21 were also involved.[147] At any particular camp, one type has had a tendency to persist and to cause recurrent outbreaks, though sometimes joined by a second type. The principal clinical manifestation was a sore throat, such as will now be described.

There is gradual onset of illness, resembling a mild cold; then the throat becomes acutely sore on swallowing; there is fever, headache, and malaise, some degree of nasal discharge, and a cough. Later the voice becomes hoarse; the cough is irritating, substernally painful, and productive, and there may be chest involvement. On examination, the tonsils are red and swollen, and the pharynx is diffusely red, particularly if the tonsils have been removed. There may be a white exudate on the tonsils of flocculent or mucoid appearance, and this may suggest hemolytic streptococcal infection. If the cervical lymph nodes beneath the angle of the jaws are enlarged and tender, the clinical resemblance of the two conditions is close and can only be resolved by cultivation of a throat swab for streptococci. Children with tonsillitis from either cause may vomit at the onset. They may later complain of earache and become deaf from extension of the infection to the middle ear.

Adenovirus infection involving the lung has been described in military groups and in babies. X-ray opacities are found, though physical signs are slight, and *Mycoplasma* atypical pneumonia may be suspected.[148] Lang et al.[149] described 25 cases of adenovirus pneumonia due to type 21 with serious pulmonary sequelae in Polynesian babies, 3 to 18 months of age, in Auckland, New Zealand.

The clinical span of adenovirus infections in the community is much broader than that glimpsed from outbreaks in the military, in residential schools, or in children infected in swimming baths; victims in all of these suffer from pharyngitis, perhaps with conjunctivitis. In the British M.R.C./P.H.L.S. study, adenoviruses were recovered from 2 and 4% respectively of the patients in general practice and in hospitals.[150] No adenoviruses were recovered from hospitalized children with lower-respiratory-tract involvement. However, in general practice, upper- and lower-respiratory syndromes were equally concerned. Tables 4 and 5 show the findings in the M.R.C./P.H.L.S. studies. In Washington, D.C., a series of 18,000 children with illnesses studied as inpatients or as outpatients at the Children's Hospital yielded adenoviruses more commonly than a comparable group in Britain.[151] In 11% of babies and children with respiratory illnesses, adenoviruses were found by throat or anal swab, and 6.7% of children with non-respiratory illnesses also yielded the viruses. Perhaps the use of anal swabs may have increased the numbers of isolations, because types 1, 2, 3, and 5 predominated, and types 1, 2, and 5 may be found in the stools of healthy children. The clinical illnesses observed ranged from common colds to pneumonia.

The epidemic of adenovirus type 7 in England and Wales and 1972 and 1973 was studied by the Public Health Laboratory Service.[152] In babies under 1 year of age, gastrointestinal illnesses were at least half as frequent as respiratory ones, but the latter predominated in infants and older children. Pyrexia, conjunctivitis, and pharyngitis were commonly found. Some neurological effects were also seen.

The use of adenovirus vaccines in military groups is beyond the scope of this chapter. Top et al.[153] have shown that an oral live-adenovirus vaccine containing types 4 and 7 will reduce the prevalence of the homologous viruses. Chanock et al.[154] described the preparation and use of vaccine in enteric-coated capsules. In spite of the absence of evidence of virus spread in servicemen, Stanley and Jackson[155] found that the vaccine virus spread from a vaccinated subject to an unvaccinated consort. Many problems concerning safety arise in connection with such a live vaccine, and a safer inactivated preparation, though sought, has yet to be evolved.

H. Herpesvirus Infection

Herpes simplex virus type 1 was first associated with respiratory disease by Morgan and Finland,[156] who recovered virus from the lung of a patient with atypical pneumonia. In 1966, Herout et al.[157] examined autopsy material for evidence of eosinophilic inclusion bodies. They found these in a 13-month-old baby originally diagnosed as a case of pneumonia associated with measles, but who died with purulent tracheobronchitis. Not only were there eosinophilic inclusions in nuclei of epithelial cells of the trachea, but herpesvirus was recovered from lung suspensions inoculated into tissue cultures, rabbits, and hens' eggs. In the same year, Albanese et al.[158] recovered a strain of herpesvirus from nasal washings from a child with mild upper-respiratory-tract disease. This strain failed to induce colds in volunteers inoculated with cultured virus even in large doses. The authors[158] concluded that the virus was unrelated to the child's respiratory symptoms.

These two studies underline the difficulty in accepting this common virus as a pathogenic agent of respiratory illnesses. The virus was recovered from 1% of the children serving as controls during the British M.R.C./P.H.L.S. studies.[88] It was also found in 1 to 2% of throat swabs from children or adults with respiratory disease. Volunteers inoculated by Lindgren et al.[159] with various respiratory viruses, including Coxsackievirus type A21 and rhinoviruses, yielded herpesvirus in 5% of subsequent mouth, nasal, or throat swabs, thus showing the ability of true respiratory virus infections to activate latent herpesvirus in the person concerned. In spite of this, it must be accepted that primary infection with herpesvirus may sometimes be encountered in adults and that in them it may produce an ulcerative condition of the pharynx, tonsils, and mouth.[160] Such illnesses would undoubtedly be confused with pharyngitis or tonsillitis caused by adenoviruses. Tables 4 and 5 give the findings in the M.R.C./P.H.L.S. studies.

VI. PROSPECTS FOR THE PREVENTION OF RESPIRATORY INFECTIONS

A. Immunization

Of the possible ways to prevent infection, that of immunization has until recently seemed to offer the best hope. Now that the complexity of the etiology of respiratory virus infections has been revealed, however, the prospects for prevention by immunization have undoubtedly receded. Immunization succeeds best when the natural infection against which protection is desired creates long-lasting immunity and when the antigenic structure of the virus concerned in the disease is stable. Neither of these conditions holds for the multiplicity of respiratory virus infections. Indeed, one must begin by pointing to the lack of feasibility of immunization against more than a small percentage of even rhinovirus infections, because of the enormous number of serotypes of these viruses. Similarly, the enteroviruses, with the exception of the polioviruses, are antigenically too diverse for practical immunization, and in any case, the respiratory infections caused by the Coxsackie viruses and Echoviruses are relatively unimportant. Much thought has been given to the immunization of children against the parainfluenza and RS viruses, and some practical consequences have followed. Inactivated polyvalent vaccines have been prepared and have been shown to produce an antibody response in the serum.[161-163] There has been little enthusiasm for such vaccines in recent years, and this may be related to doubts concerning the ability of inactivated vaccines to trigger an immunological response in nasal secretory immunoglobulin,[117] such as occurs in natural infections.

Efforts to devise an inactivated monovalent vaccine against RS virus, which exhibits little antigenic diversity, were checked by the ill-fated experiences of apparently enhanced illness in babies thus immunized, who were later exposed to natural infection.[130] Attenuated strains of RS virus (ts mutants) have been developed by Gharpure et al.[164] Thus far, these have been tried in only small numbers of children and infants, and they have sometimes provoked unwelcome clinical reactions in the youngest subjects.[165] The problem of immunizing babies against an infection experienced within a few weeks of birth is formidable and as yet unsolved.

The adenoviruses have already been shown to be chiefly important as causes of epidemics in institutional groups of adults and children. The exploitation in military groups of an oral adenovirus vaccine contained in a capsule, which prevents contact with the oropharyngeal mucosa, has been successful in the U.S.A.[166-168] The ability of certain adenoviruses to produce tumors in experimental animals has checked the use of live vaccines in the community, and the alternative of inactivated whole-virus vaccines does not appear to have been developed since studies were made in the early 1960s. Possible use of subunit adenovirus vaccines has been opened up by developments in the purification of antigenic components.[169] But even if such preparations are effective, their use in man would probably be limited to special population groups.

The prevention of influenza by means of inactivated or live attenuated virus vaccines thus remains the most important achievement of immunization against respiratory virus infections. It would be inappropriate to discuss this subject in detail in this chapter, and readers must refer to accounts such as those of Stuart-Harris and Schild.[5] Suffice it to say that both types of vaccines have been used widely in many countries, though persons subject to special risk from influenza rather than the general community have been immunized. Where mass immunization of the adult community has been practiced, there has been little evidence of control of spread of epidemic disease. The problem of the immunization of children with influenza vaccines remains largely unsolved, because inactivated whole-virus vaccines provoke fever in infants, and live viruses attenuated for adults cause clinical illness in children. Thus, influenza vaccines remain only a partial answer to the problem of the containment of influenza. Still, they do afford temporary protection of individuals against the disease, provided that an alteration in the antigens of the epidemic viruses is matched by a change in the formula of the vaccines. This dictum holds both for inactivated and for live vaccines, though some immunity results from use of an antigenically heterologous strain in the vaccine, except when the virus encountered in nature belongs to a different subtype.

B. Chemoprophylaxis

It is likely that chemotherapeutic compounds active against the respiratory viruses will be more effective when used prophylactically than therapeutically. This is the case for amantadine against influenza A virus both in experimental situations and in man. In the latter, certainly, disorders that follow a short period of incubation from the acquisition of infection give little time for the operation of a chemotherapeutic compound. Prophylaxis, on the other hand, though effective in the abortion of infection, necessitates prolonged exhibition of the drug. Such prolonged use of any chemical compound not only enhances the risk of toxic effects in man, but may further the emergence of viruses with an altered sensitivity to the drug as a result of propagation in partially protected persons. Virus infections that spread in any particular community for a relatively short period of time – for instance, the influenza viruses – are also more appropriate subjects for chemoprophylaxis than those due to agents that cause infections throughout an entire season.

It may be appropriate to rank the respiratory viruses according to the relative need for development of active chemotherapeutic compounds. The rhinoviruses take first place in such an order for two reasons – because of the frequency with which their infections cause illnesses, and because the multiplicity of their serotypes renders immunization an academic subject rather than a practical possibility.

Next in order come the influenza viruses, not because immunization is unsuccessful, but because it has serious limitations. It is impossible to visualize that immunization of entire populations will ever be successful against a virus with such remarkable power to change its antigens. On the other hand, chemoprophylaxis possesses no such drawbacks in relation to the antigens at least of type A virus. A second major need for effective chemotherapy against influenza is due to the ability of the virus to cause fatal illnesses and epidemics, with resulting disturbance to community life on a grand scale.

Third come the childhood infections with parainfluenza and RS viruses. They rank lower than the first two, because they possess fewer antigenic subtypes, and immunization against them may ultimately prove successful. Though most infections by these viruses are benign, the toll of ultimate damage to the child's respiratory tract is unknown and may in some individuals be serious. Chemoprophylaxis in small babies could, nevertheless, be difficult to initiate and sustain, because of the prolonged periods of prevalence of the viruses in the community.

The adenoviruses rank fourth, on the basis of the relative infrequence of their infections with clinical illness. The tantalizing possibilities of immunization with new antigenic materials also argue against the urgency of a need for chemoprophylaxis. Yet, because these viruses cause epidemics in institutional communities, their infections might be regarded as ideal subjects for chemoprophylaxis. It would certainly be far easier to test a candidate drug against epidemic adenovirus infection in localized communities than against many other respiratory viruses with wider prevalence in the general community.

Finally, there seems little point in attempting to rank the coronaviruses and the enteroviruses on the order of need. Immunization appears as yet impracticable, and the precise order of magnitude of their role as respiratory pathogens is far from clear; it is, therefore, better to reserve judgement. Perhaps it would have been wiser indeed not to attempt this exercise in the first place, but it is hoped that it may be viewed as an epilog to the foregoing sections.

REFERENCES

1. **Andrewes, Sir C. and Pereira, H. G.**, *Viruses of Vertebrates*, 3rd ed., Ballière Tindall, London, 1972.
2. **Dowdle, W. R., Galphin, J. C., Coleman, M. T., and Schild, G. C.**, A simple double immunodiffusion test for typing influenza viruses, *Bull. W.H.O.*, 51, 213, 1974.
3. **Almeida, J. D. and Brand, C. M.**, A morphological study of the internal component of influenza virus, *J. Gen. Virol.*, 27, 313, 1975.
4. W.H.O. Report, A revised system of nomenclature for influenza viruses, *Bull. W.H.O.*, 45, 119, 1971.
5. **Stuart-Harris, C. H. and Schild, G. C.**, *Influenza – the Viruses and the Disease*, Edward Arnold Publ., London, 1976.
6. **Chakraverty, P.**, Antigenic relationships between the neuraminidases of influenza B virus, *Bull. W.H.O.*, 46, 473, 1972.
7. **Schild, G. C., Pereira, M. S., Chakraverty, P., Coleman, M. T., Dowdle, W. R., and Chang, W. K.**, Antigenic variants of influenza B virus, *Br. Med. J.*, 4, 127, 1973.
8. **Curry, R. L., Brown, J. D., Baker, F. A., and Hobson, D.**, Serological studies with purified neuraminidase antigens of influenza B virus, *J. Hyg.* (Cambridge), 72, 197, 1974.
9. **Laver, W. G. and Webster, R. G.**, Studies on the origin of pandemic influenza. III. Evidence implicating duck and equine influenza viruses as possible progenitors of the Hong Kong strain of human influenza, *Virology*, 51, 383, 1973.
10. **Schild, G. C., Pereira, H. G., and Schettler, C. H.**, Neuraminidase in avian influenza A virus antigenically related to that of human A0 and A1 subtypes, *Nature*, 222, 1299, 1969.
11. **Vogel, J. and Shelokov, A.**, Adsorption-haemagglutination test for influenza virus in monkey kidney tissue culture, *Science*, 126, 358, 1957.
12. **Compans, R. W. and Caliguiri, L. A.**, Isolation and properties of an RNA polymerase from influenza virus-infected cells, *J. Virol.*, 11, 441, 1973.
13. **Bishop, D. H. L., Obijeski, J. F., and Simpson, R. W.**, Transcription of the influenza ribonucleic acid genome by a virion polymerase, *J. Virol.*, 8, 66, 1971.
14. **Dourmashkin, R. R. and Tyrrell, D. A. J.**, Electron-microscopic observations on the entry of influenza virus into susceptible cells, *J. Gen. Virol.*, 24, 129, 1974.
15. **Mills, J., van Kirk, J., Hill, D. A., and Chanock, R. M.**, Evaluation of influenza virus mutants for possible use in a live-virus vaccine, *Bull. W.H.O.*, 41, 599, 1969.
16. **Beare, A. S. and Hall, T. S.**, Recombinant influenza A viruses as live vaccines for man, *Lancet*, 2, 1271, 1971.
17. **Lecomte, J., Boudreault, A., and Pavilanis, V.**, Selection and in vitro properties of inhibitor-non-sensitive influenza (H3N2) strains, *Can. J. Microbiol.*, 20, 491, 1974.
18. **Cook, M. K., Andrews, B. E., Fox, H. H., Turner, H. C., James, W. D., and Chanock, R. M.**, Antigenic relationships among the "newer" myxoviruses (parainfluenza), *Am. J. Hyg.*, 69, 250, 1959.

19. **Chanock, R. M. and Coates, H. V.**, Myxoviruses – a comparative description, in *Newcastle Disease Virus, An Evolving Pathogen*, Hanson, R. P., Ed., University of Wisconsin Press, Madison, 1964, 279.
20. **Craighead, J. E. and Brennan, B. J.**, Cytopathic effects of parainfluenza virus type 3 in organ cultures of human respiratory tract tissue, *Am. J. Pathol.*, 52, 287, 1968.
21. **Chanock, R. M., Parrott, R. H., Johnson, K. M., Kapikian, A. Z., and Bell, J. A.**, Myxoviruses: parainfluenza, *Am. Rev. Respir. Dis.*, 88(2), 152, 1963.
22. **Stark, J. E., Heath, R. B., and Peto, S.**, A study of the antibodies against parainfluenza viruses in children's sera, *Arch. Gesamte Virusforsch.*, 14, 160, 1964.
23. **Gardner, P. S. and McQuillin, J.**, *Rapid Virus Diagnosis and Application of Immunofluorescence*, Butterworth, London, 1974.
24. **Coates, H. V., Alling, D. W., and Chanock, R. M.**, An antigenic analysis of respiratory syncytial virus isolates by a plaque-reduction neutralization test, *Am. J. Epidemiol.*, 83, 299, 1966.
25. **Joncas, J., Berthiaume, L., and Pavilanis, V.**, The structure of the respiratory syncytial virus, *Virology*, 38, 493, 1969.
26. **Rosenquist, B. D.**, Isolation of respiratory syncytial virus from calves with acute respiratory disease, *J. Infect. Dis.*, 130, 177, 1974.
27. **Kogon, A., Spigland, I., Frothingham, T. E., Elveback, L., Williams, C., Hall, C. E., and Fox, J. P.**, The virus watch program in metropolitan New York families. VII. Observations on viral excretion, seroimmunity, intrafamilial spread and illness association in Coxsackie and Echovirus infections, *Am. J. Epidemiol.*, 89, 51, 1969.
28. **Vernon, E.**, Coxsackievirus infections in the United Kingdom, 1958–62, *Mon. Bull. Min. Health Public Health Lab. Serv.*, 23, 210, 1964.
29. **Lennette, E. H., Fox, V. L., Schmidt, N. J., and Culver, J. O.**, The Coe virus: an apparently new virus recovered from patients with mild respiratory disease, *Am. J. Hyg.*, 68, 272, 1958.
30. **Johnson, K. M., Bloom, H. H., Mufson, M. A., and Chanock, R. M.**, Acute respiratory disease associated with Coxsackie A-21 virus infection. I. *J. Am. Med. Assoc.*, 179, 112, 1962.
31. **Buckland, F. E., Bynoe, M. L., Rosen, L., and Tyrell, D. A. J.**, Inoculation of human volunteers with Echovirus type 20, *Br. Med. J.*, 1, 397, 1961.
32. **Saliba, G. S., Franklin, S. L., and Jackson, G. G.**, Echo-11 as a respiratory virus: quantitation of infection in man, *J. Clin. Invest.*, 47, 1303, 1968.
33. **Peckham, C. S.**, Echovirus infections in England and Wales, 1959–63, *Mon. Bull. Min. Health Public Health Lab. Serv.*, 23, 217, 1964.
34. **Kono, R., Sasagawa, A., Ishii, K., Ochi, M., Sugiura, S., Matsumiya, H., Uchida, Y., Kameyama, K., Kaneko, M., and Sakurai, N.**, Pandemic of new type of conjunctivitis, *Lancet*, 1, 1191, 1972.
35. **Higgins, P. G. and Scott, R. J. D.**, The isolation of enteroviruses from cases of acute conjunctivitis, *J. Clin. Pathol.*, 26, 706, 1973.
36. **Kapikian, A. Z., Almeida, J. D., and Stott, E. J.**, Immune electron microscopy of rhinoviruses, *J. Virol.*, 10, 142, 1972.
37. **Kapikian, A. Z. (Chairman)**, Collaborative study report. Rhinoviruses: a numbering system, *Nature*, 213, 761, 1967.
38. **Pelon, W.**, Classification of the "2060" virus as Echo 28 and further study of its properties, *Am. J. Hyg.*, 73, 36, 1961.
39. **Kapikian, A. Z. (Chairman)**, Collaborative study report. Rhinoviruses: extension of the numbering system, *Virology*, 43, 524, 1971.
40. **Gwaltney, J. M., Jr.**, Medical reviews: Rhinoviruses, *Yale J. Biol. Med.*, 48, 17, 1975.
41. **Minor, T. E., Dick, E. C., Peterson, J. A., and Docherty, D. E.**, Failure of naturally acquired Rhinovirus infections to produce temporal immunity to heterologous serotypes, *Infect. Immun.*, 10, 1192, 1974.
42. **Hayflick, L. and Moorhead, P. S.**, The serial cultivation of human diploid cell strains, *Exp. Cell Res.*, 25, 585, 1961.
43. **Tyrrell, D. A. J. and Bynoe, M. L.**, Cultivation of a novel type of common cold virus in organ cultures, *Br. Med. J.*, 1, 1467, 1965.
44. **Tyrrell, D. A. J. and Blamire, C. J.**, Improvements in a method of growing respiratory viruses in organ cultures, *Br. J. Exp. Pathol.*, 48, 217, 1967.
45. **Tyrell, D. A. J.**, Rhinoviruses, in *Monographs in Virology*, Vol. 2, Springer-Verlag, Vienna, 1968, 67.
46. **Koliais, S. I. and Dimmock, N. J.**, Rhinovirus RNA polymerase: products and kinetics of appearance in human diploid cells, *J. Virol.*, 14, 1035, 1974.
47. **Sethi, S. K. and Schwerdt, C. E.**, Studies on the biosynthesis and characterization of Rhinovirus ribonucleic acid, *Virology*, 48, 221, 1972.
48. **Dimmock, N. J.**, Biophysical studies of a rhinovirus. Extraction and assay of infectious ribonucleic acid, *Nature*, 209, 790, 1966.
49. **Stott, E. J. and Killington, R. A.**, Rhinoviruses, *Annu. Rev. Microbiol.*, 26, 503, 1972.
50. **Hamparian, V. V., Conant, R. M., and Thomas, D. C.**, Rhinovirus Reference Laboratory Report to National Institute of Allergy and Infectious Diseases, Washington, D. C., 1970.
51. **Monto, A. S. and Cavallaro, J. J.**, The Tecumseh study of respiratory illness. IV. Prevalence of Rhinovirus serotypes, 1966–1969, *Am. J. Epidemiol.*, 96, 352, 1972.

52. **Gwaltney, J. M., Jr., Hendley, J. O., Simon, G., and Jordan, W. S., Jr.**, Rhinovirus infections in an industrial population. 1. The occurrence of illness, *N. Engl. J. Med.*, 275, 1261, 1966.
53. **Fox, J. P., Cooney, M. K., and Hall, C. E.**, The Seattle virus watch. V. Epidemiologic observations of rhinovirus infections, 1965–1969, in families with young children, *Am. J. Epidemiol.*, 101, 122, 1975.
54. **Hamre, D. and Procknow, J. J.**, A new virus isolated from the human respiratory tract, *Proc. Soc. Exp. Biol. Med.*, 121, 190, 1966.
55. **Almeida, J. D. and Tyrrell, D. A. J.**, The morphology of three previously uncharacterized human respiratory viruses that grow in organ cultures, *J. Gen. Virol.*, 1, 175, 1967.
56. **McIntosh, K., Kapikian, A. Z., Hardison, K. A., Hartley, J. W., and Chanock, R. M.**, Antigenic relationships among the coronaviruses of man and between human and animal coronaviruses, *J. Immunol.*, 102, 1109, 1969.
57. **McIntosh, K., Kapikian A. Z., Turner, H. C., Hartley, J. W., Parrott, R. H., and Chanock, R. M.**, Seroepidemiologic studies of coronavirus infection in adults and children, *Am. J. Epidemiol.*, 91, 585, 1970.
58. **Bradburne, A. F. and Somerset, B. A.**, Coronavirus antibody titres in sera of healthy adults and experimentally infected volunteers, *J. Hyg.* (Cambridge), 70, 235, 1972.
59. **Kaye, H. S. and Dowdle, W. R.**, Seroepidemiologic survey of coronavirus (strain 229 E) infections in a population of children, *Am. J. Epidemiol.*, 101, 238, 1975.
60. **Norrby, E.**, The structural and functional diversity of adenovirus capsid components, *J. Gen. Virol.*, 5, 221, 1969.
61. **Wadell, G.**, *Structural and Biological Properties of Capsid Components of Human Adenoviruses*, David Brobey Publ., Stockholm, 1970.
62. **Boyer, G. S., Leuchtenberger, C. and Ginsberg, H. S.**, Cytological and cytochemical studies of HeLa cells infected with adenoviruses, *J. Exp. Med.*, 105, 195, 1957.
63. **Craighead, J. E.**, Cytopathology of adenoviruses types 7 and 12 in human respiratory epithelium, *Lab. Invest.*, 22, 553, 1970.
64. **Rowe, W. P., Baum, S. G., Pugh, W. E., and Hoggan, M. D.**, Studies of adenovirus SV_{40} hybrid viruses. I. Assay system and further evidence for hybridization, *J. Exp. Med.*, 122, 943, 1965.
65. **Rowe, W. P. and Baum, S. G.**, Studies of adenovirus SV_{40} hybrids. II. Defectiveness of the hybrid particles, *J. Exp Med.* 122, 955, 1965.
66. **Morgan, C., Rosenkranz, H. S., and Mednis, B.**, Structure and development of viruses as observed in the electron microscope. X. Entry and uncoating of Adenovirus, *J. Virol.*, 4, 777, 1969.
67. **Wigand, R. and Schmieder, J.**, Inhibition of Adenovirus multiplication by metabolic inhibitors, *Arch. Gesamte Virusforsch.*, 42, 324, 1973.
68. **Russell, W. C. and Becker, Y.**, A maturation factor for Adenovirus, *Virology*, 35, 18, 1968.
69. **Ginsberg, H. S.**, Biological and physical properties of the adenoviruses, *Ann. N.Y. Acad. Sci*, 67, 383, 1957.
70. **Burnett, J. P. and Harrington, J. A.**, Infectivity associated with simian adenovirus Type SA7 DNA, *Nature*, 220, 1245, 1968.
71. **Nicolson, M. O. and McAllister, R. M.**, Infectivity of human Adenovirus-1 DNA, *Virology*, 48, 14, 1972.
72. **Pereira, H. G., Allison, A. C., and Niven, J. S. F.**, Fatal infection of new-born hamsters by an adenovirus of human origin, *Nature*, 196, 244, 1962.
73. **Connor, J. D.**, A generalised infection of new-born hamsters with type 2 Adenovirus, *Proc. Soc. Exp. Biol. Med.*, 133, 655, 1970.
74. **Black, P. H.**, The oncogenic DNA viruses: a review of in vitro transformation studies, *Annu. Rev. Microbiol.*, 22, 391, 1968.
75. **Ross, J. G. and Potter, C. W.**, A possible causal factor in intussusception in infancy, *Lancet*, 1, 81, 1961.
76. **Knox, E. G., Court, S. D. M., and Gardner, P. S.**, Aetiology of intussusception in children, *Br. Med. J.*, 2, 692, 1962.
77. **Fox, J. P., Brandt, C. D., Wassermann, F. E., Hall, C. E., Spigland, I., Kogon, A., and Elveback, L. R.**, The virus watch program in New York families. VI. Observations of adenovirus infections, virus excretion patterns, antibody response, efficiency of surveillance, patterns of infection and relation to illness, *Am. J. Epidemiol.*, 89, 25, 1969.
78. **Dingle, J. H., Badger, G. F., and Jordan, W. S., Jr.**, *Illness in the Home*, Press of Case Western Reserve University, Cleveland, 1964.
79. **Miyamoto, K. and Morgan, C.**, Structure and development of viruses as observed in the Electron Microscope. XI. Entry and uncoating of herpes simplex virus, *J. Virol.*, 8, 910, 1971.
80. **Nahmias, A. J. and Roizman, B.**, Infection with herpes simplex viruses 1 and 2, *N. Engl. J. Med.*, 289, 667, 719, 781, 1973.
81. **Bastian, F. O., Rabson, A. S., Yee, C. L. and Tralka, T. S.**, *Herpes hominis* virus: isolation from human trigeminal ganglion, *Science*, 178, 306, 1972.
82. **Baringer, J. R.**, Recovery of herpes simplex virus from human sacral ganglions, *N. Engl. J. Med.*, 291, 828, 1974.
83. **Hope-Simpson, R. E.**, Common upper respiratory diseases, *R. Soc. Health J.*, 78, 593, 1958.
84. **Sutton, R. N. P., Lane, C. A., Stenhouse, A. C., Schild, G. S., and Oxford, J. S.**, A family study of respiratory illness, *J. R. Coll. Gen. Pract.*, 14, 32, 1967.
85. Symposium on Acute Respiratory Virus Diseases, *Postgrad. Med. J.*, 49, 748, 1973.
86. **Miller, D. L.**, Collaborative studies of acute respiratory disease in patients seen in general practice and in children admitted to hospital. Aims, field methods and morbidity rates, *Postgrad. Med. J.*, 49, 749, 1973.

87. **Hope-Simpson, R. E. and Higgins, P. G.,** A respiratory virus study in Great Britain, in *Progress in Medical Virology,* Vol. 11, Melnick, J. L., Ed., Phiebig, White Plains, N.Y., 1969, 354.
88. **Poole, P. M. and Tobin, J. O'H.,** Viral and epidemiological findings in M.R.C./P.H.L.S. surveys of respiratory disease in hospital and general practice, *Postgrad. Med. J.,* 49, 778, 1973.
89. **Andrewes, C. H.,** Adventures among viruses. III. The puzzle of the common cold, *N. Engl. J. Med.,* 242, 235, 1950.
90. **Dowling, H. F., Jackson, G. G., Spiesman, I. G., and Inouye, T.,** Transmission of the common cold to volunteers under controlled conditions. III. Effect of chilling of the subjects upon susceptibility, *Am. J. Hyg.,* 66, 59, 1958.
91. **Monto, A. S., Cavallaro, J. J., and Keller, J. B.,** Seasonal patterns of acute infection in Tecumseh, Michigan, *Arch. Environ. Health,* 21, 408, 1970.
92. **Sutton, R. N. P.,** Acute respiratory disease and climate, in *Biometeorology II,* Tromp, S. W. and Weihe, W. H., Eds., Pergamon Press, Oxford, 1966, 41.
93. **Monto, A. S. and Johnson, K. M.,** A community study of respiratory infections in the tropics, *Am. J. Epidemiol.,* 86, 78, 1967.
94. **Hemmes, J. H., Winkler, K. C., and Kool, S. M.,** Virus survival as a seasonal factor in influenza and poliomyelitis, *Nature,* 188, 430, 1960.
95. **Buckland, F. E. and Tyrrell, D. A. J.,** Loss of infectivity on drying various viruses, *Nature,* 195, 1063, 1962.
96. **Tyrrell, D. A. J.,** in *Common Colds and Related Diseases,* Williams & Wilkins, Baltimore (Edward Arnold, London), 1965, 39.
97. **Monto, A. S. and Ullman, B. M.,** Acute respiratory illness in an American community, *J. Am. Med. Assoc.,* 227, 164, 1974.
98. **Stuart-Harris, C. H.,** in *Influenza and Other Virus Infections of the Respiratory Tract,* 1st ed., Williams & Wilkins, Baltimore (Edward Arnold, London), 1953, 121.
99. Commission on Acute Respiratory Disease, Acute respiratory disease among new recruits, *Am. J. Public Health,* 36, 439, 1946.
100. **Ginsberg, H. S., Badger, G. F., Dingle, J. H., Jordan, W. S., Jr., and Katz, S.,** Etiologic relationship of the R1-67 agent to "acute respiratory disease (ARD)", *J. Clin. Invest.,* 34, 820, 1955.
101. **Stuart-Harris, C. H., Andrewes, C. H., and Smith, W.,** A Study of Epidemic Influenza with Special Reference to the 1936–37 Epidemic, Medical Research Council Special Report Series, No. 228, H. M. Stationery Office, London, 1938.
102. **Greenwood, M., Hill, A. B., Topley, W. W. C., and Wilson, J.,** Experimental Epidemiology, Medical Research Council Special Report Series., No. 209, H. M. Stationery Office, London, 1936.
103. **Tyrrell, D. A. J., Buckland, R., Rubenstein, D., and Sharpe, D. M.,** Vaccination against Hong Kong influenza in Britain, 1968–69, *J. Hyg.* (Cambridge), 68, 359, 1970.
104. **Fox, J. P., Elveback, L. R., Spigland, I., Frothingham, T. E., Stevens, D. A., and Huger, M.,** The virus watch program: a continuing surveillance of viral infections in metropolitan New York families. The overall plan etc., *Am. J. Epidemiol.,* 83, 389, 1966.
105. **Fox, J. P., Hall, C. E., Cooney, M. K., Luce, R. E., and Kronmal, R. A.,** The Seattle virus watch. II. Objectives, study population, etc., *Am. J. Epidemiol.,* 96, 270, 1972.
106. Report to the Medical Research Council, A collaborative study of the aetiology of acute respiratory infections in Britain, 1961–64, *Br. Med. J.,* 2, 319, 1965.
107. **Court, S. D. M.,** The definition of acute respiratory illnesses in children, *Postgrad. Med. J.,* 49, 771, 1973.
108. **Hall, C. E., Cooney, M. K., and Fox, J. P.,** The Seattle virus watch. IV. Comparative epidemiologic observations of infections with influenza A and B viruses, 1965–1969, in families with young children, *Am. J. Epidemiol.,* 98, 365, 1973.
109. **Grist, N. R.,** Influenza virus infections, *Postgrad. Med. J.,* 49, 807, 1973.
110. **Stark, J. E., Heath, R. B., and Curwen, M. P.,** Infection with influenza and parainfluenza viruses in chronic bronchitis, *Thorax,* 20, 124, 1965.
111. **Stenhouse, A. C.,** Viral antibody levels and clinical status in acute exacerbations of chronic bronchitis: a controlled prospective study, *Br. Med. J.,* 3, 287, 1968.
112. W.H.O. Respiratory Disease Survey in Children, a serological study, *Bull. W.H.O.,* 37, 363, 1967.
113. **Assaad, F. and Cockburn, W. C.,** A seven-year study of W.H.O. virus laboratory reports on respiratory viruses, *Bull. W.H.O.,* 51, 437, 1974.
114. **Clarke, S. K. R.,** Parainfluenza virus infections, *Postgrad. Med. J.,* 49, 792, 1973.
115. **Tyrrell, D. A. J., Bynoe, M. L., Petersen, K. B., Sutton, R. N. P., and Pereira, M. S.,** Inoculation of human volunteers with parainfluenza viruses types 1 and 3 (HA2 and HA1), *Br. Med. J.,* 2, 909, 1959.
116. **Kapikian, A. Z., Chanock, R. M., Reichelderfer, T. E., Ward, T. G., Huebner, R. J., and Bell, J. A.,** Inoculation of human volunteers with parainfluenza virus type 3, *J. Am. Med. Assoc.,* 178, 537, 1961.
117. **Smith, C. B., Purcell, R. H., Bellanti, J. A., and Chanock, R. M.,** Protective effect of antibody to parainfluenza type 1 virus, *N. Engl. J. Med.,* 275, 1145, 1966.
118. **Bloom, H. H., Johnson, K. M., Jacobsen, R., and Chanock, R. M.,** Recovery of parainfluenza viruses from adults with upper respiratory illness, *Am. J. Hyg.,* 74, 50, 1961.
119. **Tyrrell, D. A. J. and Bynoe, M. L.,** Studies on parainfluenza type 2 and 4 viruses obtained from patients with common colds, *Br. Med. J.,* 1, 471, 1969.

120. Monto, A. S., Tecumseh study of respiratory illness. V. Patterns of infection with the parainfluenza viruses, *Am. J. Epidemiol.*, 97, 338, 1973.
121. Parrott, R. H., Kim, H. W., Brandt, C. D., and Chanock, R. M., Respiratory syncytial virus in infants and children, *Prev. Med.*, 3, 473, 1974.
122. Gardner, P. S., Turk, D. C., Aherne, W. A., Bird, T., Holdaway, M. D., and Court, S. D. M., Deaths associated with respiratory-tract infection in childhood, *Br. Med. J.*, 4, 316, 1967.
123. Downham, M. A. P. S., Gardner, P. S., McQuillin, J. J., and Ferris, J. A. J., Role of respiratory viruses in childhood mortality, *Br. Med. J.*, 1, 235, 1975.
124. Gardner, P. S., Respiratory syncytial virus infections, *Postgrad. Med. J.*, 49, 788, 1973.
125. Kravetz, H. M., Knight, V., Chanock, R. M., Morris, J. A., Johnson, K. M., Rifkind, D., and Utz, J. P., Respiratory syncytial virus. III. Production of illness and clinical observations in adult volunteers, *J. Am. Med. Assoc.*, 176, 657, 1961.
126. Hamre, D. and Procknow, J. J., Virologic studies on common colds among young adult medical students, *Am. Rev. Respir. Dis.*, 88, Part 2, 277, 1963.
127. Stenhouse, A. C., Rhinovirus infection in acute exacerbations of chronic bronchitis: a controlled prospective study, *Br. Med. J.*, 3, 461, 1967.
128. Mills, J. V., van Kirk, J. E., Wright, P. F., and Chanock, R. M., Experimental respiratory syncytial virus infection of adults, *J. Immunol.*, 107, 123, 1971.
129. Cooney, M. K., Fox, J. P., and Hall, C. E., The Seattle virus watch. VI. Observations of infections with and illness due to parainfluenza, mumps and respiratory syncytial viruses and *Mycoplasma pneumoniae*, *Am. J. Epidemiol.*, 101, 532, 1975.
130. Kim, H. W., Canchola, J. G., Brandt, C. D., Pyles, G., Chanock, R. M., Jensen, K., and Parrott, R. H., Respiratory syncytial virus disease in infants despite prior administration of antigenic inactivated vaccine, *Am. J. Epidemiol.*, 89, 422, 1969.
131. Kapikian, A. Z., Mitchell, R. H., Chanock, R. M., Shvedoff, R. A., and Stewart, C. E., An epidemiologic study of altered clinical reactivity to respiratory syncytial (RS) virus infection in children previously vaccinated with an inactivated RS virus vaccine, *Am. J. Epidemiol.*, 89, 405, 1969.
132. Chanock, R. M., Kapikian, A. Z., Mills, J., Kim, H. W., and Parrott, R. H., Influence of immunological factors in RS virus disease, *Arch. Environ. Health*, 21, 347, 1970.
133. Gardner, P. S., McQuillin, J., and Court, S. D. M., Speculation on pathogenesis in death from RS virus infection, *Br. Med. J.*, 1, 327, 1970.
134. Pereira, M. S. and Pereira, H. G., COE virus properties and prevalence in Great Britain, *Lancet*, 2, 539, 1959.
135. Bloom, H. H., Johnson, K. M., Mufson, M. A., and Chanock, R. M., Acute respiratory disease associated with Coxsackie A21 virus infection, II, *J. Am. Med. Assoc.*, 179, 120, 1962.
136. Parsons, R., Bynoe, M. L., Pereira, M. S., and Tyrrell, D. A. J., Inoculation of human volunteers with strains of COE virus isolated in Britain, *Br. Med. J.*, 1, 1776, 1960.
137. Lang, D. J., Cate, T. R., Couch, R. B., Knight, V., and Johnson, K. M., Response of volunteers to inoculation with haemagglutinin-positive and haemagglutinin-negative variants of Coxsackie A_{21} virus, *J. Clin. Invest.*, 44, 1125, 1965.
138. Higgins, P. G., Rhinovirus infections, *Postgrad. Med. J.*, 49, 802, 1973.
139. Cate, T. R., Couch, R. B., Fleet, W. F., Griffith, W. R., Gerone, P. J., and Knight, V., Production of tracheobronchitis in volunteers with rhinovirus in a small-particle aerosol, *Am. J. Epidemiol.*, 81, 95, 1965.
140. Minor, T. E., Dick, E. C., De Meo, A. N., Ouellette, J. J., Cohen, M., and Reed, C. E., Viruses as precipitants of asthmatic attacks in children, *J. Am. Med. Assoc.*, 227, 292, 1974.
141. Gregg, I., The role of viral infection in asthma and bronchitis, in *Symposium on Viral Diseases*, Vol. 46, Proudfoot, A. T., Ed., Royal College of Physicians, Edinburgh, 1975, 82.
142. Bradburne, A. F., Bynoe, M. L., and Tyrrell, D. A. J., Effects of a "new" human respiratory virus in volunteers, *Br. Med. J.*, 3, 767, 1967.
143. Kaye, H. S., Marsh, H. B., and Dowdle, W. R., Seroepidemiologic survey of coronavirus (strain OC43) related infections in a children's population, *Am. J. Epidemiol.*, 94, 43, 1971.
144. McIntosh, K., Chao, R. K., Krause, H. E., Wasil, R., Mocega, H. E., and Mufson, M. A., Coronavirus infection in acute lower-respiratory-tract disease of infants, *J. Infect. Dis.*, 130, 502, 1974.
145. Monto, A. S. and Lim, S. K., The Tecumseh study of respiratory illness. VI. Frequency of and relationship between outbreaks of coronavirus infection, *J. Infect. Dis.*, 129, 271, 1974.
146. Commission on Acute Respiratory Diseases, Clinical patterns of undifferentiated and other acute respiratory diseases in army recruits, *Medicine*, 26, 441, 1947.
147. Dudding, B. A., Top, F. H., Jr., Winter, P. E. Buescher, E. L., Lamson, T. H., and Leibowitz, A., Acute respiratory disease in military trainees. The adenovirus surveillance program, 1966–71, *Am. J. Epidemiol.*, 97, 187, 1973.
148. Bryant, R. E. and Rhoades, E. R., Clinical features of adenoviral pneumonia in Air Force recruits, *Am. Rev. Respir. Dis.*, 96, 717, 1967.
149. Lang, W. R., Howden, C. W., Laws, J., and Burton, J. F., Bronchopneumonia with serious sequelae in children and evidence of adenovirus type 21 infection, *Br. Med. J.*, 1, 73, 1969.

150. **Pereira, M. S.**, Adenovirus infections, *Postgrad. Med. J.*, 49, 798, 1973.
151. **Brandt, C. D., Kim, H. W., Vargosko, A. J., Jeffries, B. C., Arrobio, J. O., Rindge, B., Parrott, R. H., and Chanock, R. M.**, Infections in 18,000 infants and children in a controlled study of respiratory-tract disease. 1. Adenovirus pathogenicity in relation to serologic type and illness syndrome, *Am. J. Epidemiol.*, 90, 484, 1969.
152. Public Health Laboratory Service, Epidemiology – Adenovirus Type 7, *Br. Med. J.*, 3, 498, 1973.
153. **Top, F. H., Jr., Dudding, B. A., Russell, P. K., and Buescher, E. L.**, Control of respiratory disease in recruits with types 4 and 7 adenovirus vaccines, *Am. J. Epidemiol.*, 94, 142, 1971.
154. **Chanock, R. M., Gutekunst, R. R., Ludwig, W., Edmondson, W. P., Gundelfinger, B. F. and Kasel, J. A.**, Immunization against type 4 adenovirus by selective enteric infection. in *First International Conference on Vaccines against Viral and Rickettsial Diseases of Man*, Scientific Publication No. 147, Pan American Health Organization, Washington, D.C., 1967, 85.
155. **Stanley, E. D. and Jackson, G. G.**, Spread of enteric live adenovirus type 4 vaccine in married couples, *J. Infect. Dis.*, 119, 51, 1969.
156. **Morgan, H. R. and Finland, M.**, Isolation of herpesvirus from a case of atypical pneumonia and erythema multiforme exudativum with studies of four additional cases, *Am. J. Med. Sci.*, 217, 92, 1949.
157. **Herout, V., Vortel, V., and Vondráčková, A.**, Herpes simplex involvement of the lower respiratory tract, *Am. J. Clin. Pathol.*, 46, 411, 1966.
158. **Albanese, M., Bynoe, M. L., and Tyrrell, D. A. J.**, Studies of a strain of herpesvirus isolated from a case of upper respiratory disease, *Arch. Gesamte Virusforsch.*, 18, 356, 1966.
159. **Lindgren, K. M., Douglas, R. G., Jr., and Couch, R. B.**, Significance of *Herpesvirus hominis* in respiratory secretions of man, *N. Engl. J. Med.*, 278, 517, 1968.
160. **Glezen, W. P., Fernald, G. W., and Lohr, J. A.**, Acute respiratory disease of university students with special reference to the etiologic role of *Herpesvirus hominis*, *Am. J. Epidemiol.*, 101, 111, 1975.
161. **Sweet, B. H., Tytell, A. A., Potash, L., Weibel, R. E., Stokes, J., Jr., Drake, M. E., Woodhour, A. F., and Hilleman, M. R.**, Respiratory virus vaccines. III. Pentavalent respiratory syncytial-parainfluenza- *Mycoplasma pneumoniae* vaccine, *Am. Rev. Respir. Dis.*, 94, 340, 1966.
162. **Woodhour, A. F., Sweet, B. H., Tytell, A. A., Potash, L., Stokes, J., Jr., Weibel, R. E., Metzgar, D. P., and Hilleman, M. R.**, Respiratory virus vaccines. IV. Heptavalent respiratory syncytial-parainfluenza-*Mycoplasma pneumoniae*-influenza vaccine in institutionalized persons, *Am. Rev. Respir. Dis.*, 94, 350, 1966.
163. **Leagus, M. E., Weibel, R. E., Mascoli, C. C., Stokes, J., Jr., and Hilleman, M. R.**, Respiratory virus vaccines. VI. Serologic responses to heptavalent vaccine among young children in families, *Am. Rev. Respir. Dis.*, 95, 838, 1967.
164. **Gharpure, M. A., Wright, P. F., and Chanock, R. M.**, Temperature-sensitive mutants of respiratory syncytial virus, *J. Virol.*, 3, 414, 1969.
165. **Kim, H. W., Arrobio, J. O., Pyles, G., Brandt, C. D., Camargo, E., Chanock, R. M., and Parrott, R. H.**, Clinical and immunological responses of infants and children to administration of low-temperature-adapted respiratory syncytial virus, *Paediatrician*, 48, 745, 1971.
166. **Gutekunst, R. R., White, R. J., Edmondson, W. P., and Chanock, R. M.**, Immunization with live type 4 adenovirus: determination of infectious virus dose and protective effect of enteric infection, *Am. J. Epidemiol.*, 86, 341, 1967.
167. **Top, F. H., Jr., Grossman, R. A., Bartelloni, P. J., Segal, H. E., Dudding, B. A., Russell, P. K., and Buescher, E. L.**, Immunization with live types 7 and 4 adenovirus vaccines, I., *J. Infect. Dis.*, 124, 148, 1971.
168. **Top, F. H., Jr., Buescher, E. L., Bancroft, W. H., and Russell, P. K.**, Immunization with live types 7 and 4 adenovirus vaccines. II. *J. Infect. Dis.*, 124, 155, 1971.
169. **Couch, R. B., Kasel, J. A., Pereira, H. G., Haase, A. T., and Knight, V.**, Induction of immunity in man by crystalline adenovirus type 5 capsid antigens, *Proc. Soc. Exp. Biol. Med.*, 143, 905, 1973.
170. **Melnick, J. L., Tagawa, I., and Von Magnus, H.**, Enteroviruses, 69, 70, and 71. *Intervirol.*, 4, 369, 1974.

Chapter 2

THE BASIS OF SELECTIVITY IN DRUGS

Adrien Albert

TABLE OF CONTENTS

I. INTRODUCTION

Selectivity is the name given to a substance's built-in ability to modify certain kinds of cells without affecting any neighboring cells. Such modification can be made beneficial or adverse at will. When a new candidate drug is in the early stages of design, the provision of adequate selectivity is kept very much in mind.

Those cells that a drug is required to influence adversely are conveniently thought of as the uneconomic species, and those that must be left unchanged are the economic species. In chemotherapy and chemoprophylaxis, the invading parasite is the uneconomic species; the aim is usually to overpower this species at an appropriate time. The alternative goal of strengthening the host's defences, a related facet of selectivity, is more rarely achieved. Immunochemicals (the antigens and antibodies of vaccines and sera, respectively) can be highly discriminating, but nature has not provided them for all diseases. Hence, the subject of selectivity has been investigated mainly among synthetic drugs of low molecular weight, most of them smaller than 400 daltons. A great advantage in working with small molecules is the ease of effecting many stepwise changes in the molecule until the most selective product emerges.

In general, the more toxic an agent is, the more it is valued. This may at first seem surprising, but a high degree of selectivity requires, by definition, a powerfully toxic effect exerted on the uneconomic species; while retaining this effect, the agent should be susceptible to modification so that the toxicity to the host becomes minimal. This favorable ratio is exemplified by the most used of the general and local anaesthetics, chemotherapeutic agents (e.g., antibiotics, sulfonamides, and antimalarials), vermifuges, and insecticides.

The subject of selective toxicity has tremendous scope, embracing not only human and veterinary medicine, but the use of herbicides, fungicides, and insecticides to defend crops. A comprehensive but readable account has been given elsewhere.[1] The present chapter will review some of the major accomplishments of selectivity. It will range over the whole subject and attempt to explain the physical basis on which selectivity rests. This general discussion will be followed by an account of what has been accomplished for virus diseases. Although there have been comparatively few antiviral successes in this field, it is hoped that the more general part can indicate the attitudes and the interdisciplinary studies that have brought success in the wider context. The search for antiviral drugs has now reached the stage that antibacterial chemotherapy had attained in 1940, when only a few effective drugs were available, but several fortunate clues were on hand.

II. THE PHYSICAL AND CHEMICAL PROPERTIES THAT PROVIDE SELECTIVITY

Three principles are recognized as controllers of selectivity in biologically active agents. They may operate singly or, less frequently, in combination. The first principle, *selectivity through distribution*, utilizes substances that are accumulated principally by the uneconomic species. Such agents are often toxic to both economic and uneconomic organisms, but this property does no harm, provided that the agents are taken up principally by the latter organisms.

The second principle, *selectivity through comparative biochemistry*, uses biochemical differences that exist between the two species to injure the uneconomic organisms. The third principle, *selectivity through comparative cytology*, makes similar use of structural differences. This section will illustrate each of these three principles in turn by discussing several examples of selective agents, whose ability to differentiate will be analyzed at the level of molecular biology and the underlying physics and chemistry.

A. Selectivity through Distribution

In the first part of this century, it was discovered in France[2] that dilute (10%) sulfuric acid could safely be sprayed on crops of cereals to destroy the weeds, a result that has been repeatedly confirmed. Although sulfuric acid is as injurious to the cytoplasm of the wheat as to that of the weed, it never reaches the former. There are two helpful circumstances at work here. In the first place, the exterior of cereal plants is smooth and waxy, whereas that of the usual dicotyledenous weeds is rough and free from wax. The second factor is that the tender new shoots of cereals are protected by leaf sheaths, whereas the growing point of dicotyledons is exposed and vulnerable.[3]

In rather similar fashion, phenothiazine (Figure 1), which has long and successfully been used as an oral antihelminth in sheep, is accumulated by intestinal worms, but not by the cells that line the sheep's intestines. Yet, if injected into the sheep, phenothiazine is toxic for *both* species.[4] A truly dramatic example from human medicine is the accumulation of tetracycline antibiotics by bacteria (both Gram-negative and Gram-positive), whereas mammalian cells do not act in this way. As a result, the synthesis of proteins on bacterial ribosomes is drastically inhibited, and the organisms soon die. Yet, when isolated ribosomes are exposed to tetracyclines, there is an identical response, those of the host being just as sensitive as those of the parasite. The difference lies in the cytoplasmic membranes: that of the bacteria actually concentrates the drug against a gradient, whereas that of the host remains almost impermeable to the drug.[5] The net result is that the tetracyclines constitute one of the most selective, and hence safest, families of chemotherapeutic agents.

Even more striking than these examples is the selective partitioning of an agent between the tissues of a single organism. For instance, 5-fluorouracil (Figure 2), which is used by

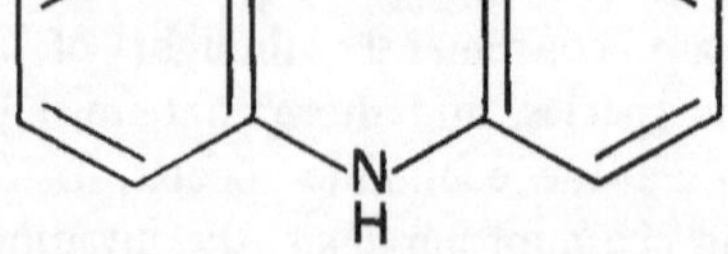

FIGURE 1. Molecular formula of phenothiazine.

dermatologists to destroy all the principal types of skin cancer, is accumulated preferentially by the malignant tissue. So selective is this agent that the patient is allowed to use the bare fingers to rub it daily into the affected area. As a result, the cancerous growth (but this only) becomes inflamed, then disintegrates, and finally is replaced by normal skin.[6] The introduction of this drug followed an observation that malignant tissues accumulate uracil more avidly than do normal tissues.

Examples of selective distribution of inorganic substances will now be given. Some 80% of a minute oral dose of iodine (10^{-12} g), given in radioactive form (^{131}I), was rapidly accumulated by the thyroid. This selective accumulation is used to destroy tumors of the thyroid gland. Similarly, ^{32}P, which is specifically taken up by the bone marrow, is used for treating polycythemia. Brain tumors accumulate about eight times as much organic boron compounds as do their surrounding tissues. When organoboranes enriched in the stable isotope ^{10}B are injected, the malignant tissue can be selectively destroyed by a beam of neutrons for which this isotope shows very strong affinity.[7] Many patients treated in this way have been able to resume their occupations and lead normal lives.

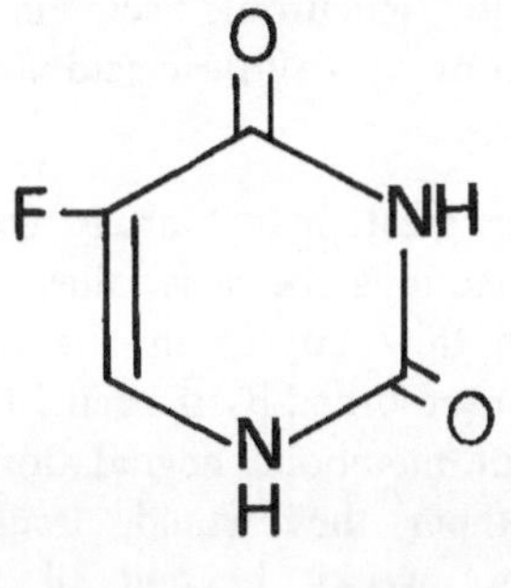

FIGURE 2. Molecular formula of 5-fluorouracil.

1. Qualitative Basis of Distribution

Striking as the above examples may appear, their physical basis becomes understandable only after the factors that govern distribution are enumerated (first qualitatively, and later quantitatively).

A drug, after administration, must cross several selectively permeable membranes to reach its receptor. Thus, the antimalarial chloroquin must penetrate the gastrointestinal membranes, then that of the erythrocyte, and then those of the parasite inside the red cell. Each membrane, because of its restrictive permeability, contributes to the total selectivity. Further discrimination is imposed by the nature of the receptor and by the losses that the biochemistry of the host is able to inflict on the drug during its passage from the site of administration to the receptor.

Figure 3 epitomizes what can happen to a drug in typical compartments (the regions isolated by the selectively permeable membranes) during the journey of the drug to the receptor. Sometimes, for the purpose of sustained release, the drug is given in masked form (often spoken of as the "pro-drug"), which can be converted to the true, active drug by enzymes normally present in the blood stream or tissues. This is that area of our subject where metabolic change leads to activation; however, most drugs are thought to act in the form in which they are administered.

The first pro-drug to be designed was hexamine,

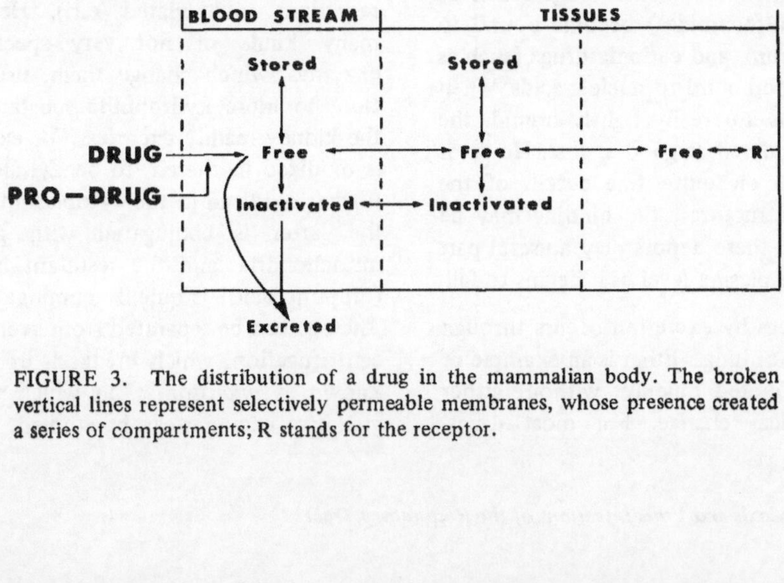

FIGURE 3. The distribution of a drug in the mammalian body. The broken vertical lines represent selectively permeable membranes, whose presence created a series of compartments; R stands for the receptor.

also known as urotropin or methenamine. It is a bland, odorless condensation product of ammonia and formaldehyde; the acidity of the urine liberates the latter, thus confining its pungent action to the region where the antiseptic action is required. This ingeniously conceived (and effective) product was introduced as long ago as 1899. When a drug is excreted (or metabolized) too fast for effective medication, it can often be administered in masked form. For example, testosterone is injected as the phenylpropionic ester dissolved in oil; the ester slowly escapes from its depot in the oil and is hydrolyzed to testosterone by the nonspecific esterases in the blood stream. Another familiar example is penicillin, injected as the insoluble salt that it forms with procaine. One of the most useful remedies for bacterial dysentery is phthaloylsulfathiazole, which has the advantage over sulfathiazole of not being absorbable from the small intestine; however, when this pro-drug has passed on to the colon, it is hydrolyzed by benign bacteria to sulfathiazole, which suppresses the pathogenic organisms.

Masking can also be used to improve absorption. Whereas 6-azauridine (Figure 4) is not absorbed from the intestine, its more liposoluble triacetyl derivative, azaribine, freely enters the blood stream, where esterases reconvert it to 6-azauridine.[8] Azarabine, which has some antiviral properties, is most used as selective medication in generalized psoriasis.[9]

As soon as a drug begins to circulate freely in the blood stream, its concentration falls, due to the operation of three factors: storage, inactivation, and excretion. Liposoluble drugs (such as the barbiturate thiopentone) become stored in deposits of fat, whereas anionic drugs (such as aspirin and the sulfonamides) become bound to the albumin of serum, and cationic drugs (such as chloroquin) become bound to nucleic acids. When these stored forms are really tightly bound, the drug may be considered as good as lost. However (and this depends on quite fine details of the drug's molecular structure), the binding may be loose; in that case, these depots play a useful part in replenishing the plasma level as it begins to fall.

The loss of drugs by excretion occurs through kidneys, bile duct, or lungs. Ether is an example of a drug that is excreted quickly, without either storage or chemical change. For most drugs,

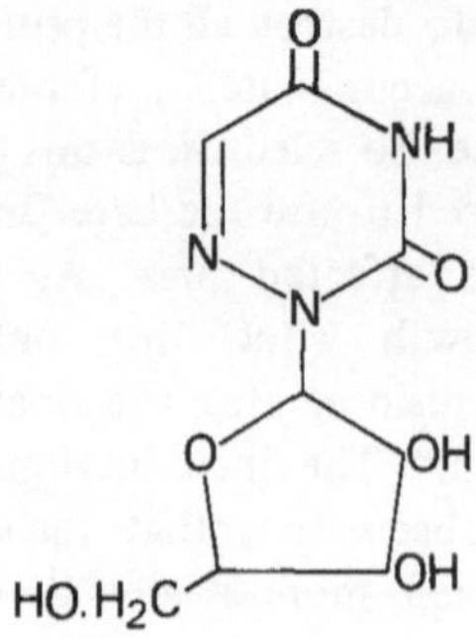

FIGURE 4. Molecular formula of 6-azauridine.

however, excretion is preceded by a metabolic change in which covalent bonds are formed or broken; this process is seldom reversible. For example, organic acids that are too weak to be ionized at the low pH of urine, and therefore would be difficult to eliminate, become conjugated with glycine and are then readily excreted by the kidney. Most amines and phenols are conjugated with glucuronic acid in the liver, kidney, or gut, or with sulfuric acid in the cell sap of several organs.

However, a great many drugs escape these conjugative reactions because they are rather liposoluble. If they appear in the initial renal filtrate, they are resorbed by the renal tubules. If a special form of metabolic degradation were not available for them, they would circulate in the body for many weeks, beyond all therapeutic control. As it happens, their dipophilic nature causes them to accumulate in a membranous organelle of the liver, known as the endoplasmic reticulum (abbreviated e.r.). Here they meet many kinds of not very specific scavenging enzymes, which change them, usually by oxidation, to more hydrophilic substances, and these the kidney readily excretes. For example, toluene is oxidized in the e.r. to benzyl alcohol, which is further oxidized in the cytoplasm to benzoic acid; the latter is conjugated with glycine in the mitochondria, and the resultant benzoyl glycine (hippuric acid) is quickly eliminated in the urine. The e.r. can be separated from liver by differential centrifugation, which breaks it up into spherules, known as *microsomes,* in which the various degradative enzymes can be studied. These enzymes probably arose (in the course of evolution) to deal with toxic substances occurring in food or produced by bacterial decomposition in the gut. Conversely, the e.r. enzymes cannot attack either the raw materials or the products of intermediary metabolism, because such substances are too hydrophilic to penetrate into the e.r. This fact affords us a useful glimpse of how selectivity is achieved in nature.

Metabolic alteration of foreign substances is sometimes referred to as *detoxification,* but many examples are known where the product is more toxic than the substrate. Thus, methanol is transformed by the body to formaldehyde, which can lead to blindness.

After surviving all these hazards of storage, excretion, and inactivation, and also penetration through other membranes into other compartments, a successful drug eventually enters the compartment that contains the receptor with which it unites to produce its characteristic response. Union between drug and receptor is occasionally covalent (penicillin and the insecticidal phosphates and carbamates provide examples), but usually it takes place by secondary (easily reversible) bonds. Thus, because the concentration of a drug steadily falls for the reasons given above, medication must be maintained to keep an effective concentration of the drug near the receptor.

Three striking properties of drugs indicate that they act on small, specific parts of cells, which were named *receptors* by Ehrlich in 1910. These properties are (1) retention of activity at high dilution (e.g., $10^{-9}M$), (2) high *chemical* specificity (even the D- and L-stereoisomers of a single substance often have different pharmacological action), and (3) the high *biological* specificity (e.g., the powerful effect that epinephrine, or adrenaline, exerts on cardiac muscle, yet not on striated muscle). Development of this idea by A. J. Clark and by others has led to the current recognition of a receptor as a small molecule or as part of a larger molecule. In some cases it has turned out to be an enzyme. For the phosphorus insecticides, the target receptor is the enzyme acetylcholinesterase, which becomes phosphorylated on the hydroxy group of the serine moiety.[10] This leads to death of the insect from overdosage of its own accumulated acetylcholine, a neurotransmitter. Similarly, the principal receptor for the antibacterial sulfonamides is an enzyme, dihydrofolate synthetase, whose normal function is to incorporate *p*-aminobenzoic acid into the molecule of dihydrofolic acid, an essential coenzyme for the biosynthesis of purines and thymine, and hence of DNA. Dihydrofolate synthetase was isolated and purified by Brown[11] in 1962.

The receptor for acetylcholine, the neurotransmitter, is not an enzyme, such as acetylcholinesterase, but a permease, i.e., a membrane-sited regulator of the permeability of inorganic ions. Recently isolated, it was found to be a highly lipophilic protein.[12]

Sometimes the receptor is not a protein, but a nucleic acid. The antibacterial aminoacridines, such as aminacrine (Figure 5), need two physical properties in order to develop biological activity:[13] (1) the molecules must carry the kind of substituent that will ensure complete ionization at pH 7; (2) the molecule must be built completely flat, with an area of not less than 39 Å. By obeying these rules, potent analogues were found in the phenanthridine, benzoquinoline, and benzacridine series also, and even in smaller nuclei (such as pyridine and quinoline) if their flat surface area was enlarged by a coplanar substituent.

That aminoacridines are accumulated by the nucleic acid of cells was revealed by their use in vital staining. The necessity for molecular flatness was explained in 1961, when it was shown that aminoacridine molecules became intercalated in DNA by stacking. This occurred between layers of purine/pyrimidine base pairs to which the acridines were attached by van der Waals forces and by ionic bonds to the phosphate anions (explaining the need for ionization). Figure 6 represents this situation diagramatically. The resultant increase in melting temperature showed that intercalation

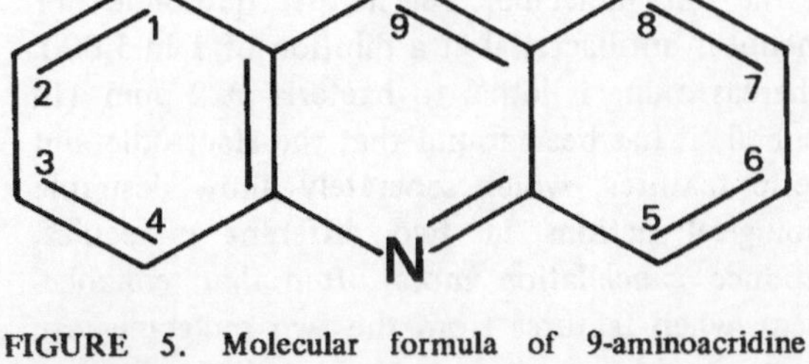

FIGURE 5. Molecular formula of 9-aminoacridine (aminacrine).

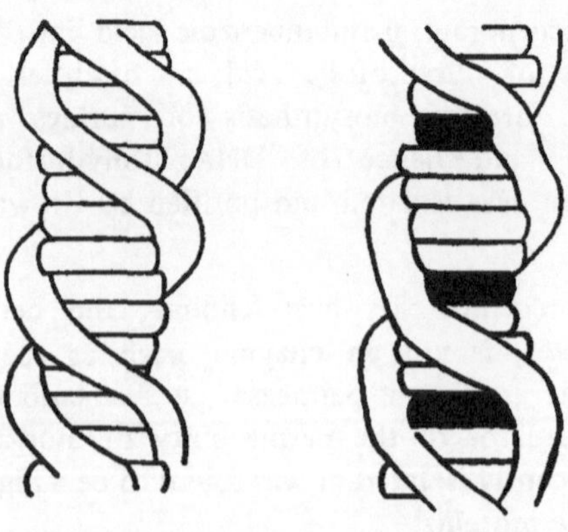

FIGURE 6. Sketches showing the secondary structure of normal DNA (left) and of DNA with intercalated molecules of proflavine (3,6-diaminoacridine). The base pairs and proflavine appear in edgewise projection, and the phosphate/deoxyribose backbone appears as a smooth coil. (From Lerman, L. S., *J. Cell. Comp. Physiol.*, 64 (Suppl. 1), 1, 1964. With permission.)

interfered with the unwinding of the DNA strands and hence with their normal functioning.[14,15]

Given these clues, it did not take long to demonstrate that aminoacridines injure bacteria by blocking the DNA template used by the enzymes synthesizing both DNA and RNA.[16] These studies with aminoacridines established that nucleic acids can be receptors. Here, in fact, a drug–receptor interaction was observed in unusually fine detail at the level of molecular biology. Molecules of DNA are large, yet quite small molecules can be receptors, as the following will show.

A study of the antimicrobial properties of 8-hydroxyquinoline (Figure 7), known for short as oxine, showed that biological activity depended on its ability to chelate, that is, to form polydentate complexes with the cations of heavy metals;[17] for example, the 1:1 ferrous complex is shown in Figure 8. Prior to this, it had been supposed that oxine owed its antibacterial properties to a combination of those of quinoline and phenol, all in the one molecule. Yet neither quinoline nor phenol is antibacterial at a dilution of 1 in 5,000, whereas oxine is lethal to bacteria at 2 ppm. (In general, it has been found that the electronic and steric features, which separately allow desirable biological actions in two different molecules, produce cancellation more often than enhancement when features from the two molecules are compressed into one.)

Once it was proved that oxine acts by chela-

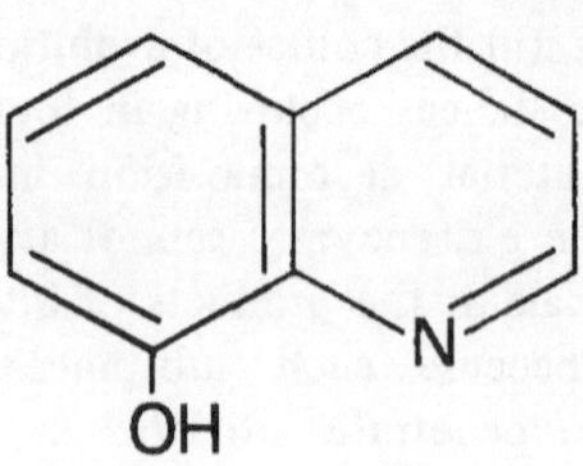

FIGURE 7. Molecular formula of oxine (8-hydroxyquinoline).

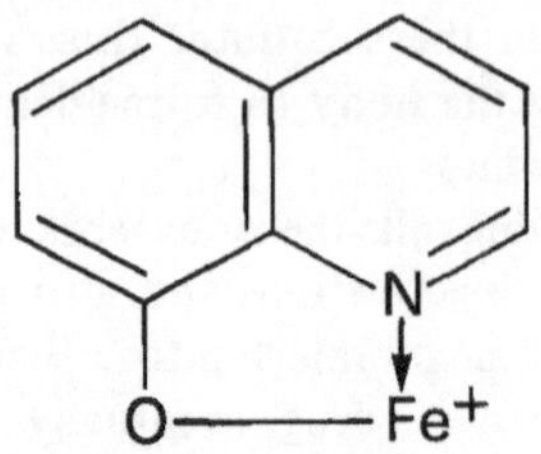

FIGURE 8. Molecular formula of the 1:1 oxine/iron complex.

tion, the following question had to be answered next: "Does it act by removing metals essential to bacterial welfare, or does it cause traces of a contaminating metal to become more toxic to the bacteria?" The latterproved to be the case, as the following results indicated. Staphylococci were completely killed in 1 hr by 0.00001 *M* oxine, but were unharmed by a solution seven times stronger; in fact, even a saturated (0.005 *M*) solution did not kill them.[18] Streptococci behaved similarly. The meaning of this phenomenon became clear when it was found to occur only in media containing traces of iron (for bacteria) or copper (for fungi). The viability of staphylococci for 24 hr in distilled water permitted decisive experiments, summarized in Table 1.

It can be seen at once from Table 1 that oxine (0.00001 *M*) is biologically inert, but becomes bactericidal in the presence of an equal quantity of iron. Clearly, the toxic agent is neither oxine nor iron, but the oxine/iron complex. However, when broth replaced water, no added iron was necessary, because it was abundantly present in the medium. It should be noted that, when the concentration of oxine was increased to 0.0013 *M*, the bactericidal action disappeared. This happens when the iron becomes saturated with oxine (as in the 2:1 ferrous complex); when sufficient extra iron was added to the broth (so that the unsaturated 1:1

TABLE 1

The Innocuousness of Oxine in the Absence of Iron: a Bactericidal Test with *Staphylococcus aureus* (pH 6 to 7; 20°)

Oxine (m*M*)	$FeSO_4$ (m*M*)	Growth on plating out after 1 hr: In glass-distilled water	In untreated meat broth
Nil	Nil	Prolific	Prolific
0.01	Nil	Prolific	Nil
Nil	0.01	Prolific	Prolific
0.01	0.01	Nil	Nil

From Albert, A., Gibson, M. I., and Rubba, S. D., *Br. J. Exp. Pathol.*, 34, 119, 1953. With permission.

complex could be re-formed), full bactericidal properties were restored.[18] It is interesting that, in the absence of external heavy metals, oxine has been found to enter the cells of bacteria and fungi without harming them.

Several other chelating antimicrobials are known that, though they have a totally different structure, reproduce the action of oxine by requiring a variable-valence metal for activity; hence, these agents all show the concentration-quenching described above. One example is dimethyldithiocarbamic acid, whose salts are among the most used of agricultural fungicides. It appears that all these chelated metal complexes act by destroying lipoic acid (Figure 9) in an oxidative chain reaction.[19,20] Lipoic acid (thioctic acid) is the essential coenzyme for the oxidative decarboxylation of pyruvic acid, and accumulation of pyruvic acid has been demonstrated.

In the previous example, the receptor is quite a small molecule: lipoic acid. Another example where a coenzyme functions as a receptor is in the lethal action (e.g., in mammals) of hydrogen cyanide; this result follows directly from the binding of this poison to the free valence of iron in the porphyrin molecule of cytochrome oxidase.

The targets of drugs are not inevitably receptors. Several chelating agents are routinely used in hospitals as antidotes for poisoning by foreign inorganic substances. These antidotes circulate in the blood stream without necessarily depleting the body's own essential metals, a tribute to the strength with which such cations are bound in nature. An example of such an antidote is calcium ethylenediaminetetracetic acid (EDTA), a most effective remedy in lead poisoning.

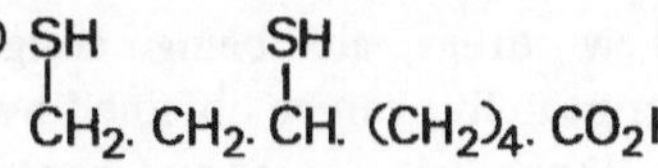

FIGURE 9. Molecular formula of thioctic acid (lipoic acid).

2. *The Permeability of Natural Membranes*

From the foregoing, it is plain that the distribution of a drug (and hence, in the present context, its selectivity) is greatly influenced by the selective qualities of natural membranes. Four main types are recognized. The most common membranes (Type 1) permit the fastest diffusion of those molecules that have some lipophilic groups (e.g., $-CH_3$), few hydrophilic atoms (e.g., oxygen and nitrogen), and no ionic charge. However, an ion can be accepted if the molecule is made more lipophilic.

The pioneer work in this field was done in Finland by the biologist Collander.[21] In recent years, Hansch and his colleagues[22] have published many partition coefficients for the solvent pair octanol/water, to establish, semiquantitatively, the relative lipophilicity of common substituents. The cardiac glycosides afford clinically interesting examples of the effect of partition coefficients on permeability. Digitoxin, the most readily accumulated glycoside in clinical use, is the most lipophilic. Related glycosides that are more hydrophilic (through having extra sugar moieties or extra hydroxyl groups in the steroid nucleus) are relatively poorly retained.[23]

The next most common type of biological membrane (Type 2) incorporates a carrier molecule, which provides what is known as facilitated diffusion. Choline, a quaternary ammonium salt, is permanently ionized, and hence cannot enter a cell by simple diffusion. However, a carrier exists for it, which seems to oscillate between the inner and outer surfaces of the membrane to pick up and then release this molecule. Foreign substances, if modeled on choline (tetramethylammonium salts, for instance), can enter cells by this mechanism. The test for this type of membrane is to show that transport ceases when the carrier is saturated, even though the concentration of the transported substance remains greater *outside* the cell. Type 3 membranes work in a similar manner, but they consume energy, as is known from a detectable increase in metabolic activity during transport. The absorption and excretion of many substances by the kidney tubules is effected in this way.

When new drugs are being designed, the requisite permeability can be obtained by stepwise increases in lipophilicity, to ensure penetration of Type 1 membranes. But greater selectivity can be obtained by making part of the drug resemble a natural substance for which a specific transport mechanism exists in particular cells having the relevant Type 2 or Type 3 membrane. Thus, the penetration of nitrogen-mustard type cancer drugs was improved by attaching various amino acids (of which phenylalanine was the most successful) to take advantage of specific amino acid uptake mechanisms.[24] Analogs of natural sugars, purines, or pyrimidines have been formulated along similar lines.

Finally, there are the less common Type 4 membranes, distinguished by the presence of pores that, as in the glomerular tuft of the kidney, can be as large as 3 nm in diameter; through these even a molecule as large as inulin (5,000 daltons) passes easily. Blood capillaries and liver parenchymal cells have Type 4 membranes.

The following biological structures in mammals offer a Type 1 membrane barrier to the passage of most foreign substances: the intestinal epithelium, the blood/brain barrier, the blood/cerebrospinal fluid interface,[25] and those areas of the renal-tubule epithelium not occupied by Type 3 patches. The high acidity of the stomach converts bases to their cations, which are poorly absorbed, and gastric absorption is correlated with lipophilicity. The central nervous system selectively accumulates lipophilic drugs such as barbiturates; these are at first found in the more vascular gray matter, but are soon transferred to the white regions, which are richer in lipids.

A direct approach to treating respiratory diseases would be to vaporize a solution of the drug as an aerosol. Aerosols consist of particles so small that the force of gravity is effectively opposed by the viscous drag of the particle, and settling is correspondingly retarded. For example, a particle 1 μm in diameter settles at the very slow rate of 0.003 cm/sec. Selectivity in distribution depends very much on particle size. Thus, particles larger than 10 μm in diameter are largely removed by the nasal passages, those about 5 μm in diameter are accumulated mainly by the bronchi and bronchioles, and those 2 μm in size settle principally in the alveolar ducts and sacs. If the diameter is still further decreased, the total accumulation becomes less. Absorption into the blood stream is more efficient from the alveolar sacs than from any other part of the pulmonary tree; hence, these sacs are not only difficult to reach with aerosols, but do not retain for very long the little that they receive.

3. Quantitative Basis of Distribution

All of the processes that govern distribution can be quantified by patient experiment. For each transition denoted by an arrow in Figure 3, a velocity constant can be determined; this branch of the subject is known as pharmacokinetics. Transport, storage, excretion, and inactivation rates are being determined for as many drugs as the specialized workers in this field can handle.

The rate of transfer of a drug across a membrane is described by the differential da/dt, where a is the change in amount (or concentration) effected in a very small interval of time (t). Most pharmacokinetic processes appear to have first-order kinetics. In these, the amount (or concentration) D of a drug remaining on one side of the membrane determines the rate of transfer across the membrane, as follows:

$$-da/dt = K_a D \qquad (1)$$

where

K_a = the absorption (or distribution, or elimination) constant.

For antibacterial sulfonamides, the requisite data are readily obtainable by giving oral doses of any one of these drugs to volunteers, followed by analysis, at intervals, of blood and urine samples. In a typical rate project of this kind (see Figure 10), it was possible to obtain one constant for the storage of the sulfonamide in serum albumin,

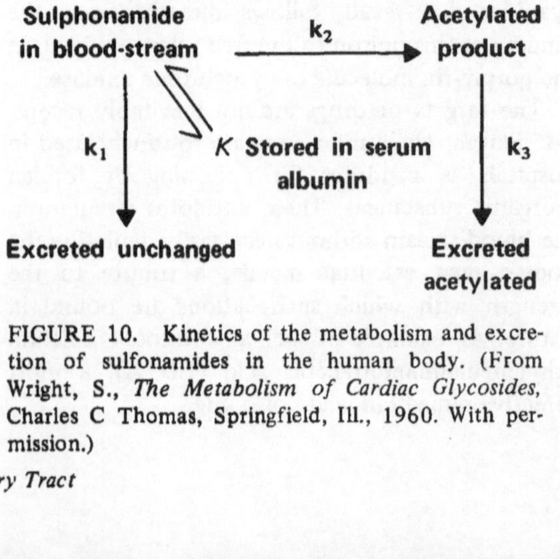

FIGURE 10. Kinetics of the metabolism and excretion of sulfonamides in the human body. (From Wright, S., *The Metabolism of Cardiac Glycosides*, Charles C Thomas, Springfield, Ill., 1960. With permission.)

as well as the release of the unchanges sulfonamide from serum albumin, another forthe renal elimination of the unchanged sulfonamide, another one for its acetylation, and one for the elimination of the acetylated product.[26] Each constant is an independent variable, characteristic of minute changes in the structure of the sulfonamide under examination. Thus, the addition, subtraction, or mere relocation of a (chemically inert) methyl group significantly alters several of these constants.

Similarly, for the many other equilibria denoted in Figure 3, constants are steadily being obtained for both established drugs and candidate drugs. In this way, drug designers are acquiring the data to control distribution, and thereby improve selectivity.

Two other benefits of great importance have come from these studies. First, the half-life ($t_{0.5}$) of the drug in the blood stream is calculated from the elimination constant by using this simple equation:

$$t_{0.5} = 0.693/K_a \quad (2)$$

It represents the time taken for half of the drug to disappear from the circulation. Knowledge of this constant gives the best dose interval, usually made equal to $t_{0.5}$.

Once the dose interval has been found, the size of the dose is chosen as follows. A priming dose is found that quickly reaches the blood level below which no beneficial effect on the patient can be observed; further doses are smaller, yet still large enough to effect a beneficial result. Such simple pharmacokinetic studies prevent the cure being delayed by underdosage or the patient being injured by overdosage.

B. Selectivity Through Comparative Biochemistry

Few differences exist between species in the way nutrients are broken down and the resulting energy is stored, but there are many significant differences in the building up of essential cell constituents and in the breaking down of foreign substances. Enzymes, as well as their substrates and coenzymes, play a large part in maintaining these differences, and so they provide foundations on which the selectivity can be based. Before discussing species differences, it may be noted that the different tissues of a single organism maintain characteristic enzymes to carry out their work. For example, three enzymes (xanthine oxidase, catalase, and D-amino acid oxidase) present in liver and kidney in abundance are absent from muscle. Moreover, arginase occurs only in the liver, alkaline phosphatase only in the intestinal mucosa, 5-nucleotidase in the testis, and α-mannosidase in the epididymis. The blood is disproportionately rich in carbonic anhydrase, and the pancreas in ribonuclease. Glutamine synthetase is abundant in brain and liver, but almost absent from all other human tissues. Large quantitative differences are found between the enzyme contents of different kinds of muscle. Thus, aconitase is much more abundant in heart than in skeletal muscle, but the reverse is true for aldolase.[27] It is evident that a pro-drug could be selectively converted to the active drug by a specific enzyme resident in the target organ.

1. Presence of Analogous Enzymes in Dissimilar Organisms

Many examples are known where a pair of enzymes occurring in two organisms are chemically dissimilar, though carrying out apparently identical functions. These are usually known as analogous enzymes (isoenzymes, a subclass, show smaller differences). Pairs of analogous enzymes are demonstrable by differences in kinetics or electrophoretic behavior or by specificity in their coenzymes, substrates, or inhibitors.

Many selective drugs operate by inhibiting one of a pair of analogous enzymes, but not the other. Thus, rifamycin acts by combining with the protein of bacterial RNA polymerase (the DNA-dependent one), thus bringing synthesis of bacterial RNA to a halt. It was found that it combines strongly (though not covalently) with this protein through its β-chain, a structure absent from the RNA polymerases of higher organisms.[28] Hence, rifamycin exhibits selectivity of the highest order, for it has no action at all on the RNA-synthesizing mechanism of mammals. All details of how the rifamycins combine with the bacterial enzyme are not yet known, but the flat naphthalene ring and several of the free hydroxyl groups are essential. Because rifamycins are expensive, their application is usually reserved for drug-resistant cases (tuberculosis and staphylococcal infections). They have been found free of toxicity to the patient.

The use of antimonials in the treatment of the severe and widespread tropical disease bilharzia (schistosomiasis) is a traditional remedy whose

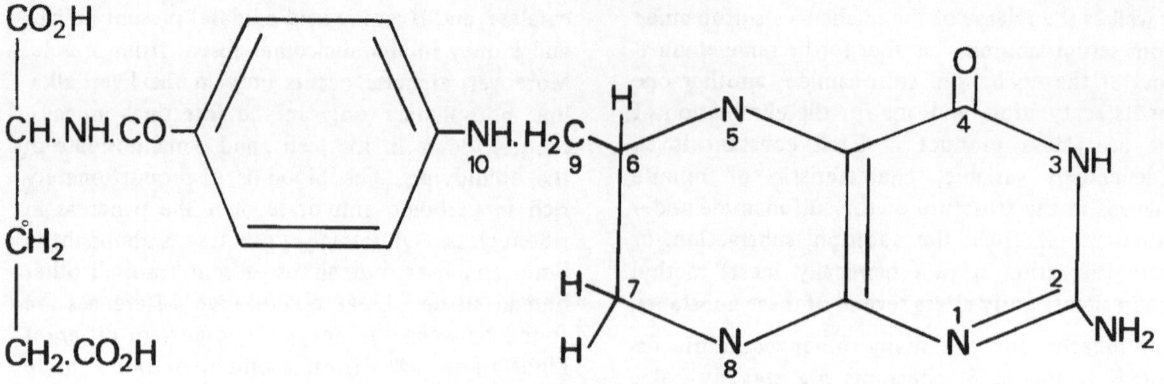

FIGURE 11. Molecular formula of tetrahydrofolic acid.

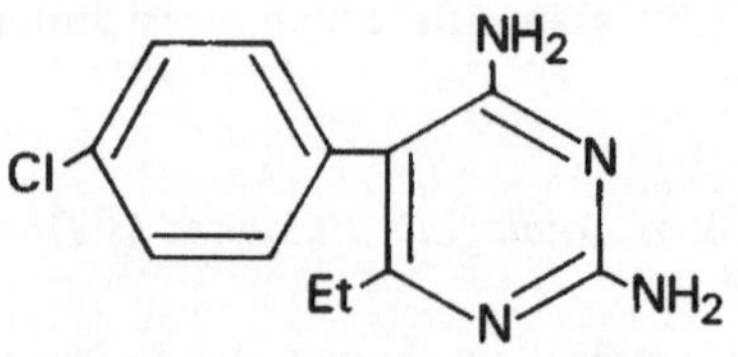

FIGURE 12. Molecular formula of pyrimethamine.

selective action has been traced to the presence of analogous enzymes – one in the host, the other in the parasites. The latter are blood flukes (trematodes), and their phosphofructokinase (the enzyme for converting fructose-6-phosphate to the diphosphate during glycolysis of carbohydrate foods) has been isolated and shown to be much more sensitive to antimony compounds than mammalian phosphofructokinase.[29] The therapeutic success of antimony depends on the selective blocking of this enzyme, because glycolysis is the main source of energy for this parasite. However, antimonials do not possess absolute selectivity, and unpleasant side reactions sometimes occur.

A much higher degree of selectivity has been built into those drugs, both antibacterial and antiprotozoal, that interfere with folic acid metabolism. Tetrahydrofolic acid (Figure 11) is important for all forms of life, for it combines with one-carbon intermediates to form coenzymes. These specifically insert a carbon atom into many vitally important molecules (such as the purines, pyrimidines, and amino acids) that the cell has to synthesize to carry on its metabolism and reproduction. For example, the coenzyme for the insertion of C-2 into the purine nucleus is 10-formyl-5,6,-7,8-tetrahydrofolic acid. Because all biosynthesis of purines follows this pathway, lack of tetrahydrofolic acid causes gross purine deficiency in the cell. A closely related coenzyme inserts the 5-methyl group into uridylic acid to give thymidylic acid, which is essential for DNA formation and cannot be biosynthesized in any other way. The sole source of tetrahydrofolic acid in nature is the reduction of natural dihydrofolic acid. This reduction is effected by the enzyme dihydrofolate reductase (the new hydrogen atoms occupy the 5- and 6-positions). When this enzyme is inhibited by appropriate drugs, the affected organisms die, many through lack of thymidylic acid, others through lack of purines.

Remissions in acute lymphatic leukemia follow treatment with the drug methotrexate, a molecule in which the oxo group of dihydrofolic acid has been replaced by an amino group, and the hydrogen atom in the 10-position by a methyl group. Methotrexate effects a lasting cure in two other types of cancer: choriocarcinoma and Burkitt's lymphoma (the latter occurs in association with the Epstein-Barr virus). However, methotrexate (and similar altered folic acids) proved useless for treating bacterial and protozoal diseases, because they were not absorbed by microorganisms.

About 1950, G. H. Hitchings started to simplify the molecule of methotrexate to find out which part inhibited the dihydrofolate reductase. The action was found to reside in the 2,4-diaminopteridine moiety, and further simplification revealed that the 2,4-diaminopyrimidine portion of the latter carried the full inhibitory activity. This knowledge was used at once to develop pyrimethamine (Figure 12), a highly effective antimalarial prophylactic.[30] The lipoidal (ethyl and chlorophenyl) groups in this molecule favor its uptake by the parasite and increase its

TABLE 2

Concentrations (× 10^8 *M*) of Antifolic Drugs Needed for 50% Inhibition of Dihydrofolate Reductase, Isolated from Various Sources[31]

Substance	Human liver	Mouse erythrocyte	*Plasmodium berghei*	*Trypanosoma equiperdum*	*Escherichia coli*
Pyrimethamine	180	100	0.05	20	2,500
Trimethoprim	30,000	100,000	7.0	100	0.5
Methotrexate	9	Not done	0.07	0.02	0.1

TABLE 3

Effect of Trimethoprim on Isolated Dihydrofolate Reductases in Terms of Concentration (× 10^8 *M*) Causing 50% Inhibition

From mammalian liver		From bacteria	
Man	30,000	*Escherichia coli*	0.5
Rat	26,000	*Staphylococcus aureus*	1.5
Rabbit	37,000	*Proteus vulgaris*	0.4

From Roth, B., Falco, E. A., and Hitchings, G. H., *J. Med. Pharm. Chem.*, 5, 1103, 1962. With permission.

adsorption on the enzyme by van der Waals forces.

As Table 2 shows, the plasmodial enzyme is inhibited by pyrimethamine at a concentration about 2,000 times lower than that inhibiting an analogous mammalian enzyme. This antimalarial concentration equaled that achieved in the tissues after the usual prophylactic dose. These data established that the extraordinarily selective action of pyrimethamine in malaria is due to the high sensitivity of the parasite's enzyme compared to that of the host.[31]

To provide strong antibacterial action in mammals, the 2,4-dihydropyrimidines were found to require a different substitution pattern of alkyl and aryl groups, as in trimethoprim (Figure 13). Table 3 shows that pyrimethamine is most selective against the malarial parasite enzyme, whereas trimethoprim is selective against the analogous enzymes from various types of bacteria.[32] Compared with the high selectivity shown by both these pyrimidines, the pteridine drug methotrexate has a much narrower range.

Table 3 illustrates how trimethoprim exerts high selectivity against the dihydrofolate reductase of bacteria (both Gram-positive and Gram-negative types) while leaving the analogous mammalian enzymes unharmed.

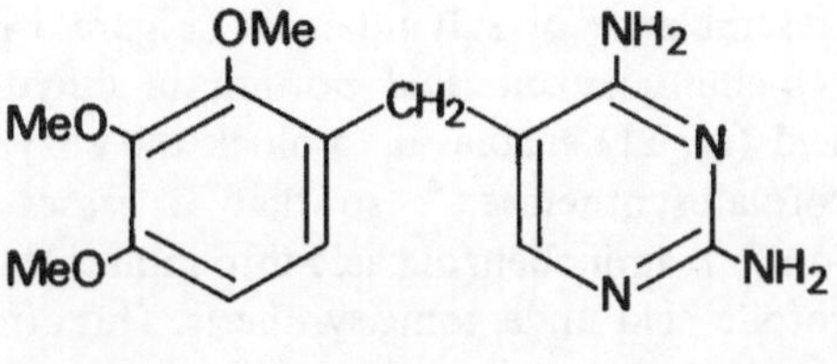

FIGURE 13. Molecular formula of trimethoprim.

It is noteworthy that these 2,4-diaminopyrimidine drugs have been discovered and developed by the exercise of pure scientific reasoning. Moreover, the therapeutic value of trimethoprim has been increased by using the drug in conjunction with a sulfonamide (sulfamethoxazole) to achieve sequential blocking, namely, inhibition of the synthesis of dihydrofolic acid as well as inhibition of its reduction.[33] The combination has been found extraordinarily effective when given orally in bacterial dysentery, bronchitis, and long-standing infections of the urinary tract.

The chance of abortion or damage to the fetus makes it necessary to use antifolic drugs cautiously during pregnancy.

2. Absence of Enzymes from One of the Species

An extreme situation is found with analogous enzymes when one member of the pair is present in relatively small amounts or is even completely absent. The high selectivity of the antibacterial sulfonamides (such as sulfadiazine) depends on two reciprocal relationships. In the first place, mammals have no enzyme for synthesizing folic acids, whereas pathogenic microorganisms are equipped with such an enzyme, known as dihydrofolate synthetase. In the second place, mammals have a permease, which enables them to absorb dihydrofolic acid from foods, whereas pathogenic microorganisms have no such facility. For these reasons, the antibacterial sulfonamides can overcome a massive bacterial infection with very little

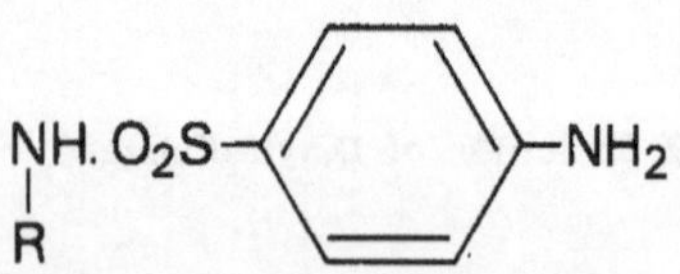

FIGURE 14. Molecular formulas of antibacterial sulfonamides. (A) Sulfanilamide, R = H. (B) R = heterocyclic substituent.

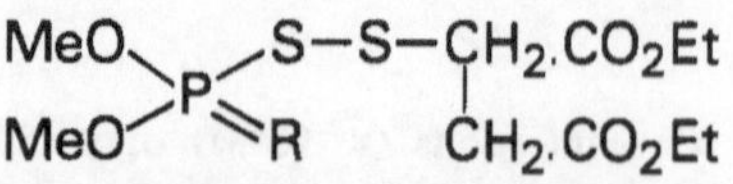

FIGURE 15. (A) Malathion, R = S. (B) Malaoxon, R = O.

inconvenience to the host. The electronic and steric resemblance of sulfanilamide (Figure 14A) to the *p*-aminobenzoic acid portion of dihydrofolic acid (cf. 11) enables it to block the enzyme dihydrofolatesynthetase,[34] so that it ceases to incorporate *p*-aminobenzoic acid into molecules of dihydrofolic acid undergoing synthesis. This effect has been demonstrated on the purified enzyme.[11] The presence of the heterocyclic substituent in the molecules of the most prescribed sulfonamides (Figure 14B) does not interfere, because the heterocyclic ring makes a sharp angle with the plane of the benzene ring of these drugs. In fact, by greatly increasing the percentage of the drug that is ionized (as anion), the heterocyclic substituent increases the resemblance of the drug to *p*-aminobenzoic acid (which is fully anionic) and hence blocks the enzyme that much more effectively. The small proportion of the dose of a sulfonamide drug that gets built into analogs of folic acids in place of *p*-aminobenzoic acid has no therapeutic significance.

A somewhat similar basis for selective action will now be described for the organophosphate insecticides, where use has been made of the weakness, or even absence, of two enzyme processes, one in the economic and one in the uneconomic species.

The toxicity of organic phosphorus compounds is directly proportional to their inhibition of the enzyme acetylcholinesterase. When they suspend the function of this enzyme, acetylcholine accumulates at nerve/muscle junctions in vertebrates and at nerve/nerve junctions (synapses) in both vertebrates and insects. Although acetylcholine is a natural neurotransmitter, an excess of it is fatal. The first organophosphate insecticides were insufficiently selective, but improvement came with increasing knowledge of the comparative biochemistry of the relevant species.

The highly selective insecticide malathion was introduced in 1950. This substance (Figure 15A) is not the actual toxicant, but insects have an enzyme that rapidly hydrolyzes it to malaoxon (Figure 15B). The latter compound phosphorylates the serine residue in acetylcholinesterase, and so brings about the death of the insect. The P–S hydrolyzing enzyme is not present in vertebrates, which are very little affected by this substance. But vertebrates are given a second selective advantage by this agent. The nonspecific esterases in their blood quickly hydrolyze the carboxylic ester group of malaoxon to a carboxylic acid group (CO_2H), thus yielding a type of molecule that the kidney rapidly eliminates. On the other hand, insects are poorly supplied with esterases. Thus, the high selectivity of this compound is achieved by two independent metabolic changes, one of them toxifying and one detoxifying.[35] Figure 16 illustrates this selectivity between cockroach and mouse. In the cockroach the level malaoxon builds up quickly after administration of malathion, and this level is slow to decline; in the mouse, on the other hand, the level of the toxic compound never becomes high, and its elimination is obviously rapid.

Many other highly selective phosphorus insecticides have since been marketed, inspired by the biochemistry underlying the usefulness of malathion, but differing in detail. Organophosphate anthelmintics are now used for killing worms in all kinds of farm animals. For example, the worm *Haemonchus contortus*, which parasitizes the sheep's gut, has an acetylcholinesterase that is irreversibly inhibited by the organophosphate known as haloxon, whereas the acetylcholinesterase of the sheep's gut is only temporarily inhibited and soon recovers.[36]

Examples of selectivity through comparative biochemistry are by no means confined to enzyme inhibition, for effective drugs have been obtained at many other sites, from blocking nucleic acids down to competition between small molecules.

C. Selectivity through Comparative Cytology

The organization of plant and animal cells into a variety of tissues makes a valuable division of labor possible, and further divisions occur in all forms of life at the subcellular level. The many, often conflicting reactions that take place simultaneously in cells require many isolated compart-

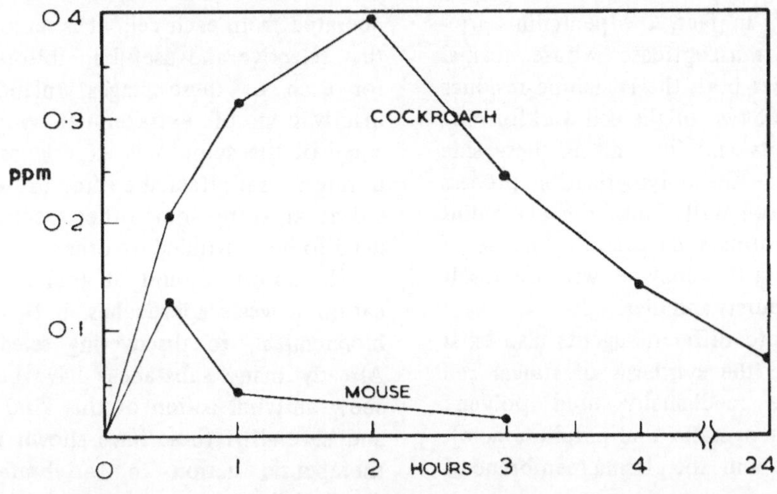

FIGURE 16. Malaoxon level after injecting 30 ppm of ^{32}P-malathion. (From Krueger, H. R. and O'Brien, R. D., *J. Econ. Entomol.*, 52, 1063, 1959. With permission.)

ments, separated by membranes of selective permeability. When any type of organelle is examined closely in the electron microscope, it is seen to differ not only from organism to organism, but even between the different tissues of one organism.

The chloroplast, the green photosynthetic organelle in plants, about 0.2 μm in diameter, consists of a double-layered lipoprotein membrane, which contains the lamellae carrying the chlorophyll molecules. Many herbicides have been devised to attack the chloroplasts specifically without harm to animals.

Vulnerability to selective toxicants is nowhere more evident than in bacteria, whose small size, compared to that of eukaryotic cells, leaves no space for a nucleus or for even one mitochondrion. In place of a nucleus, the chromosomal DNA forms a strand attached to the plasma membrane, and the latter undertakes the usual functions of a mitochondrion. Thus, the exposed position of the nuclear and mitochondrial functions in bacteria contrasts sharply with the membrane-protected situation of these functions in the cells of higher organisms. The selectivity of 9-aminoacridine against bacteria in wounds is based on the difference of accessibility of the DNA with which it reacts.

The selective action of the antibiotic chloramphenicol depends on its distinguishing between the 70*S* ribosomes of bacteria, with which it combines, and the 80*S* ribosomes of higher forms of life. In finer detail, it is now known that chloramphenicol unites with the 50*S* *subunit* of the bacterial ribosome, where it prevents protein synthesis by blocking peptidyl group transfer.[28] Erythromycin, a macrolide antibiotic that shows little toxicity to man, is strongly bound to this 50*S* subunit, but not at the site that binds chloramphenicol.

Another unique structural feature in bacteria is the cell wall, grossly porous, but able to restrain the bursting that is a constant threat imposed by the high osmotic pressure of the bacterial interior. Several antibiotics (oxamycin, penicillin, and bacitracin, among others) prevent the synthesis of new cell wall. As soon as bacteria begin to grow in the absence of new wall, they promptly burst and die. The strong framework that holds the wall together is the polymer of a carbohydrate that bears a short polypeptide side chain; there is nothing else like this in all of nature. This polypeptide ends in two residues of D-alanine, one of which is lost when cross-linking takes place.

The antibiotic oxamycin is recognized as a rigid structural analog of D-alanine. It has been shown to prevent the incorporation of D-alanyl-D-alanine into the polypeptide when a new antibacterial wall is being formed.[37]

Penicillin acts in similar, but much more effective fashion. When molecular models were built and compared, it was seen that the penicillins are structural analogs of a D-dipeptide; both nitrogen atoms and the carbon of the carboxyl group exactly coincide.[38] In fact, the penicillins irreversibly acylate a transpeptidase, whose normal function is to receive both the D-alanine residues of the side chains of two of the cell-wall-forming saccharopeptide units and (by uniting these side chains) to commence the polymerization process that leads to new cell wall. Thus, the penicillins (and the cephalosporins too) prevent linkage of the units by blocking this enzyme, with the result that the bacterium bursts and dies.

Specific, but quite different agents also exist for interfering with the synthesis of fungal cell wall, whereas the medicinally used polyene antibiotics (such as nystatin and amphoteracin), interfere specifically with the plasma membrane of fungi.

D. Conclusion

The above account of the physical basis of selectivity indicates how this property has been introduced into many agents that are used daily for the benefit of man in the chemotherapy of protozoal, bacterial, and fungal diseases, as well as in noninfectious (metabolic) diseases. Further examples are available in a brief account of this subject recently published.[39]

III. POSSIBLE SITES FOR ACTION OF ANTIVIRAL DRUGS

Antiviral chemotherapy seems to have reached the stage that antibacterial chemotherapy had attained by 1940; following the discovery of sulfonamides in 1935, only a few effective drugs were available, but clues were on hand that were to lead to the discovery of a whole wealth of new remedies.

When a virus infects a cell, several events occur that are specific to the invader, and hence offer opportunities for selective attack. First there is the moment of contact with the cell, followed by penetration of the cell's plasma membrane and by (often simultaneous) rejection of the viral coating protein. Next, if the virus has an RNA genome, RNA replicase is soon manufactured or becomes activated, but in any case, the synthesis of nucleic acid polymerases dominates the pattern of invasion at this stage. Then follows the synthesis of viral nucleic acids, structural proteins, and yet more enzymes, and eventually the assembly of these components to form the completed virus. Finally, several thousand of these virions are liberated from each cell. It is reasonable to expect that selective and useful inhibitors will be found for each of these stages, including the direct inactivation of extracellular virions. Moreover, some of the symptoms experienced by a patient during a viral attack are often caused by hypersensitization, or by some other secondary effect, and need to be controlled by other means.

The above account of viral invasion and replication suggests approaches, both distributive and biochemical, to discovering selective inhibitors. Already many substances have been found that show antiviral action of this kind in cell culture, and several of these have shown prophylactic or therapeutic action in vertebrate hosts. Some details follow.

Agents acting on the free virion have, so far, been the least successful. They include kethoxal [3-ethoxy-2-oxobutanal hydrate, $CH_3 \cdot CH(OEt) \cdot CO \cdot CH(OH)_2$], active against vaccinia and Newcastle disease, and 2-thiouracil (Figure 17), which decreases the infectivity of poliovirus by reaction with the SH group of the capsid[40] and is also a potent inhibitor of the reproduction of plant viruses.

The influenza prophylactic, amantadine, is an example of a drug acting at the site of penetration of the virion into the host cell (see Chapter 5).

However, the majority of the antivirals currently used owe their selective action to blocking an intracellular stage of viral multiplication. For example, idoxyuridine, adenine arabinoside, and 5-trifluoromethyldeoxyuridine, because of their resemblance to natural nucleosides, are able to prevent the synthesis of normal nucleic acids; guanidine and hydroxybenzylbenzimidazole inhibit the synthesis of viral RNA polymerase; rifampicin inhibits reverse transcriptase; and

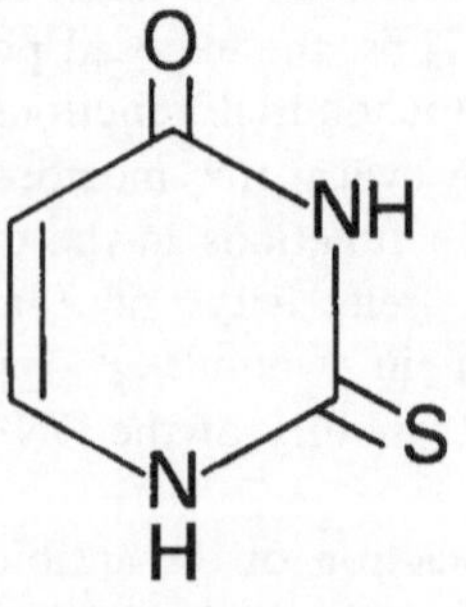

FIGURE 17. Molecular formula of 2-thiouracil.

methisazone stops the production of those proteins usually formed late in the intracellular phase. These examples of selectivity depend on such singularities as this: the enzyme that replicates a viral nucleic acid is not present in the *uninfected* cell and is quite different from the analogous polymerase of the host cell.[41] The action can be highly specific – a single pox virus particle is inactivated by as little as two molecules of isatin 3-thiosemicarbazone.

Selectivity can also be based on differences in structure. There are more than 300 immunologically distinct varieties of virus that infect man (most of them virulent, and a few tumorigenic), but most fall into eight major families characterized by distinctive structural features. The DNA or RNA that forms the core of the virion is surrounded by a protective capsid (= shell) of one, two, or more proteins. These components are often arranged in a highly ordered fashion characteristic of the particular family. Many of the viruses that infect animals are 20-sided, and hence roughly spherical, but the viruses of measles and influenza are more pleomorphic. In the influenza virus, for example, the capsid, formed of hemagglutinin and neuramidase, is embedded in lipid, and this lipoprotein envelope encloses a coiled ribonuleoprotein tube (see Chapter 5). Pox viruses, the largest known, are brick-shaped and of very complicated structure. Many viruses have an extra protein as a core around which the nucleic acid is arranged. As many as 20 kinds of protein, some of them enzymes, occur in some viruses. The nucleic acids are single-stranded in some types of virus and double-stranded in others. Coenzymes, such as dihydrofolic acid, and metal cations are commonly present as well.

The virus causing vesicular stomatitis will be described to exemplify the structural complexity of a virus infecting man. Its virions are cylinders 680 Å in diameter and 1,750 Å long, surrounded by a lipoprotein envelope from which antigen-carrying ridges (peplomers) stand out. Immediately inside the envelope is a smooth membrane, and within this lie two helices, one 500 Å and the other 175 Å in diameter, surrounding an axial canal. These helices consist of protein tubes with ribonucleic acid cores. Phages, too, have a complex structure, consisting of a capsid-DNA head, a contractile tail, a tail core, loose tail fibers, and an enzyme-containing base plate.[42]

To maximize the possibilities of the structural singularities of viruses, it is necessary to catalog the structural possibilities of the vulnerable tissues of the host (in influenza, for example, details of the nasal turbinates, pharynx, and trachea). It must also be kept in mind that the virus can undergo eventual structural and biochemical adaptation. For example, influenza virus can be cultured to form a strain that propagates in nervous tissue, causing lethal encephalitis in the embryonic mouse.[43]

Much attention has been given to another approach, namely, a search for naturally occurring inhibitors of viruses. The most important of these is interferon, a small protein that occurs in trace amounts in all mammals, with further production occurring when a virus attacks the host. It is completely nontoxic to man and has a very broad spectrum of activity (see Chapter 6, Volume II). Interferon seems to act by selectively blocking the translation of viral RNA and is host-specific. Most of the exogenous supply has come from human blood banks. Because this is not a practical source, attempts have been made to develop substances that will induce the biosynthesis of interferon in the patient. Polymers were used first, one of the best-known being the pair of nucleotides known as poly-I–poly-C (polyriboinosinic and polyribocytidylic acids, mixed and annealed). It is inactive *per os.* When injected into mice infected with encephalomyocarditis virus (a picornavirus that, in mice, develops an infection resembling disease caused by enterovirus in man), a single prophylactic dose generated enough interferon to save the animal from certain death; however, the agent became progressively less effective when used later and later, in single doses, as the infection developed. When the agent was effective, viremia was completely suppressed.[44] Used prophylactically against lactic dehydrogenase virus infections of mice, this agent brought about only a temporary decrease in viremia.[45]

Quite similar results in inducing formation of interferon can be obtained with smaller molecules, such as tilerone (Figure 18) or bisdiethylaminoethylfluorenone. When tested for viremia sup-

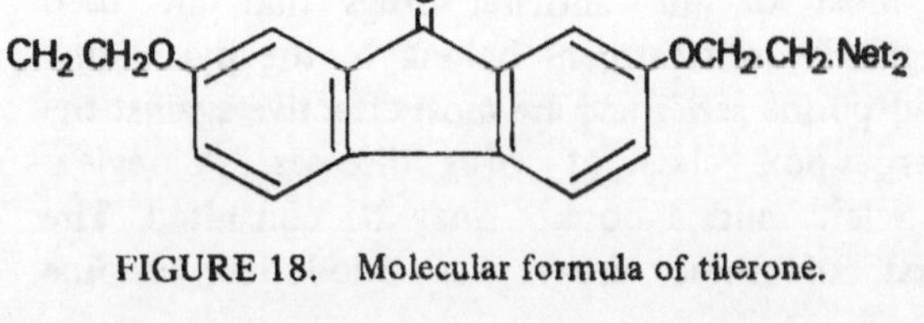

FIGURE 18. Molecular formula of tilerone.

pression in mice, tilerone proved as effective as poly-I–poly-C.[44] These smaller, aromatic molecules lend themselves to much greater variation of the details of their structure than is possible with the polymers, and hence they are being actively pursued.

IV. ANTIVIRAL AGENTS USED IN MAN

It is proposed in this section to describe briefly the chemistry, molecular biology of action (where known), and therapeutic results of antiviral agents that have at least achieved clinical trials, including several that are in regular clinical use. One major idea in assembling these data is to demonstrate in the drugs types of molecular structure and biochemical activity that, although often similar to structures and biochemical activity important in the host cell, have proved in practice to have a useful order of selectivity.

Quite the most successful of these for the chemoprophylaxis of influenza is 1-aminoadamantane (amantadine, Symmetrel®), shown in Figure 19. In double-blind trials, it has proved to be highly effective in reducing the incidence of influenza (A2 strains) during an epidemic.[46] Adamantane is a cage-like molecule (tricyclodecane, $C_{10}H_{16}$), and the amino derivative is a strong base (pK_a = 10.14 at 37°).[47] Some closely related derivatives have shown similar protective action; see Chapter 3, Volume II for details of the prophylactic trials. Amantadine is not directly virucidal. It seems to act by inhibiting the stripping of the protein capsid from the virion as the latter enters the cell.[48]

Several 1-phenoxymethyl-3,4-dihydroisoquinolines, found active in vitro against a wide range of viruses, were eventually used in man[49] without mishap, but they have not yet found a place in current therapy. One of them, known as UK 2361 (1-*p*-methoxyphenoxymethyl-3,4-dihydroisoquinoline), has shown prophylactic action against influenza B in man. Volunteers given 1.5 g orally each day and challenged after 1 day with intranasal living B virus showed only half of the control group's incidence of illness.[50]

Most of the antiviral drugs that are used regularly in treatment belong to the pyrimidine and purine series and are most effective against the herpes-pox class of virus diseases. A review article[51] and a book[52] may be consulted. The first of these drugs was 5-iododeoxyuridine

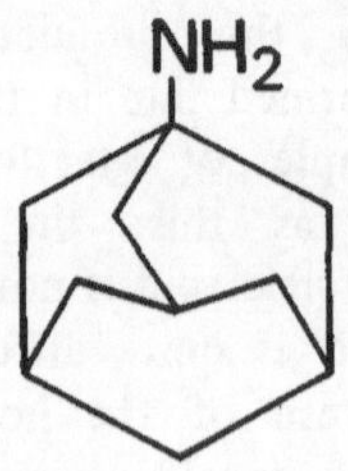

FIGURE 19. Molecular formula of 1-aminoadamantane (amantadine).

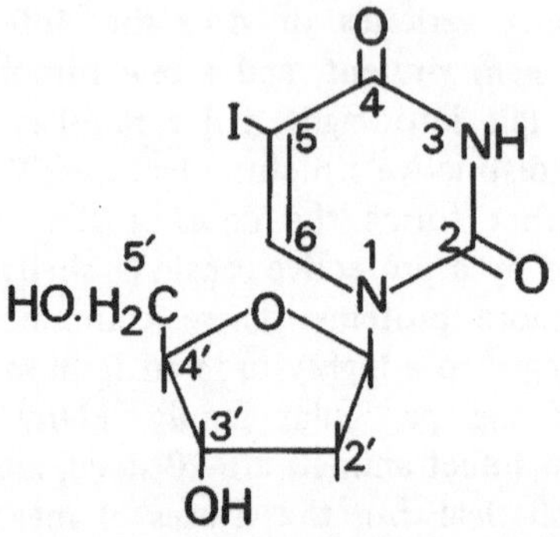

FIGURE 20. Molecular formula of 5-iododeoxyuridine (idoxuridine).

(idoxuridine), shown in Figure 20, which, after successful trials in the rabbit eye,[53] has become the standard treatment for the serious, even blinding, eye disease known as herpes simplex keratitis.[54] Acting as a thymine analog, the drug exerts several actions on the pathway to thymidylate synthesis, but its principal action follows its bioconversion to the triphosphate, which is incorporated into viral DNA and leads to miscoding.

Other successful clinical uses of idoxuridine are in herpes zoster, where a topical application of a 35% solution in dimethyl sulfoxide usually effects a cure in only 4 days.[55] In cold sores, in herpes simplex encephalitis,[56] in recurrent herpes simplex, and in early varicella[57] idoxuridine has given very good results. This drug is inactivated when taken orally, and it is seldom given parenterally, because it is incorporated, to a small extent, into mammalian DNA, and mutagenesis is feared. Recently, a strong inhibitor of herpes simplex virus has been found that has no toxicity to mammalian cells. This was made by replacing the 5′-hydroxy group in IUDR with an amino group, giving 5′-amino-2′,5′-dideoxy-5-iodouridine (AIU).[57a]

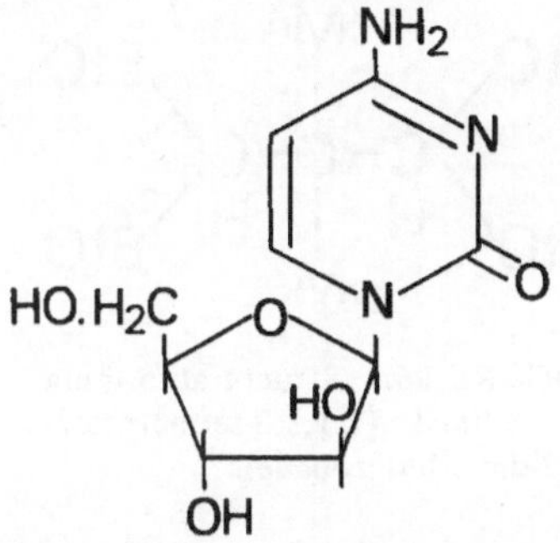

FIGURE 21. Molecular formula of cytarabine (Ara-C).

Cytarabine (Figure 21), also known as Ara-C, is a clinically important pyrimidine. Given intravenously, it brings dramatic improvement and subsequent cure in primary gingivostomatitis (caused by herpes simplex virus) in children and adults. It has also given excellent results in generalized herpes encephalitis, provided that treatment is begun as soon as the pathologist confirms the diagnosis. It is very effective in curing the rare, generalized infection that can follow vaccination; vaccinia dermal lesions, which occur a little more often, are simply treated with topical idoxuridine, without loss of immunity occurring.[52]

Cytarabine acts by preventing viral DNA synthesis and function as a result of its being an analog of cytosine deoxyriboside, an excess of which can reverse its action. In man, a few side effects have been found in therapeutic doses, and this has led to some preference for a related purine known as Ara-A (Figure 22), also called vidarabine or 9-β-D-arabinofurosyladenine. Its highly insoluble nature has necessitated the laborious process of intravenous infusion, but it seems to be effective and free of side reactions.

Recurrent herpes genitalis is often treated by painting the lesions with a solution of an aminoacridine, such as proflavine (Figure 23), followed by short exposure to visible light. In this way, free radicals, generated from the aminoacridine, inactivate the virus.[58] Other nitrogenous tricyclic families, such as the phenazines (e.g., neutral red) and phenothiazines, also show this effect, though not as strongly.

Remarkable ability to prevent infection by the virus of smallpox, among people who have been in intimate contact with fatal cases, is shown by methisazone (Figure 24), i.e., *N*-methylisatin

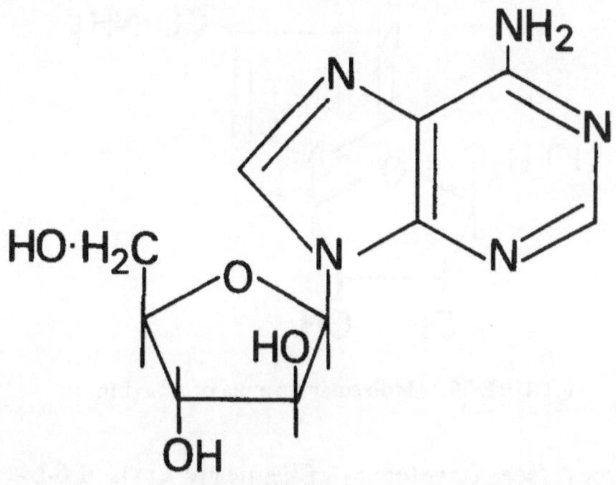

FIGURE 22. Molecular formula of vidarabine (Ara-A, 9-β-D-arabinofurosyladenine).

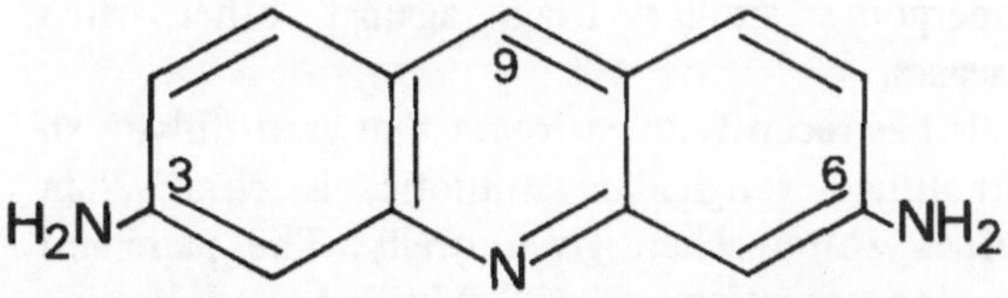

FIGURE 23. Molecular formula of proflavine.

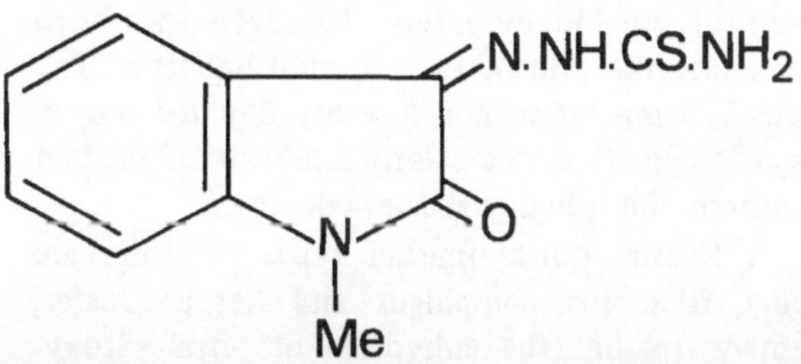

FIGURE 24. Molecular formula of methisazone (*N*-methylisatin β-thiosemicarbazone).

β-thiosemicarbazone. During an outbreak in Madras, the drug was given orally to 1,100 unvaccinated people who had been in close contact with cases of smallpox. Only three cases, all mild, occurred among this treated group, whereas in a similar number of controls 78 contracted smallpox, and 12 died of it.[59] Although the drug gives protection when administered too late in the incubation for vaccination to be effective, it cannot cure established infection. Regrettably, severe nausea is a common side effect. For further information, see Reference 51, p. 35.

Ribavirin (Figure 25), also marketed as Virazole®, has broad-spectrum antiviral activity, in vitro and in vivo, against both DNA and RNA

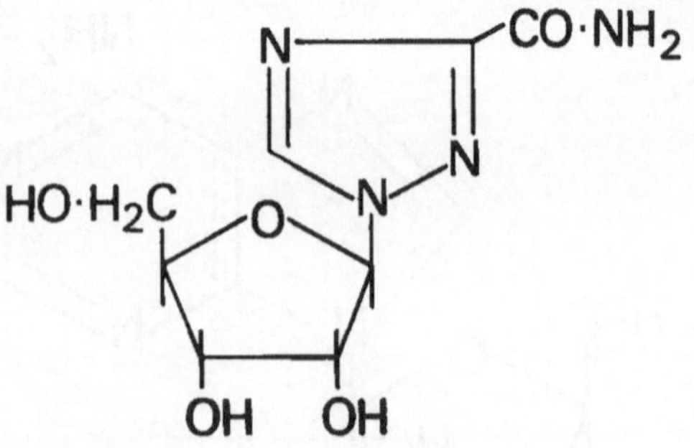

FIGURE 25. Molecular formula of ribavirin.

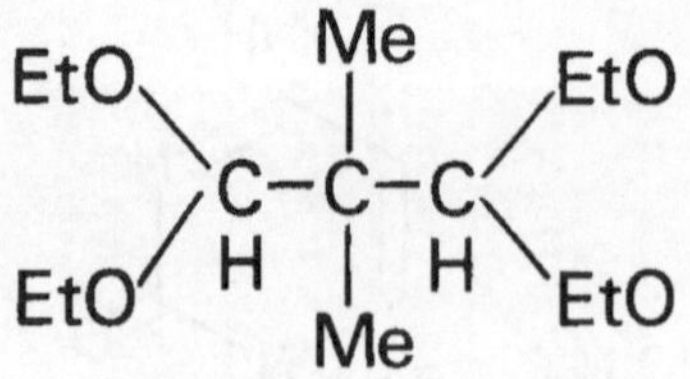

FIGURE 26. Structural formula of adimal (1,1,3,3-tetraethoxy-2,2-dimethylpropane).

types (see Chapter 5). Chemically it is 1-β-D-ribofuranosyl-1,2,4-triazole-3-carboxamide, a stable, water-soluble nucleoside.[60] It has proved successful in clinical trials of hepatitis[61] and is undergoing similar trials against other virus diseases.

It has recently been found that griseofulvin, an established fungicidal antibiotic, is curative in herpes zoster when given orally. The pain and burning sensation quickly subsided, and healing occurred in about 7 days.[62]

The alkaloid emetine, long used in the treatment of amebic dysentery, has been shown to accelerate the rate of cure in viral hepatitis. The dose is 3 mg intravenously every day for only 4 days.[63] Emetine is a strong inhibitor of protein synthesis and is highly species-selective.[64]

In Russia, quite unusual types of drugs are being tried for pemphigus and herpes zoster, namely, oxolin (the anhydride of tetrahydroxy-tetrahydronaphthalene) and adimal (1,1,3,3-tetraethoxy-2,2-dimethylpropane);[65] the latter is shown in Figure 26.

The first success in the difficult, but economically important subject of plant antiviral chemotherapy was the use of the antibiotic blasticidin-S on bean and tobacco plants. It proved effective when applied up to 12 hr after experimental infection and acted by selectively inhibiting synthesis of the viral 30*S* RNA without affecting the host's ribosomal RNA.[66]

V. EVIDENCE OF SELECTIVITY IN OTHER ANTIVIRAL AGENTS

This section briefly mentions some chemically different antiviral agents that are, at present, undergoing laboratory tests. The alkaloid camptothecin, obtained from a Tibetan shrub, prevents fowl plague replication. It specifically inhibits "late" functions of the influenza virus in infected cells. It acts by cleaving DNA in a selectively repairable manner, with the result that the synthesis of *viral* RNA is particularly inhibited.[67]

Methotrexate, although inactive against bacteria, selectively inhibits the reproduction of oncornaviruses in mouse embryonic tissue cells by inhibiting the biosynthesis of thymidylic acid.[68]

Interesting antiviral agents are being found among the carbohydrates. For instance, 2-deoxy-D-glucose inhibits the replication and cytopathic effects of three viruses pathogenic to man, namely, respiratory syncytial, parainfluenza-3, and measles viruses. It acts by preventing the normal insertion of sugars to form the glycoprotein envelope of these viruses. Electron micrographs show a morphological change. However, at near-toxic doses, this deoxyglucose did not alter the course of influenza-A infection in mice or of parainfluenza-3 infection in hamsters.[69]

Glucosamine, another antiviral carbohydrate, was tested (in pyruvate- and fructose-based media) on fowl plague and Semliki forest viruses. It strongly inhibited the synthesis of viral RNA while hardly affecting the host's RNA and protein syntheses. This excellent selectivity is accomplished by reducing the host's pool of uridine triphosphate to a level where the virus cannot use it for RNA synthesis (presumably the host's RNA polymerase has a higher affinity for UTP).[70] The sulfur-containing antibiotic gliotoxin inhibits viral reverse transcriptase, and some of the easily synthesized analogs are equally potent.[71]

The metabolism of viruses, during replication, is intimately bound up with the metabolic processes of the invaded cell, so that agents that cause the death of viruses may be thought likely to injure the harboring host cells. However, enough examples of selective and useful virus inhibitors have been brought to light to encourage the belief that there will be many more such drugs when enough details have been uncovered (e.g., by viral biochemists) to facilitate more selective inhibition.

RECENT RESULTS

Many nonamino derivatives of adamantane have a better selectivity than amantadine (e.g., the alcohols, ketones, esters, and nitriles) when tested in vitro against Newcastle disease virus on a monolayer culture of chick embryo fibroblasts.[72] The same test also demonstrated increased potency and selectivity in 3-amino-4-homoisotwistane, an isomer of amantadine.[73]

Phosphoacetic acid, which acts by blocking viral DNA polymerase with high selectivity, is currently undergoing clinical trials against herpes infections.

REFERENCES

1. **Albert, A.,** *Selective Toxicity,* 5th ed., Chapman & Hall, London (John Wiley & Sons, New York), 1973.
2. **Rabaté, E.,** *La Destruction des Mauvaises Herbes,* Libraire de l'Académie de l'Agriculture, Paris, 1927.
3. **Blackman, G.,** Weed control in cereals by chemical methods, *Agriculture,* 53, 16, 1946.
4. **Lazarus, M. and Rogers, W. P.,** The mode of action of phenothiazine as an anthelmintic. The uptake of ^{35}S-labelled phenothiazine by the tissues of nematode parasites and their hosts, *Aust. J. Sci. Res. Ser. B,* 4, 163, 1951.
5. **Franklin, T. J.,** The inhibition of incorporation of leucine into protein by cell-free systems from rat liver and *Escherichia coli* by chlorotetracycline, *Biochem. J.,* 87, 449, 1963; Uptake of tetracycline by membrane preparations from *Escherichia coli, Biochem. J.,* 123, 267, 1971.
6. **Williams, A. C. and Klein, E.,** Experiences with local chemotherapy and immunotherapy in premalignant and malignant skin lesions, *Cancer,* 25, 450, 1970.
7. **Hatanaka, H. and Sano, K.,** A revised boron-neutron therapy for malignant brain tumours, *Z. Neurol.,* 204, 309, 1973.
8. **Welch, A. D.,** Some metabolic approaches to cancer chemotherapy, *Cancer Res.,* 21, 1475, 1961.
9. **McDonald, C. J. and Calabresi, P.,** Psoriasis and occlusive vascular disease, *J. Clin. Invest.,* 53, 52, 1974.
10. **Schaffer, N. K., May, S. C., and Summerson, W. H.,** Serine phosphoric acid from diisopropylphosphoryl derivative of eel cholinesterase, *J. Biol. Chem.,* 206, 201, 1954.
11. **Brown, G. M.,** The biosynthesis of folic acid. II. Inhibition by sulfonamides, *J. Biol. Chem.,* 237, 536, 1962.
12. **de Robertis, E. and Schacht, J.,** *Neurochemistry of Cholinergic Receptors,* North-Holland Publ. Co., Amsterdam, 1974.
13. **Albert, A., Rubbo, S. D., and Burvill, M. I.,** The influence of chemical constitution on antibacterial activity. Part IV. A survey of heterocyclic bases, with special reference to benzoquinolines, phenanthridines, benzacridines, quinolines, and pyridines, *Br. J. Exp. Pathol.,* 30, 159, 1949.
14. **Lerman, L. S.,** Structural considerations in the interaction of DNA and acridines, *J. Mol. Biol.,* 3, 18, 1961.
15. **Lerman, L. S.,** Acridine mutagens and DNA structure, *J. Cell. Comp. Physiol.,* 64 (Suppl. 1), 1, 1964.
16. **Hurwitz, J., Furth, J. J., Malamy, M. H., and Alexander, M.,** The role of deoxyribonucleic acid in ribonucleic acid synthesis. III. The inhibition of the enzymatic synthesis of ribonucleic acid and deoxyribonucleic acid by actinomycin D and proflavin, *Proc. Natl. Acad. Sci. U.S.A.,* 48, 1222, 1962.
17. **Albert, A., Rubbo, S. D., Goldacre, R. J., and Balfour, B. G.,** The influence of chemical constitution on antibacterial activity. Part III. A study of 8-hydroxyquinoline (oxine) and related compounds, *Br. J. Exp. Pathol.,* 28, 69, 1947.
18. **Albert, A., Gibson, M. I., and Rubbo, S. D.,** The influence of chemical constitution on antibacterial activity. Part VI. The bactericidal action of 8-hydroxyquinoline (oxine), *Br. J. Exp. Pathol.,* 34, 119, 1953.
19. **Sijpestijn, A. K. and Janssen, M. J.,** Mode of action of dialkyldithiocarbamates on moulds and bacteria, *Antonie van Leeuwenhoek J. Microbiol. Serol.,* 25, 422, 1959.
20. **Sijpestijn, A. K.,** personal communication.
21. **Collander, R.,** The permeability of plant protoplasts to non-electrolytes, *Trans. Faraday Soc.,* 33, 985, 1937.
22. **Hansch, C., Steward, A. R., Anderson, S. M., and Bentley, D.,** The parabolic dependence of drug action upon lipophilic character as revealed by a study of hypnotics, *J. Med. Chem.,* 11, 1, 1968.
23. **Wright, S.,** *The Metabolism of Cardiac Glycosides,* Charles C Thomas, Springfield, Ill., 1960.
24. **Bergel, F.,** Design of alkylating agents for selectivity of action, *Ann. N.Y. Acad. Sci.,* 68, 1238, 1958.
25. **Schanker, L. S.,** Mechanisms of drug absorption and distribution, *Annu. Rev. Pharmacol.,* 1, 29, 1961.
26. **Nelson, E. and O'Reilly, I.,** Kinetics of sulfisoxazole acetylation and excretion in humans, *J. Pharmacol. Exp. Ther.,* 129, 368, 1960.
27. **Dixon, M. and Webb, E.,** *Enzymes,* 2nd ed., Longmans, London, 1964.

28. **Gale, E. F., Cundliffe, E., Reynolds, P. E., Richmond, M. H., and Waring, M. J.,** *The Molecular Basis of Antibiotic Action,* John Wiley & Sons, New York, London, 1972.
29. **Bueding, E. and Mansour, J. M.,** The relationship between inhibition of phosphofructokinase activity and the mode of action of trivalent organic antimonials on *Schistosoma mansoni, Br. J. Pharmacol. Chemother.,* 12, 159, 1957.
30. **Falco, E. A., Goodwin, L. G., Hitchings, G. H., Rollo, I. M., and Russell, P. B.,** 2,4-Diaminopyrimidines – a new series of antimalarials, *Br. J. Pharmacol. Chemother.,* 6, 185, 1951.
31. **Burchall, J. J. and Hitchings, G. H.,** Inhibitor-binding analysis of dihydrofolate reductases from various species, *Mol. Pharmacol.,* 1, 126, 1965; **Ferone, R., Burchall, J. J., and Hitchings, G. H.,** *Plasmodium berghei* dihydrofolate reductase. Isolation, properties, and inhibition by antifolates, *Mol. Pharmacol.,* 5, 49, 1969; **Jaffe, J. J. and McCormack, J. J.,** Dihydrofolate reductase from *Trypanosoma equiperdum.* 1. Isolation, partial purification, and properties, *Mol. Pharmacol.,* 3, 359, 1967.
32. **Roth, B., Falco, E. A., and Hitchings, G. H.,** 5-Benzyl-2,4-diaminopyrimidines as antibacterial agents. I. Synthesis and antibacterial activity in vitro, *J. Med. Pharm. Chem.,* 5, 1103, 1962.
33. **Bushby, S. R. M. and Hitchings, G. H.,** Trimethoprim, a sulphonamide potentiator, *Br. J. Pharmacol. Chemother.,* 33, 72, 1968.
34. **Brown, G. M.,** Dihydrofolate synthetase, *Methods Enzymol.,* 18B, 771, 1971.
35. **Krueger, H. R. and O'Brien, R. D.,** Relation between metabolism and differential toxicity of malathion in insects and mice, *J. Econ. Entomol.,* 52, 1063, 1959.
36. **Lee, R. M. and Hodsden, M. R.,** Cholinesterase activity in *Haemonchus contortus* and its inhibition by organophosphorus anthelmintics, *Biochem. Pharmacol.,* 12, 1241, 1963.
37. **Strominger, J. L., Threnn, R. H., and Scott, S. S.,** Oxamycin: a competitive antagonist of the incorporation of D-alanine into a uridine nucleotide in *Staphylococcus aureus, J. Am. Chem. Soc.,* 81, 3803, 1959.
38. **Tipper, D. J. and Strominger, J. L.,** Mechanism of action of penicillins: a proposal based on their structural similarity to acyl-D-alanyl-D-alanine, *Proc. Natl. Acad. Sci. U.S.A.,* 54, 1133, 1965.
39. **Albert, A.,** *The Selectivity of Drugs,* Chapman & Hall, London (John Wiley & Sons, New York), 1975, 64pp.
40. **Steele, F. M. and Black, F. L.,** Inactivation and heat stabilization of poliovirus by 2-thiouracil, *J. Virol.,* 1, 653, 1967.
41. **Tamm, I. and Eggers, H. J.,** Biochemistry of virus reproduction, *Am. J. Med.,* 38, 678, 1965.
42. **Kozloff, L., Lute, M., and Henderson, K.,** Viral invasion. I. Rupture of thiolester bonds in the bacteriophage tail, *J. Biol. Chem.,* 228, 511, 1957.
43. **Gamboa, E. T., Harter, D. H., Wolf, A., Benitez, H. H., Takahata, N., Duffy, P. E., and Hsu, K. C.,** Replication of neurotropic influenza virus in organotypic cultures of embryonic mouse hypothalamus, *J. Neuropathol. Exp. Neurol.,* 33, 571, 1974.
44. **Stringfellow, D. A., Overall, J. C., and Glasgow, L. A.,** Interferon inducers in therapy of infection with encephalomyocarditis virus in mice. I. Effect of single doses of polyriboinosinic–polyribocytidylic acid and tilerone hydrochloride on viral pathogenesis, *J. Infect. Dis.,* 130, 470, 1974.
45. **Johnson, M. L. and duBuy, H. G.,** Effect of a potent interferon inducer on acute and chronic lactic dehydrogenase virus viremia, *Infect. Immun.,* 11, 113, 1975.
46. **Galbraith, A. W., Oxford, J. S., Schild, G. C., and Watson, G. I.,** Protective effect of aminoadamantane on influenza A2 infections in the family environment, *Ann. N.Y. Acad. Sci.,* 173, 29, 1970.
47. **Perrin, D. D. and Hawkins, I.,** Dissociation constant of the 1-aminoadamantane cation, *Experientia,* 28, 880, 1972.
48. **Kato, N. and Eggers, H. J.,** Inhibition of uncoating of fowl plague virus by 1-adamantanamine hydrochloride, *Virology,* 37, 632, 1969.
49. **Brammer, K. W., McDonald, C. R., and Tate, M. S.,** Antiviral properties of 1-phenoxymethyl-3,4-dihydro- and 1,2,3,4-tetrahydroisoquinolines, *Nature London,* 219, 515, 1968.
50. **Beare, A. S., Bynoe, M. L., and Tyrrell, D. A. J.,** Prophylaxis of influenza with a synthetic isoquinoline, *Lancet,* 1, 843, 1968.
51. **Bauer, D. J.,** Introduction to antiviral chemotherapy, in *Chemotherapy of Viral Diseases* (International Encyclopedia of Pharmacology and Therapeutics), Section 61, Vol. 1, Pergamon Press, Elmsford, N.Y., 1972, 1.
52. **Juel-Jensen, B. E. and MacCallum, F. O.,** *Herpes Simplex, Varicella, and Zoster,* Heinemann, London, 1973.
53. **Perkins, E. S., Wood, R. M., Sears, M. L., Prusoff, W. H., and Welch, A. D.,** Antiviral activities of several iodinated pyrimidine deoxyribonucleosides, *Nature London,* 194, 985, 1962; **Kaufman, H. E., Maloney, E. D., and Nesburn, A. B.,** Comparison of specific antiviral agents in herpes simplex keratitis, *Invest. Ophthalmol.,* 1, 686, 1962.
54. **Maxwell, E.,** Treatment of herpes keratitis with 5-iodo-2′-deoxyuridine (IDU): a clinical evaluation of 1,500 cases, *Am. J. Ophthalmol.,* 56, 571, 1963.
55. **Juel-Jensen, B. E.,** Results of the treatment of zoster with idoxuridine in dimethyl sulfoxide, *Ann. N.Y. Acad. Sci.,* 173, 74, 1970.
56. **Tomlinson, H. and MacCallum, F. O.,** The effect of iododeoxyuridine on herpes simplex virus encephalitis in animals and man, *Ann. N.Y. Acad. Sci.,* 173, 20, 1970.
57. **Kibrick, S. and Katz, A. S.,** Topical idoxuridine in recurrent herpes simplex, with a note on its effect in early varicella, *Ann. N.Y. Acad. Sci.,* 173, 83, 1970.

57a. **Cheng, Y. C., Goz, B., Neenan, J. P., Ward, D. C., and Prusoff, W. H.,** Selective inhibition of herpes simplex virus by 5′-amino-2′,5′-dideoxy-5-iodouridine, *J. Virol.,* 15, 1284, 1975.

58. **Kaufman, R. H., Gardner, H. L., Brown, D., Wallis, C., Rawls, W. E., and Melnick, J. L.,** Herpes genitalis treated by photodynamic inactivation of the virus, *Am. J. Obstet. Gynecol.*, 117, 1144, 1973.
59. **Bauer, D. J., St. Vincent, L., Kempe, C. H., and Downie, A. W.,** Prophylactic treatment of smallpox contacts with N-methylisatin β-thiosemicarbazone, *Lancet*, 2, 494, 1973.
60. **Sidwell, R. W., Huffman, J. H., Khare, G. P., Allen, L. B., Witkowski, J. T., and Robins, R. K.,** Broad-spectrum antiviral activity of virazole: 1-β-D-ribofuranosyl-1,2,4-triazole-3-carboxamide, *Science*, 177, 705, 1972.
61. **Zuniga, C. B., Almeida, C., Iervolino, A. C. L., Castro, I. O., and Galvao, P. A. A.,** Acao do 1-beta-D-ribofuranosyl-1,2,4-triazol-3-carboxamide no tratamento das hepatites virais agudas, *Rev. Assoc. Med. Bras.*, 20, 386, 1974 (in Portuguese, with English summary).
62. **Chaturvedi, U. C. and Mathur, A.,** Griseofulvin in herpes zoster, *Br. Med. J.*, 3, 438, 1975.
63. **Fusillo, A.,** Effect of small doses of emetine in the therapy of virus diseases. New specific therapeutic method in viral hepatitis, *Minerva Med.*, 64, 129, 1973.
64. **Grollman, A. P.,** Inhibitors of protein biosynthesis. V. Effects of emetine on protein and nucleic acid biosynthesis in HeLa cells, *J. Biol. Chem.*, 243, 4089, 1968.
65. **Samsanov, V. A., Nikitina, M. N., and Mizonova, T. P.,** Therapeutic effectiveness of oxolin and adimal in dermatoses of viral and presumably viral origin, *Vestn. Dermatol. Venerol.*, 12, 58, 1974 (in Russian, with English summary).
66. **Hirai, T., Hirashima, A., Itoh, T., Takahashi, T., Shimomura, T., and Hayashi, Y.,** Inhibitory effect of blasticidin S on tobacco mosaic virus multiplication, *Phytopathology*, 56, 1236, 1966.
67. **Kelly, D. C., Avery, R. J., and Dimmock, N. J.,** Camptothecin: an inhibitor of influenza virus replication, *J. Gen. Virol.*, 25, 427, 1974.
68. **Shobukov, V. M. and Galagov, G. A.,** Effect of folic acid and methotrexate on reproduction of Rauscher leukemia, *Bull. Exp. Biol. Med. USSR*, 77, 801, 1974 (English translation).
69. **Hodes, D. S., Schnitzer, T. J., Kalica, A. R., Camargo, E., and Chanock, R. M.,** Inhibition of respiratory syncytial, parainfluenza-3, and measles viruses by 2-deoxy-D-glucose, *Virology*, 63, 201, 1975.
70. **Scholtissek, C., Rott, R., and Klenk, H.-D.,** The different mechanisms of the inhibition of the multiplication of enveloped viruses by glucosamine, *Virology*, 63, 191, 1975.
71. **Ottenheijm, H. C. J., Kerkhoff, G. P. C., Bijen, J. W. H. A., and Spande, T. F.,** A three-step synthesis of a gliotoxin analogue with anti-reverse transcriptase activity, *J. Chem. Soc. Chem. Commun.*, p. 768, 1975.
72. **Aigami, K., Inamoto, Y., Takaishi, N., Hattori, K., Takatsuki, A., and Tamura, G.,** Biologically active polycycloalkanes. I. Antiviral adamantane derivatives, *J. Med. Chem.*, 18, 713, 1975.
73. **Aigami, K., Inamoto, Y., Takaishi, N., and Fujikura, Y.,** Biologically active polycycloalkanes. II. Antiviral 4-homoisotwistanes, *J. Med. Chem.*, 19, 536, 1976.

Chapter 3

INFLUENZA IN MAN AND ANIMALS: EPIDEMIOLOGY, SURVEILLANCE, AND IMMUNOPROPHYLAXIS

G. C. Schild

TABLE OF CONTENTS

INTRODUCTION

The influenza A and B viruses are major and continuing causes of human respiratory disease. There is little doubt that for at least the past several hundred years influenza has exerted an enormous epidemiological impact in all areas of the world. During the present century, world-wide outbreaks, pandemics, due to influenza A have occurred with an irregular periodicity of some 10 to 30 years and have been associated with high levels of morbidity and mortality. Less widespread outbreaks of influenza A infection occur between pandemics. However, the total impact of such interpandemic epidemics on morbidity and mortality over several years may be greater than that which occurs in pandemic years.[1] The influenza B virus produces epidemics which are less widespread and less frequent than those due to influenza A and which are not consistently associated with raised mortality.[2] While influenza A epidemics typically involve all age groups, mortality occurs largely in elderly persons and those who are at high risk due to preexisting chronic disease of the respiratory or circulatory system; secondary bacterial pneumonia is a major factor in fatal infections (see Chapter 1).[3] However, acute deaths due to primary viral pneumonia do occur but are relatively infrequent, and their occurrence in previously healthy young persons, as noted in the Asian influenza outbreaks of 1957, is particularly disturbing.[3] In contrast, influenza B characteristically produces outbreaks of a more localized nature, especially in children and institutionalized populations.[1,3] In countries with developed economies, economic loss due to influenza is an important factor[4] (see Section II).

Influenza has justifiably been called "the last great plague"[5] since it is the only pandemic disease of man remaining at a time when many infectious diseases, including smallpox and poliomyelitis, caused by viruses have been eliminated or reduced to insignificant proportions as a result of improved standards of public health and hygiene, extensive vaccination programs, and, in the case of bacterial and parasitic disease, chemoprophylaxis or chemotherapy.

An important factor in the success of influenza virus as a human pathogen is its antigenic adaptability. The high degree of antigenic variation exhibited by influenza A and, to a lesser extent, influenza B is unique among human viruses. The study of antigenic variation is an important aspect in attempts to understand the epidemiological behavior of the disease. This feature of the virus also presents serious problems for control by immunoprophylaxis. Antigenic variation is discussed in detail in Section II.

The ultimate control of influenza by vaccination or chemoprophylaxis depends on a better understanding of the biology and epidemiology of the virus and the nature of immunity. Prospects for such control are currently increasing with expanding knowledge of the nature of the virus and how to manipulate it in the laboratory. In particular, the technology now exists for the preparation of inactivated vaccines with a high degree of purity and antigenicity; also, rapid progress is being made toward the development of live attenuated vaccines. Prospects for the control of influenza by vaccines are discussed in Section VII.

Effective world-wide surveillance of influenza, particularly for the early detection of new variants, is another essential aspect in attempting to understand the epidemiology of influenza and to mount effective vaccination or, when compounds become available, chemoprophylactic programs (see Section V).

Influenza A infections occur in many different species of animals. Some animal influenza viruses show antigenic similarities to human influenza A viruses. Animal influenza viruses may provide useful experimental model systems relevant to the study of human influenza but their main significance is the possibility that animal influenza A viruses may be important as potential progenitors for new pandemic viruses of man. This possibility has stimulated much interest in animal influenza viruses, and extensive investigations of the ecology of influenza in animals and comparative studies on the genetic and antigenic character of animal and human influenza viruses (see Section VI) are currently in progress.

II. THE VIRUS AND ITS ANTIGENIC CHARACTERISTICS

A. Structure and Antigenic Composition

The influenza viruses, classified as orthomyxoviruses,[6] are lipid-containing, enveloped viruses in which the genetic material is single-stranded RNA. The RNA is present in the virus particle as eight distinct species which are probably monocistonic in function[8] (Chapter 4); molecular weights range from 2×10^5 to 1×10^6 daltons.[7] This fragmented nature of the influenza virus genome has important implications for the understanding of the intrinsic antigenic variability of the influenza virus and the emergence of new pandemic virus strains. The virus RNA is "negative" stranded (nonmessage polarity), and the virus contains RNA-dependent RNA transcriptase to facilitate conversion to message RNA upon infection of host cells (see Chapter 4). Influenza viruses possess the property of hemagglutination of erythrocytes and contain the enzyme neuraminidase. Both functions are associated with specific surface antigens of the virus and are useful characteristics in laboratory detection, identification, and assay of the viruses.[9]

Figure 1 diagrammatically illustrates the structure of the virus which exists as irregularly shaped particles 75 to 110 mm in diameter, bound by an outer lipid membrane. The envelope glycoproteins hemagglutinin (HA) and neuraminidase (NA) take the form of "spikes" covering the virus surface and are attached by one end at the lipid membrane. The morphology of these structures has been described in detail by Laver and Valentine,[10] and further details are given in Chapter 5. Internally, the particle contains several other components; the two major ones are the ribonucleoprotein (NP), which is closely associated with the helical RNA genome of the virus, and the matrix protein (M), which underlies the lipid membrane and may form a matrix surrounding the NP-RNA complex. The antigenic nature of the HA, NA, and NP components of the virus has been well established.[9,11,12] More recently, it was shown that the hitherto unrecognized M protein is an additional virion antigen.

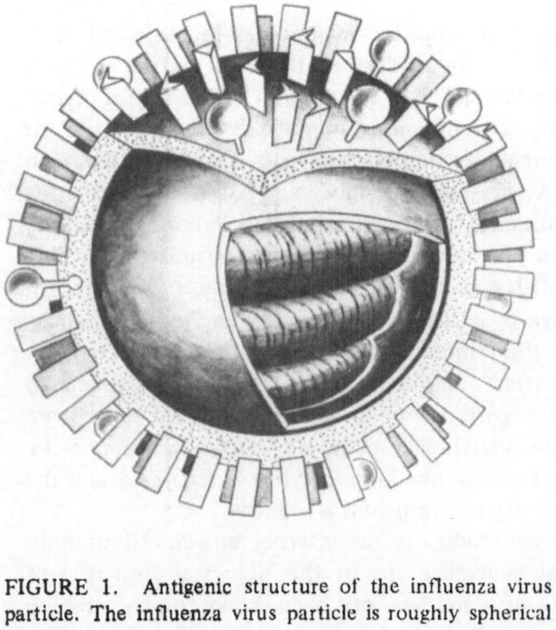

FIGURE 1. Antigenic structure of the influenza virus particle. The influenza virus particle is roughly spherical or sometimes filamentous in shape and some 75 mm in diameter. Four major protein antigens (and two or three minor proteins) constitute the particle. Two major proteins, hemagglutinin (HA) and neuraminidase (NA), are on the virus surface and are inserted in the lipid bilayer (Li) which bounds the particle. Internally there are two major proteins, the matrix protein (M) and the nucleoprotein (NP).

The polypeptide composition of highly purified influenza virus preparations has now been determined by polyacrylamide gel electrophoresis studies of purified virus preparations.[14-16] Eight polypeptides (Table 1) have been identified in the virus particle, with molecular weights ranging from 25,000 to 96,000 daltons (see also Chapter 5). The relationship of five of the viral polypeptides to known structural and antigenic components of the virus, as indicated in Table 1, has been established in several laboratories.[13-15] The HA is synthesized as a single protein with a molecular weight of approximately 80,000 daltons, but it may be cleaved into two species of glycoprotein designated HA-1 and HA-2 with molecular weights of 58,000 and 28,000 daltons,[14,17,18] respectively. The total molecular weight of the functional HA subunit is 210,000 daltons;[19] thus, it contains two or more molecules of each of the component polypeptides. The NA antigen is associated with a single glycoprotein species with a molecular weight of 60,000 daltons.[14] The molecular weight of the functional NA subunit is 240,000 daltons,[20] and it appears to contain four polypeptide molecules per subunit (Chapter 5). The NP antigen is associated with a single polypeptide with a molecular weight of approximately 53,000 for influenza A viruses and 60,000 for influenza B viruses.[14,21,22] The M antigen, which is the smallest but most abundant polypeptide of the virus, has a molecular weight of 25,000 daltons.[13,14,22] P-1, P-2, and P-3 represent minor internal components of the virus;[13,16,22] their antigenic character remains unknown but these components are probably associated with the RNA and are involved in the RNA-dependent RNA transcriptase activity of the virus.[23] In addition to the structural virus polypeptides, at least two other virus-coded polypeptides (NS1 and NS2) are synthesized in infected

TABLE 1

Proteins and Antigens of the Influenza Virus

Protein designation	Mol wt × 10^3 daltons	Location and function, antigenicity
P_1, P_2, P_3	85–96	Minor internal components of the virus; probably associated with polymerase activity; antigenicity unknown
HA[a]	75–80	Hemagglutinin – antigenically variable ("shift" and "drift") surface glycoprotein surface antigen; antibody to HA neutralizes virus infectivity and is the major protective antibody following infection or immunization; antibody levels provide the best index of immunity to infection
NA[a]	60–65	Neuraminidase – antigenically variable surface glycoprotein; antibody does not neutralize virus infectivity directly but inhibits virus release from infected cells and contributes to immunity following infection or immunization
NP	53	Nucleoprotein – antigenically invariable internal antigen provides basis for classification of viruses into Types A and B; antibody not protective, apparently no role in immunity
M	25	Matrix protein – antigenically invariable internal antigen; antigenically distinct for influenza A and B viruses; antibody not protective, apparently no role in immunity

[a]Essential components of subunit vaccines (see Section VII).

host cells[8] (Table 1) but do not become incorporated into virus particles.

B. Properties of Antibody to Specific Virus Components

If directed against the HA of homologous or antigenically closely related[24] virus, antibody to HA is efficient in neutralizing virus infectivity and in inhibiting the agglutination of erythrocytes (hemagglutination inhibition, HI). Specific antibody to NA is capable of inhibiting the enzyme activity of the virus on high molecular weight substrates such as fetuin, but not on low molecular weight substrates (Chapter 6). Neuraminidase inhibition (NI) is due to the steric effects of antibody hindering the access of enzyme to substrate. Although anti-NA is not generally associated with virus neutralization, it may neutralize infectivity if present in high concentrations.[25] Antibody to NA also inhibits the release of virus from infected host cells when constantly present in the cell environment.[26,27] Such inhibitory activity on virus shedding may be independent of the ability of the antibody to neutralize the enzymic activity of neuraminidase or the substrate on the membranes of infected host cells.[27] Antibodies to the two surface antigens provide an index of immunity to influenza in man and experimental animals.[28,29] Protection of animals by administration of passive antibody to HA or NA strongly suggests that circulating antibody to the antigens is an important mechanism of immunity.[30,34] The mechanisms by which anti-HA and anti-NA protect appear to differ. In man or experimental animals, antibody to hemagglutinin imparts solid immunity to infection; in contrast, antibody to neuraminidase is associated with partial immunity.[30,31] It has been shown that individuals possessing anti-NA may be infected, but shedding of virus is restricted and the severity of symptoms is reduced.[31,32]

Antibodies to the internal antigens NP or M do not neutralize any of the biological properties of the virus and have not been shown to play a role in immunological protection.[13,33,34] Cross immunity between influenza A viruses with completely unrelated surface HA and NA antigens has been observed in experimental animal systems.[35,36] This could not be attributed to production of interferon, virus interference, or nonspecific immunological effects since a related phenomenon was not observed between influenza A and B viruses; further studies are required to determine the mechanism involved.

The internal antigens of the influenza virus are antigenically invariable and form the antigenic basis of classification of viruses in influenza A and influenza B.[37] The NP antigens of all influenza A viruses are apparently of the same immunological specificity as their M antigens.[13,38] Influenza B viruses also possess common NP and M antigens which are distinct from those of influenza A.

Little is known about the number, position, or chemical nature of the specific antigenic determinants of the various structural antigens of influenza virus. Most of the available information is on the HA subunit, which has been shown to have at least two groups of determinants.[39,40] Two different populations of antibody molecules can be distinguished in anti-HA sera; one population reacts specifically with antigenically variable (strain-specific, SS) determinants of the molecule, and others react with determinants which were antigenically stable within an HA subtype (cross reacting, CR) (see Section II.C). These determinants do not relate to the HA-1 and HA-2 polypeptides. Antibody to the SS determinants of HA is efficient in reacting with sites of the HA molecule critical for virus neutralization, whereas antibody to CR determinants is relatively inefficient in neutralizing infectivity.[41] Immunization of thymus-deprived (TXBM) mice showed a thymus dependence of antibody formation against both determinants.[39] However, the thymus dependence of antibody elicited by the CR determinants is overcome by repeated inoculations of HA, whereas the dependence of antibody formation to the SS determinants is less readily overcome.[39] Further investigation is required to determine the significance of SS and CR determinants in studies of antigenic "drift" in the HA antigen and to determine the properties of antibodies to these determinants in naturally acquired immunity following vaccination.

C. Antigenic Variation

The antigenic varieties of human influenza A virus prevalent from 1918 to the present are shown in Table 2. HA and NA, the surface antigens of the virus, are antigenically variable for influenza viruses Type A and B; from time to time, these antigens undergo independent antigenic changes. At entirely unpredictable intervals of 10 to 39 years (Table 2), complete changes have occurred in the antigenic character of the hemagglutinin and (in 1957) simultaneously in the neuraminidase antigen of the prevalent influenza A viruses. Such changes are known as antigenic "shift." Influenza A viruses bearing such new surface antigens spread rapidly, causing world-wide outbreaks of disease. At more frequent intervals, minor changes (termed antigenic "drift") take place in the HA and NA antigens. Although each episode of "drift" is in itself minor, the effects are additive and over a period of several years result in a virus showing a considerable degree of antigenic difference from the parent strain (Table 1l). In antigenic "shift" both CR and SS determinants of the HA antigen undergo change, while only SS determinants change in antigenic "drift" as the CR determinant enables the identification of the HA as being within a particular subtype, e.g., H2 or H3.*

There is circumstantial evidence indicating that the causative virus of the 1918 pandemic was antigenically close to the swine (Hsw1N1)* influenza virus which has been endemic in swine populations in the U.S. since 1930. Viruses of the swine-influenza antigenic type have recently been isolated from military recruits in the U.S.,[42] and this has caused considerable speculation in the U.S. and elsewhere about the possibility of swine (Hsw1N1) virus becoming the prevalent human influenza A virus in the future (see Section VI).

Influenza A viruses undergo both "shift" and "drift," while influenza B viruses only undergo infrequent antigenic "drift." Antigenic "drift" occurs as a result of the natural selection of spontaneously occurring mutants under the pressure of increasing levels of immunity in the population. The mechanism of "shift" remains unknown. However, it seems unlikely that "shift" is due to genetic mutation of the previously prevalent strain. Recent findings[43] indicate that new "shifted" strains may arise directly from the reservoirs of influenza A virus which exist in nature in lower mammals and birds; alternatively, they may be due to genetic recombination between human or animal influenza A viruses, resulting in a reassortment of their genetic material (RNA fragments) and a sharing of their biological and antigenic characteristics.

*For WHO system of nomenclature of influenza viruses, see Reference 37.

TABLE 2

Antigenic Subtypes of Hemagglutinin and Neuraminidase of Influenza of Man

Period of prevalence of subtype	Reference strain	HA and NA subtypes	Variants showing antigenic "drift"	Type of antigenic variation and pandemic periods	
1918–1920+	A/Swine/15/30[a]	Hsw1N1		"SHIFT"	Great pandemic of 1918–20 probably caused by swine/30 virus
1933–1957	A/PR/8/34[b]	H0N1	A/Weiss/43 (H0N1)	"drift"	
	A/FM1/47	H1N1	A/England/1/51 (H1N1)		
			A/Denver/1/57 (H1N1)		
1957–1968	A/Singapore/57	H2N2	A/England/12/64 (H2N2)	"SHIFT"	Asian pandemic of 1957–58
			A/Tokyo/3/67 (H2N2)	"drift"	
1968–?	A/Hong Kong/68	H3N2	A/England/42/72 (H3N2)	"SHIFT"	Hong Kong pandemic of 1968–69
			A/Port Chalmers/1/73 (H3N2)	"drift"	
			A/Victoria/3/75 (H3N2)		
1976	A/New Jersey/76	Hsw1N1		"SHIFT"?	Possible future pandemic virus

[a] Viruses antigenically close to the swine influenza virus (A/Swine/Iowa/1930 ($H_{sw1}N1$) are thought to have caused the 1918 pandemic on the basis of indirect evidence since virus isolation from man was not achieved before 1933.

[b] Although designated Hsw1, H0, and H1, the hemagglutins of strains isolated in man from 1933 to 1957 and of A/Swine/Iowa/1930 are of a single antigenic subtype.

Antigen "shift" occurred in 1957, resulting in the appearance of the "Asian" (H2N2) pandemic virus, and again in 1968, resulting in the Hong Kong (H3N2) pandemic. The 1957 "shift" involved both the HA and NA antigens, while the 1968 "shift" involved only the HA antigen. Both "shifts" were associated with pandemics of influenza. However, the mortality associated with the 1957 pandemic was greater than that encountered in 1968; this was probably due to the protective effect of antibody to the unchanged NA antigen in the case of the 1968 epidemics.

The enormous impact of the pandemic which occurred in 1918 to 1919 is now legendary.[44] It seems likely that the onset of the outbreaks in 1918 coincided with the antigenic "shift" in the influenza A virus and that this subtype of virus persisted from 1918 until 1956, undergoing only progressive antigenic "drift" during this period (Table 2). Indirect circumstantial evidence based on studies of the sera of persons alive before 1918 indicates that the 1918 pandemic virus is closely related to influenza A viruses of swine (Hsw1N1).

Both "Asian" and "Hong Kong" viruses have undergone progressive antigenic "drift" from 1957 to 1967 and 1968 to 1976, respectively. The high frequency of antigenic "drift" shown by the "Hong Kong" virus is remarkable and considerably greater than the "Asian" virus.[45] Successive variants of "Hong Kong" appeared in 1971 (A/Hong Kong/71), 1972 (A/England/42/72), 1973 (A/Port Chalmers/1/73), 1974 (A/Scotland/840/74), and 1975 (A/Victoria/3/75).[46] With the exception of the 1971 strain, all of these variants became prevalent and produced an epidemic impact.

With the appearance of completely new antigenic variants (antigenic "shift") in 1957 and 1968, the formerly prevalent strains rapidly disappeared from most of the world over a period of a few months, persisting only in a few isolated foci. There is currently no explanation for this dramatic disappearance of former epidemic viruses from man. However, evidence (mentioned elsewhere in this chapter) indicates that former human strains may persist in animal hosts and possibly emerge after several decades as new pandemic viruses.

III. THE HISTORY OF INFLUENZA EPIDEMICS AND PANDEMICS

Although the influenza virus was not isolated until 1933, there is circumstantial evidence (discussed below) as to the causative influenza viruses of major outbreaks which occurred in 1889 to 1890, 1898 to 1900, and 1918 to 1919. The isolation and characterization of the influenza virus by Smith et al.[47] led to rapid progress in the identification of viruses and the study of their epidemiological behavior, founding the epidemiological study of the disease on a sound scientific basis. Since 1933 recurrent epidemics of influenza A or influenza B infection have been identified in all countries where virological investigations have been carried out.

In countries from which epidemiological information is available, years in which no influenza outbreaks are recorded are rare. Most outbreaks of influenza and all pandemics are due to the influenza A virus. Influenza B outbreaks are four to six times less frequent than those caused by influenza A virus[2,48] and are relatively localized, producing their major impact in children. Raised mortality is not a constant feature of influenza B epidemics, probably because of the relatively young age group affected. Figure 2 shows the relatively frequent occurrence of influenza A outbreaks in the U.S. from 1934 to 1970 and the lower frequency of influenza B epidemics. Excess mortality associated with influenza A epidemics was of a considerably higher order than that for influenza B with the exception of 1935 to 1936, when high mortality rates were apparently associated with influenza B epidemics.[2]

A. Early History

It is difficult to establish with certainty the past history of influenza epidemics, even in the developed areas of the world. The interpretation of historical accounts of influenza-like diseases requires caution because of the abundance of epidemics of disease of diverse etiology in medieval times, some of which may have had a clinical resemblance to influenza. Descriptions of extensive outbreaks of disease appear in the literature from as early as the 12th century,[49] apparently showing close clinical and epidemiological similarities to present-day outbreaks and reportedly associated with high mortality rates. Outbreaks of a widespread nature occurred in 1781 to 1782, 1803, 1833, 1837, 1847 to 1848, 1889 to 1890, 1898 to 1900, and 1918 to 1919. There is little doubt that the outbreaks of 1847 to 1848 and 1889 to 1890 were pandemic in nature,

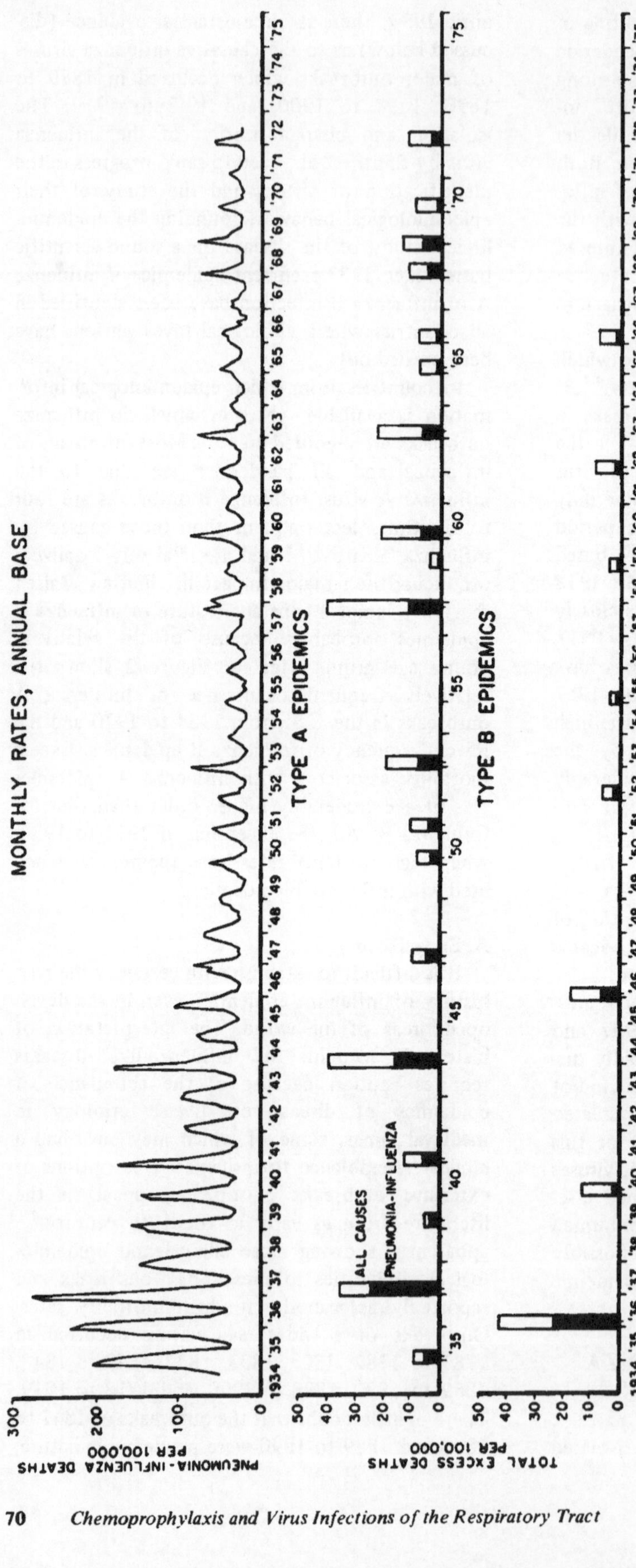

FIGURE 2. Pneumonia-influenza death rates by month and excess mortality during epidemic periods in the U.S., 1934 to 1970. (After Dowdle et al.[2])

as were the subsequent outbreaks of 1918 to 1919, 1957 to 1958, and 1968 to 1970.

Available information on these early outbreaks of influenza-like illness has been reviewed by Hirsch,[49] Creighton,[50,51] and Burnet and White.[52] The 1781 outbreak appeared in Asia and spread west to Europe by 1782. This outbreak was reported to have affected persons of middle age rather than the young or elderly, but the mortality rate was not high. The 1847 pandemic spread from the east through Russia, reaching the U.K. in the winter of 1847 to 1848. This pandemic was reported to have been associated with high rates of mortality in the elderly. The 1889 to 1890 pandemic also spread from the east through Russia to Europe and subsequently to the U.S. during the autumn and winter, producing outbreaks with high mortality levels. In Europe the following 4 years were also marked by waves of severe influenza, possibly caused by the same pandemic virus.

B. The 1918 to 1919 Pandemic

The years 1918 to 1919 saw the most severe pandemic of influenza yet recorded, the so-called "Spanish Influenza" pandemic. Although much anecdotal information[53] is available on the distribution and severity of this pandemic, its place of origin is uncertain.[54] However, infection probably spread from Asia to Europe. The pandemic occurred in several waves during a period of little over 1 year. The first, in the spring of 1918, was relatively mild; the attack rate was 20 to 40% in individuals up to 50 years old and considerably lower in the aged. The second wave, in the autumn of 1918, was exceptionally severe and produced enormous mortality rates in the 20- to 40-year-old age group. A less severe wave occurred in early 1919. The total mortality for this pandemic is estimated to be 15 to 50 million deaths, and closer estimates are impossible to ascertain. Estimates of mortality by continents are shown in Table 3. As described above, there is circumstantial information suggesting that this pandemic was caused by the classical swine influenza virus (Hsw1N1) or a virus antigenically close to it. First, the initial appearance in 1918 of swine influenza in pigs in the U.S.[55] coincided with the onset of human pandemics. Second, antibody to the swine virus is present in a high proportion of individuals who were alive in 1918.[56,57] Thus, it seems likely that the swine influenza virus isolated by Shope[58] in 1930 was transmitted from man to pigs in 1918 and has since remained as an endemic infection of swine in the U.S. The recent 1976 isolation of swine influenza virus from man in the U.S. (described above) may be of considerable potential importance, signaling the early stages of a new Hsw1N1 pandemic era.

TABLE 3

Estimated Mortality During the 1918–1919 Pandemic

Continent	Influenza mortality 1918–19
North and Central America	1,100,000
South America	300,000
Europe	2,200,000
Australia and Oceana	900,000
Africa	1,300,000
Total for all continents	21,500,000

From Collier, R., *The Plague of the Spanish Lady*, Macmillan, London, 1974. With permission.

C. The "Asian" Influenza Pandemic and the Period 1957 to 1968

The "Asian" (H2N2) influenza virus first appeared in central China in February 1957.[59] This virus possessed hemagglutinin and neuraminidase antigens showing complete antigenic differences (antigenic "shift") from the formerly prevalent H1N1 strain; this probably represented the first antigenic shift in the influenza A virus since the pandemic of 1918 (see Table 2). This new virus spread rapidly to other areas in Asia (including Hong Kong, Taiwan, and Singapore), producing severe outbreaks. As early as June 1957, outbreaks occurred in several European countries (Czechoslovakia, Netherlands, U.K.), the Middle East, and the U.S. In the southern hemisphere epidemics reached South Africa and South America by July. By August the epidemics in the Far East had subsided, but by late September the U.S., northern areas of South America, and all of Europe suffered rapidly evolving outbreaks associated with high rates of morbidity and mortality. The peak of the epidemic in the U.S. and Europe occurred in October 1957. Virtually every country of the world had been infected by the end of November 1957.[60]

The "Asian" virus produced high attack rates during the 1957 to 1958 pandemic period, varying from 20 to 80% in different countries. Highest rates of attack were in the younger age groups of the populations and in areas of high population density (reviewed by Stuart-Harris and Schild).[48] In the U.K.[61] and U.S.,[62] approximately 50% attack rates were recorded in the 5- to 15-year-old group.

It is of considerable interest that in 1957, before the circulation of the "Asian" virus, some 30% of serum samples from the aged persons (75 to 85 years old) were found to contain antibody to the "Asian" strain. This has been taken to indicate that the "Asian" virus of 1957 represented a reappearance of a virus which had circulated some 68 to 70 years before. As mentioned above, pandemics occurred in the period 1889 to 1890 which might have been produced by an earlier emergence of the "Asian" virus.

During the following 10 years (1958 to 1968), many influenza outbreaks due to the influenza A (H2N2) virus occurred. In general, these had a small impact on mortality and morbidity in comparison to the first waves of the "Asian" epidemic. In 1959 significant outbreaks occurred in the U.S.S.R., but not elsewhere. In 1960 and 1963 widespread influenza occurred in many countries. Again in 1965 widespread outbreaks occurred in Europe and the U.S.S.R. During the next 2 years, outbreaks were mild and sporadic, but in the winter of 1967 to 1968 major influenza epidemics were documented in the U.K. and other European countries and in Taiwan and Australia. During this period, progressive antigenic "drift" occurred in the HA and NA antigens of the prevalent strains. These were typified by the A/England/12/64 (H2N2) strains which caused the 1965 outbreak and the A/Tokyo/3/67 (H2N2) strain which was involved in the outbreak during the winter of 1967 to 1968.

D. The "Hong Kong" Pandemic and the Period 1968 to 1976

Although its true geographic origin is not known, the "Hong Kong" (H2N2) virus was first isolated during influenza outbreaks in Hong Kong in July 1968.[63] It has been suggested that outbreaks of respiratory disease on the Chinese mainland some weeks earlier were caused by the same virus. By early autumn, the virus spread rapidly in the Far and Middle East, India, Japan, and Australia.[64] However, only the U.S. suffered major outbreaks with increased mortality during the winter of 1968 to 1969.[65] In other countries, although it was prevalent in the populations, the strain produced a relatively mild epidemiological impact. However, the full-blown effect of the Hong Kong pandemic did not occur until the next winter (1969 to 1970) in Europe, Asia, and most other areas of the world.

In the U.S. Hong Kong influenza produced more than 51 million illnesses during the epidemic of 1968 to 1969, an incidence of 25%.[2,4] The disruptive effect of the epidemic is evident in that it caused excessive school absenteeism involving the loss of almost 35 million days from school (16% of the total number of days of school missed by all school children during the entire school year). Individuals then employed missed more than 66 million days from work, a figure which represents 22% of total annual days lost from work due to all acute conditions. An estimated 27,495 excess deaths were attributed to the epidemic. The mortality rate in the population at large was 14.0 per population of 100,000. Kavet[4] has estimated that in 1968 to 1969 influenzal disease cost an overall $3.5 billion in the U.S.

The differing chronology of the outbreaks in the U.S. and the U.K. is shown by the excess mortality data for these countries in the successive winters of 1968 to 1969 and 1969 to 1970 (Figure 3) but it is completely unexplained.

Table 4 shows the claims for sickness benefit due to influenza among insured persons in the U.K. The Hong Kong outbreak of 1968 to 1970 produced some 25 million lost working days during the period December 1969 to March 1970. This was approximately 7% of the total lost working days for all causes during the whole year. In general, the Hong Kong pandemic was less severe in terms of mortality and morbidity than the Asian pandemic in 1957. The antigenic changes in the virus may provide a clue to its epidemiological behavior. While the HA antigen (H3) was completely different from that of the Asian virus (H2), the neuraminidase was of the same antigenic subtype (N2). The Hong Kong virus showed only partial antigenic "shift;" thus, the common NA antigen of the Asian and Hong Kong virus may have contributed to a cross immunity between these strains.

Recurrent influenza outbreaks have been associated with H3N2 viruses since 1970. Only the period 1970 to 1971 was free from influenza activity. Successive antigenic variants of the Hong

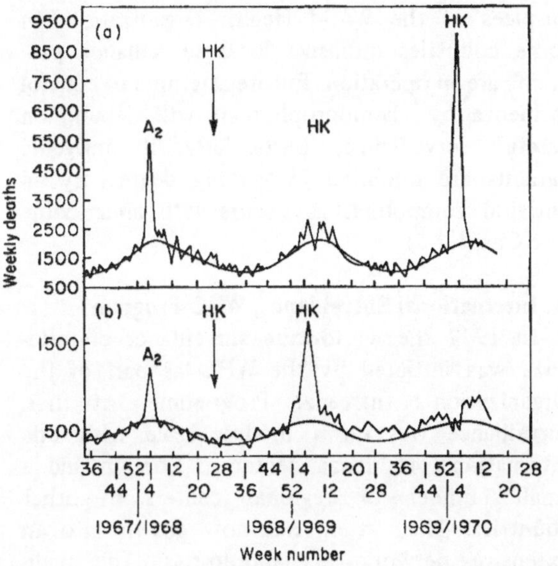

FIGURE 3. (a) Pneumonia, influenza, and bronchitis deaths in England and Wales. (b) Pneumonia and influenza deaths in 122 U.S. cities. A_2 indicates epidemic due to A/Tokyo/67(H2N2) virus. HK indicates epidemics due to A/Hong Kong/68(H3N2) virus. HK↓ indicates time of first appearance of Hong Kong/68 virus. (From Oxford, J. S. and Williams, J. D., Eds., *Chemotherapy and Control of Influenza*, Academic Press, London, 1975. With permission.)

TABLE 4

Sickness Benefit Claims Due to Influenza and All Cases Among Insured Persons in the U.K.

	Days claimed (millions)	
June to May	**All causes**	**Influenza**
1962–63	289	13.9
1963–64	287	9.4
1964–65	299	9.6
1965–66	312	16.7
1966–67	301	6.8
1967–68	328	15.5
1968–69	329	15.7
1969–70	342	25.7[a]
1970–71	314	6.6
1971–72	307	10.4
1972–73	320	13.6
Mean	312	13.8

Data from official U.K. statistics.

[a]Hong Kong/68 (H3N2) pandemic period.

Kong virus typified by the strains A/England/42/72 (H3N2), A/Port Chalmers/73 (H3N2), A/Scotland/42/72 (H3N2), and A/Victoria/3/75 (H3N2) have produced outbreaks of almost annual incidence in many countries. The most recent variant (A/Victoria/75) was first isolated in the autumn of 1975 in Papua New Guinea from outbreaks associated with high mortality.[46] This variant produced severe outbreaks of disease in Europe and in the U.S. during the winter of 1975 to 1976, causing an excess mortality rate higher than any since the 1968 to 1970 pandemic period.[66]

It is interesting that the studies of Masurel[83] indicate that a proportion of individuals born before 1898 appeared to possess antibody to Hong Kong virus before its appearance in 1968. This has been taken to indicate that the Hong Kong virus might have produced the severe outbreaks which occurred in 1898 to 1890 before losing prevalence to reemerge in 1968.

IV. ENVIRONMENTAL AND NONSPECIFIC HOSTS FACTORS IN INFLUENZA

Influenza epidemics occur in the cold seasons in both the northern and southern hemispheres, essentially between December and April in the northern hemisphere and May to September in the south; thus, the disease appears to be under climatological control. Exceptions to the winter prevalence of epidemics in the northern temperate zone may occur with the emergence of new pandemic strains, as was seen in 1957 when summer and early autumn outbreaks occurred due to the Asian virus. In tropical areas influenza may coincide with the wet season. Schild and McGregor[67] noted widespread outbreaks of Hong Kong influenza in tropical West Africa (Gambia) in autumn of 1968 and again in autumn of 1971. In equatorial areas, outbreaks may not be of clear-cut seasonal occurrence; for example, in Southest Asia the influenza A virus may be isolated the whole year round but, typically, with a peak incidence of isolations and disease in June and July.

The way in which climatological conditions affect the epidemiological behavior of the virus has not been clearly established. An analysis of the situation must consider that certain factors may affect the susceptibility of the host to infection, to develop disease symptoms, and to shed virus, while others may influence the viability and infectivity of the virus itself. Hope-Simpson[68] and Davey and Reid[69] found a positive correlation in successive winters between the incidence of influenza activity and the coldness of the winter. Reid[70] suggested that rapid fluctuation in temperature values may be more important than actual values. The effect of humidity has not been well established. Hemmes et al.[71] found that the influenza virus itself survives best in aerosols at low relative humidities. The effects of temperature and humidity on aspects of human physiology which may influence influenza virus replication are unknown. However, it should be emphasized that severe influenza epidemics are truly global in distribution and, therefore, environmental effects must have a relative rather than absolute controlling influence on influenza epidemiology.

Another factor which may be important is the population distribution at different seasons. At certain times of the year the requirement for heating in the temperatue zones causes relative crowding of persons into heated zones. The school year also has an effect on local population density. The efficient spread of the virus during the winter in temperate countries may be a result of several of the above factors (low exterior temperature, low interior humidity, and high local population density). There is evidence indicating that spread of infection is generally more efficient in densely populated urban areas than rural or remote areas of low population density. A low socioeconomic status contributes to high population density and may thus influence the impact of influenza epidemics. Assaad and Reid found that the lower socioeconomic groups in Glasgow suffered a higher mortality risk from influenza than higher groups. Other nonspecific factors include the distribution in the population of noninfluenzal microbial factors such as organisms which produce secondary bacterial pneumonia (*Pneumococcus, Staphylococcus, Haemophilus*). Such organisms may themselves have a season-dependent distribution in the population. A more detailed review of nonspecific factors in influenza epidemics is given by Kilbourne.[72]

V. MECHANISMS OF INFLUENZA SURVEILLANCE

For influenza, international surveillance is particularly important because of the constantly changing antigenic character of the prevalent virus and the association of major changes (antigenic "shifts") with pandemics of disease. Continuous international surveillance is conducted under the auspices of the World Health Organization. In some countries intensive local surveillance programs are in operation. Future attempts to control influenza by chemoprophylaxis will depend on careful surveillance, particularly if antigenic variants are inhibited in varying degress by an antiviral compound, as occurs with amantadine (see Chapter 5).

A. International Surveillance, WHO Program

In 1947 the world-wide surveillance of influenza was initiated by the WHO as part of the Organization's Influenza Programme. At first, surveillance was on a modest scale with one international influenza center in London and a small number of regional centers in other countries. The system has now grown into an extensive network of laboratories. The main purpose of the program is the early detection and isolation of new antigenic variants of influenza virus and the collection of epidemiological data leading to a fuller understanding of the epidemiology and natural history of the disease. The early and accurate information available to the WHO on the behavior of the virus provides the basis for recommendations given to National Authorities enabling the preparation of appropriate public health and prophylactic measures for combating the disease.[9] The activities of the program are co-ordinated by the WHO in Geneva. There are two international influenza laboratories, the WHO Collaborating Center for Reference and Research on Influenza in London and the WHO Collaborating Centre for Influenza for the Americas in Atlanta, Georgia, together with 97 national influenza laboratories in 69 different countries.

When new influenza virus isolates showing significant antigenic change are detected, they are made available by the international centers to competent laboratories, including organizations producing influenza vaccines. The characterization of new viruses is carried out in the two international centers in London and Atlanta. The national centers are annually provided with a kit of reagents enabling them to identify current subtypes. Isolates not readily identifiable by reference reagents are of interest as possible variants, and national influenza centers isolating such viruses immediately submit them to one of the international centers for rapid identification. The WHO program involves the collection and distribution of epidemiological information from national authorities. This comes from various national influenza centers, national epidemiological services, or the ministries of health of different collaborating countries. During the influenza season in the part of the world concerned, a report is sent weekly to the WHO virus unit in Geneva, where the information is collected and published in the WHO Weekly Epidemiological Record. Although the program has attained a considerable degree of success in meeting its objectives, influenza epidemics remain largely unpredictable.

The World Health Organization, in recognition of the possibility that animal sources of influenza may play a role in the emergence of new human epidemic viruses, also coordinates ecological studies of animal influenza in various countries, notably the U.S., U.S.S.R., and several European countries. Two WHO collaborating centers for studies of animal influenza have been set up in the U.S.

B. National Surveillance

The basis of national surveillance is the existence of a laboratory service able to diagnose influenza infection and isolate the causative viruses. In many countries there are already entirely adequate laboratories for this purpose. In some countries there is also a system of notification which, in cooperation with the laboratory, provides information on the incidence and severity of influenza. If an effective inhibitor of influenza virus becomes available, then accurate notification of influenza circulating in the community is important so that persons need only receive the compound for a defined period of time while at special risk (Chapter 3, Volume II). A method which has proved useful in some countries is the setting up of a network of "spotter" physicians in clinics who sending specimens to the diagnostic laboratory from suspected clinical cases. However, in many countries plans for regular and uniform collection and dissemination of epidemiological information on influenza have yet to be developed.

Epidemiological data from which a useful indication of the prevalence of influenza can be obtained in the country concerned are as follows.

1. Mortality Rates

The weekly number of deaths from influenza and its complications and from all causes in the whole country (or in selected areas) is assessed. The concept of excess mortality due to influenza[73] or pneumonia and influenza[74] is of considerable value and assessed by comparing mortality rates during epidemics with those which occured in the same population at the same period of the year in noninfluenza years. The strikingly different mortality impacts of influenza epidemics in different countries during the same years has been emphasized by Assaad et al.[73] who compared excess mortality data from 13 different countries. Data for the U.K., U.S., and U.S.S.R. from 1963 to 1973 are shown in Figure 4.

2. Morbidity Rates

The indices of influenza activity are as follows:

1. The weekly number of cases of influenza, or acute respiratory disease including influenza, from national sources or from selected practices or hospitals.
2. The weekly number of absences due to acute respiratory disease or to all causes from sickness benefit claims or from reports from selected groups, factories, administrative services, military units, schools, and universities and the weekly number of emergency hospital admissions.

In the case of an outbreak, the clinical features of the illness (estimated attack rates in various age groups of the population) and an account of locations involved and duration of illness and its age distributions should be reported.

The basis of the collection of epidemiological information in the U.K. is described below. Indirect methods for the detection of influenza[75,76] in the U.K. can be summarized as follows:

1. Sickness benefit claims are assessed and cover all of the adult working population.
2. Total mortality and mortality due to respiratory diseases and their complications are determined on the basis of an examination of death certificates sent to the Registrar General.
3. Rates of emergency admissions to hospitals are recorded weekly.
4. Industrial and school absenteeism give valuable information.
5. Returns of new consultations for acute respiratory infection from general practitioners in 100 practices under the auspices of the Royal College of General Practitioners are particularly

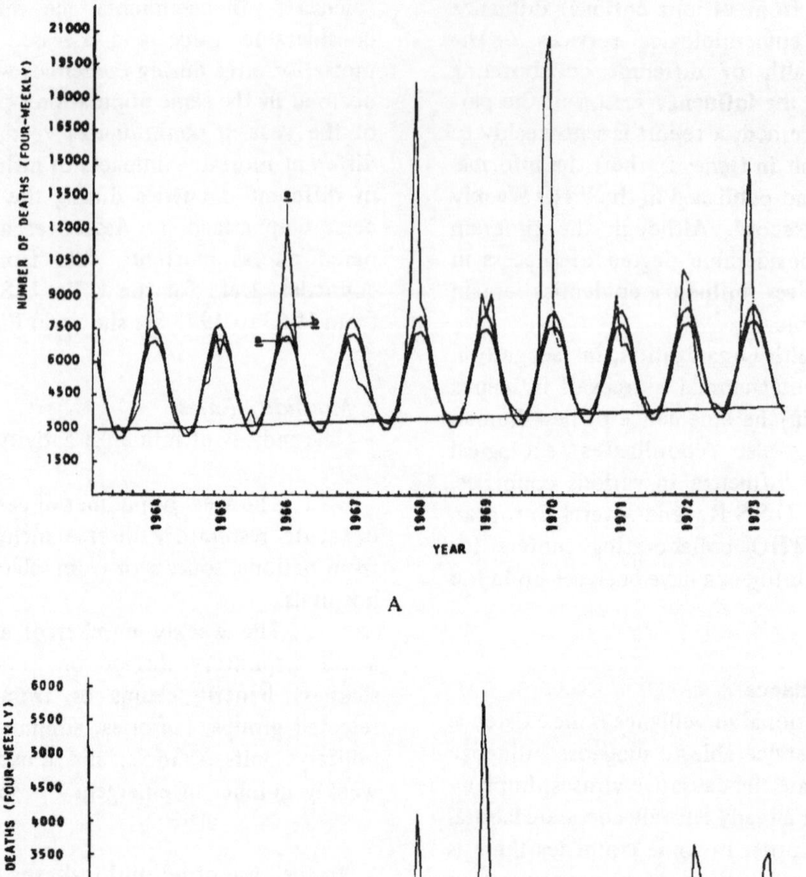

FIGURE 4. Influenza mortality rates in the U.K., U.S., and U.S.S.R. (A) Deaths from influenza, pneumonia, and bronchitis in England and Wales from 1963 to 1973. Curves a, b, and c represent, during a 4-week period, (a) the expected seasonal curve obtained from deaths in years of low mortality when no sizable influenza outbreaks occurred; (b) the expected seasonal curve based on deaths in all years under study, except those in which very high peaks occurred; and (c) the actual number of deaths. (B and C) Curves of the actual and expected seasonal number of deaths from influenza and pneumonia in the U.S. and U.S.S.R., respectively. (From Assaad, F., Cockburn, W. C., and Sundaresan, T. K., *Bull. WHO*, 49, 219, 1973. With permission.)

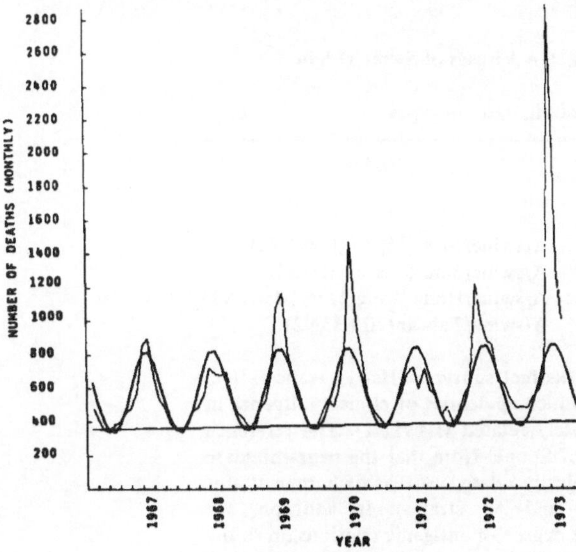

FIGURE 4C

useful in that they reflect experience in all ages of the population.

6. Some 44 regional Public Health Laboratory Service virological units in the U.K.[76] routinely return laboratory evidence of influenza isolation to the Central Public Health Laboratory Service or the Communicable Diseases Unit in Scotland.

7. Excess mortality is used as an index.

In the U.S., influenza surveillance activities are coordinated by the Center for Disease Control (CDC) at Atlanta, Georgia. The CDC collects current epidemiological data, disseminates information to state health officers and evaluates it, and facilitates the exchange of information with the WHO. Dowdle et al.[2] recently reviewed in detail the natural influenza surveillance program in operation in the U.S. The major emphasis of this surveillance has been the concept of excess mortality in 122 large cicites.

The above methods provide accurate efficacy data when any large-scale attempt to control influenza by either chemo- or immunoprophylaxis is implemented.

VI. ANIMAL INFLUENZA VIRUSES AND THEIR POSSIBLE ASSOCIATION WITH HUMAN DISEASE

Influenza A viruses produce outbreaks of disease in swine, horses, and many species of wild and domestic birds. These viruses may also be recovered from sporadic cases or from asymptomatic infections. In contrast, influenza B viruses infect only man and have not been recovered from animal or avian sources. The fact that influenza B does not undergo antigenic "shift" may be related to the apparent absence of animal reservoirs of this virus. The antigenic subtypes of influenza A viruses from swine, equine, and avian sources are shown in Tables 5, 6, and 7, respectively. The ecology antigenic spectrum and biology of animal influenza viruses have recently been reviewed.[77,78]

Classical swine influenza viruses exist as a single antigenic subtype (Hsw1N1) (Table 5). Swine influenza viruses were first isolated by Shope[58] in 1930, but the epidemic swine influenza has been recognized in the U.S. since 1918.[55] Viruses of this type are endemic in swine herds in the U.S. They repeatedly produce annual outbreaks of disease among swine herds in the U.S. in the autumn and winter, but have only been sporadically isolated elsewhere (Central Europe, Hong Kong, and Sweden). Among Hsw1N1 viruses isolated from 1930 to the present, there is evidence of only a minor degree of antigenic "drift."

The swine population of the U.K. has apparently been free of swine influenza infection since 1939, when the last strain was isolated.[79] However, after the emergence of the human Hong Kong (H3N2) virus in man in 1968, this virus was found to produce frequent infections in swine in the U.K. and elsewhere.[80,81] The increasing frequency of antibody to the human Hong Kong strain in swine sera collected in the U.K. in 1969 and later is illustrated in Figure 5. The Hong Kong virus apparently produces only asymptomic infections, in contrast to the influenza-like illness produced by the classical swine virus of Shope. It is not certain whether this virus has become endemic in swine or whether most infections in swine occur mainly as a result of primary introduction of the virus from human sources.

Two antigenically distinct subtypes (Table 6) of equine influenza virus (Heq1Neq1 and Heq2Neq2) appear to be distributed internationally, producing frequent epidemics of respiratory disease. These viruses were first isolated in 1956 and 1963, respectively, but may have been prevalent long before this time. There is evidence indicating a significant degree of antigenic "drift" among equine Heq2Neq2 viruses isolated since 1963.[11]

TABLE 5

Antigenic Classification of Influenza A Viruses of Swine Origin

Hemagglutinin and neuraminidase subtypes		
HA subtype	NA subtype	Strains
Hsw1	N1	A/swine/Iowa/15/30 (Hsw1N1)[a]
Hsw1	N1	A/swine/Manitoba/67 (Hsw1N1)
Hsw1	N1	A/swine/Hong Kong/1/74 (Hsw1N1)
H3	N2	A/swine/Taiwan/70 (H3N2)

[a]Strains isolated from swine fall into two distinct subtypes, Hsw1N1 and H3N2. The "classical" swine influenza viruses produce epidemics of clinical influenza in swine herds and are all antigenically closely related (Hsw1N1). The reference strain of this subtype is the 1930 isolate of Shope. Note that the neuraminidase N1 of classical influenza viruses is antigenically related to that of human strains isolated between 1932 and 1956 (A0 and A1 strains). In addition, the hemagglutinin antigen (Hsw1) shows some degree of antigenic relationship to the (H0) of the human strains isolated from 1932 to 1946 (A0 strain). In contrast, A/Hong Kong/1/68-like strains of subtype H3N2 have been isolated from sporadic, asymptomatic infections in swine.

TABLE 6

Antigenic Classification of Influenza A Viruses of Equine Origin

Hemagglutinin and neuraminidase subtypes		
HA subtype	NA subtype	Reference strains
Heq1	Neq1	A/equine/Prague/1/56 (Heq1Neq1)
Heq2	Neq2	A/equine/Miami/1/63 (Heq2Neq2)

TABLE 7

Antigenic Subtypes of Hemagglutinin and Neuraminidase of Avian Influenza A Viruses

HA subtype	Reference strains	NA subtype	Reference strains
Hav1	A/FPV/Dutch/27 (Hav1Neq1)*	N1	A/chicken/Brescia/1902 (Hav1N1)
Hav2	A/chicken/Germany "N"/49 (Hav2N1)	N1	A/chicken/Scotland/59 (Hav5N1)
Hav3	A/duck/England/56 (Hav3Nav1)	N2	A/turkey/Massachusetts/65 (Hav6N2)
Hav4	A/duck/Czech(56 (Hav4Nav1)	Neq1	A/chicken/Germany "N"/49 (Hav2Neq1)
Hav5	A/tern/S. Africa/61 (Hav5Nav2)	Neq1	A/FPV/Dutch/27 (Havl Neql)
Hav6	A/turkey/Mass/67 (Hav6N2)	Neq2	duck/Ukraine/1/63 (Hav7Neq2)
Hav7	A/duck/Ukraine/1/63 (Hav7Neq2)	Nav1	A/duck/England/56 (Hav3Nav1)
Hav8	A/turkey/Ontario/6118/68 (Hav8Nav4)	Nav2	A/tern/S. Africa/61 (Hav5Nav2)
		Nav3	A/turkey/England/63 (Hav1Nav3)
		Nav4	A/turkey/Ontario/6118/68 (Hav8Nav4)
		Nav5	A/shearwater/Australia/72 (Hav6Nav5)

The detailed antigenic characteristics of avian influenza viruses and their antigenic relationships with human influenza A viruses have been fully described elsewhere.[78] In brief, the hemagglutin in antigens of avian viruses has been divided into eight groups, Hav1 to Hav8. Five antigenic varieties of neuraminidase, Nav1 to Nav5, are found uniquely among avian viruses, while other

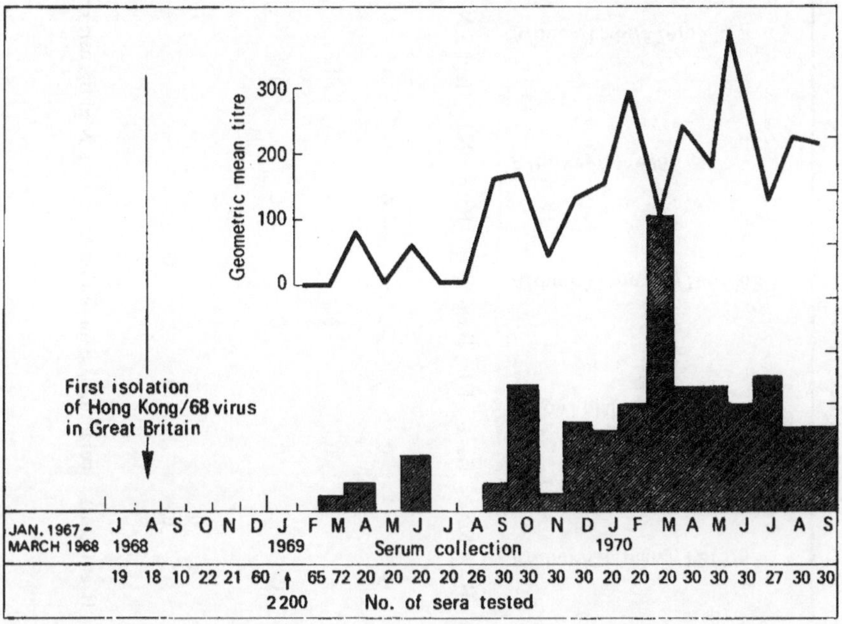

FIGURE 5. Proportion of swine sera collected in the U.K. during various periods from 1967 to 1970 with HI antibody titers of 1:40 or greater for A/Hong Kong/68 virus and geometric mean HI titer of positive sera. (From Harkness, J. W., Schild, G. C., Lamont, P. H., and Brand, C. M., *Bull. WHO*, 46, 709, 1972. With permission.)

avian strains contain neuraminidase antigens also found in human, swine, and equine viruses. It is of particular interest that among avian strains a given HA antigen is not always associated with the same NA antigen and a large number of different combinations of HA and NA antigen substances exist (see Reference 80).

The classic fowl plague virus which produces a fatal epidemic disease of chickens was isolated early in this century before swine or human influenza viruses were first cultivated, but it was only subsequently identified as an influenza A virus.[82] Subsequently, numerous antigenic varieties of avian influenza A virus have been isolated, and new subtypes are still being identified (Table 7). Avian influenza viruses have been found in each country where they have been seriously sought. Included among avian influenza A viruses are viruses with widely different degrees of virulence for chickens, turkeys, and ducks. These viruses that are known to be prevalent in domestic poultry have also been isolated from several species of wild birds including several species of duck, tern (*Sterna* sp.), shearwater (*Puffinus pacificus*), puffin (*Fratercula artica*), and wagtail (*Motacilla* sp.). Infection in wild birds has been detected frequently in the U.S., U.S.S.R., several European countries, and South Africa (see References 77 and 78). It should be emphasized that the ecological investigation of influenza in birds and mammals has been fragmentary in the past. Frequent isolations of virus are now being made from several species, particularly from wild birds, and knowledge of the ecology of the virus in nonhuman hosts is increasing rapidly on an international scale.

Influenza A viruses from nonhuman sources are not antigenically identical to those isolated from man since 1933, strains containing HA and /or NA antigens showing clear antigenic relationships to those of human viruses are found among them (Table 8, see also Stuart-Harris and Schild).[78] The only exception is the Hong Kong virus which, after it appeared in man in 1968, was found to be capable of crossing species barriers to infect swine and, occasionally, domestic and wild birds[78] without producing apparent symptoms in these hosts.

The earliest recognition of epidemic swine influenza in the U.S. (1918) merits special comment. At the time of the onset of the severe pandemics in the human population in the U.S., a veterinarian in the Midwest, Dr. J. Koen, noted the sudden appearance of an influenza-like illness among swine herds. According to Koen,[55] this was a previously unrecognized disease in swine and its

TABLE 8

The Major Antigenic Relationships Between Human Type A Influenza Viruses and Those of Lower Animals and Birds

Human Type A influenza viruses		Animal or avian influenza virus and antigenic designation																			
		Swine origin		Equine origin				Avian origin													
		A/swine/Iowa/15/30		A/equine/Prague/1/56		A/equine/Miami/1/63		A/chicken/Scotland/59		A/tern/S. Africa/1/61		A/duck/Germany/1215/63		A/fowl plague/Rostock/34		A/duck/Germany/1868/68		A/turkey/Mass/65		A/duck/Italy/574/66	
Human epidemic era and period of prevalence of antigen	H and N antigen subtype or group	Hsw1	N1	Heq1	Neq1	Heq2	Neq2	Hav5	N1	Hav5	Nav2	H_2	Nav2	Hav1	N1	Hav6	N1	Hav6	N2	Hav2	N2
Hemagglutinin antigens																					
A0–A1 1933–1956	H0–H1	+		0		0		+		+		0		0		0		0		0	
Asian (or A2) 1957–1968	H2	0		0		0		0		0		+		0		0		0		0	
Hong Kong 1968–?	H3	0		0		+		0		0		0		0		0		0		0	
Neuraminidase antigens																					
A0–A1	N1		+		0		0		+		0		0		+		+		0		0
Asian and Hong Kong 1968–?	N2		0		0		0		0		0		0		0		0		+		+

Note: + indicates evidence of antigenic relationship demonstrated by hemagglutination-inhibition tests (for H antigens), neuraminidase-inhibition tests (for N antigens), or immunodiffusion tests.

temporal coincidence with the human disease was emphasized. Since 1918 outbreaks of swine influenza have occurred annually in the fall in the Midwestern area of the U.S. In 1930, even before the isolation of influenza from man, Shope and Lewis (at the Rockerfeller Institute, Princeton, N.J.)[58] identified the etiological agent of swine influenza as a virus. Only subsequently was the virus identified as an influenza A strain.

Recent studies have shown that, although it possesses some minor antigenic differences, the swine influenza virus belongs to the same subtypes of HA and NA antigen as the human influenza virus strains which were isolated in 1933 to 1956, i.e., HON1 and H1N1 (Table 2). The earliest suspicion that swine influenza virus may have an epidemiological association with the pandemic of 1918 was based on the early work of Andrewes and his colleagues.[56] It was found that persons who in 1935 were old enough to have experience of infection in the 1918 pandemic possessed antibody to the swine virus, and younger persons did not. This finding has been confirmed in recent studies, and in 1976, essentially, persons 50 years or older possess such antibodies, representing a group of the population who even half a century later still appear to retain the immunological memory of their earliest experiences with influenza. The term "original antigen sin" has been used to describe this phenomenon.

Swine influenza was of only speculative significance for man until 1976. In mid January of that year an outbreak of influenza occurred among young military recruits who were recently enrolled in the army camp Fort Dix, New Jersey. Twelve isolates of influenza virus were identified, and six of them were antigenically identical to the A/Victoria/75 virus which was then predominant in many areas of the U.S. and Europe. The remaining six strains were identified as swine influenza virus.[42] Therefore, the outbreak at Fort Dix, which ended some 6 weeks later, appeared to be of mixed etiology, involving two distinct influenza A viruses. An immediate serological survey indicated that some 500 recruits in the camp were infected with the swine influenza virus. Intensive surveillance had not detected evidence for the cases of swine influenza infection in the camp or in the general population up to the end of August 1976. Nevertheless, the isolation of swine influenza-like viruses was considered by some authorities in the U.S. to be of great potential significance as the possible beginning of a new pandemic era. Preparations have been made to vaccinate the entire adult population in the U.S. against swine influenza during the autumn and winter of 1976 to 1977.

As will be described, evidence, based on the distribution of antibody in the sera of older persons, indicates that the Asian (H2N2) virus which appeared in 1957 produced widespread outbreaks some 67 years previously in 1890. Similarly, the Hong Kong (H3N3) virus may have circulated around 1900 before it reemerged in 1968. Based on the acceptance of this historical evidence, cyclical reemergence of influenza A viruses with a 60- to 70-year periodicity, it might be expected that the swine (Hsw1N1) virus would become the prevalent influenza A virus at the end of the present decade or in the 1980s. This hypothesis is of interest in relation to the significance of the outbreak at Fort Dix.

The major significance of animal influenza A virus is that it might provide a reservoir of genetic material from which new human pandemic strains may emerge, possibly by a process involving genetic recombination between human and animal viruses. Because of this possibility, it is important that intensive and continuous ecological surveillance of influenza viruses in nonhuman hosts be carried out to determine the distribution and antigenic spectrum of these viruses. Such studies are currently in progress as part of the WHO influenza program.

VII. INFLUENZA PROPHYLAXIS

Attempts to control influenza by immunological methods have involved inactivated and live influenza vaccines. In Europe and the U.S., inactivated vaccines remain the principal type of vaccine used. Live vaccines have been used mostly in the U.S.S.R., Eastern European countries, and, possibly, the Chinese Peoples Republic. However, clinical studies of live vaccines have been in progress for several years, and the developments in this field have enabled the first licensing and use of these vaccines to take place in the U.K. and a number of other European countries during 1975. Any further use of chemoprophylaxis is most satisfactorily considered against a background of our present success at control by vaccination.

A. Inactivated Vaccines

Several National Authorities recommend these vaccines for annual administration to those at high risk from influenza including elderly and debilitated persons, essential public health workers, and industrial employees. To be effective, vaccines must contain influenza A or B viruses identical or closely related to (in terms of the HA and NA antigens) the prevalent influenza viruses against which it is required to protect; vaccination must be renewed annually or at least biannually. With vaccines of appropriate antigenic composition and potency, clinical protection rates of 60 to 90% have been recorded.[84] Vaccines are routinely prepared by cultivation of the virus in the allantoic cavity of the fertile hen's egg, with the subsequent purification of the virus and inactivation of its infectivity by formalin or β-propiolactone.

The vaccines used routinely or in clinical studies can be divided into the following categories.

1. Vaccines containing inactivated whole virus particles.
2. So-called split vaccines, containing disrupted virus particles from which unwanted viral components (internal proteins, RNA) have not been removed. Ether or various detergents have been used to disrupt the virus in the manufacture of these vaccines.
3. "Subunit" vaccines, containing purified envelope subunits (HA and NA) from which unwanted viral components have been removed.
4. With any of the above types of vaccines, an adjuvant may be used to increase the antigenicity. Suitable adjuvants may be oily or simple inorganic substances such as aluminum hydroxide or aluminum oxide. An adjuvant, A65, based on peanut oil is licensed for use in the U.K.[85]

Recent technical developments in the production of inactivated vaccines include the use of continuous-flow zonal ultracentrifugation for the purification of the vaccine virus[86] and the routine use of high-yielding genetic recombinant influenza viruses from which vaccine can be prepared.[87] Zonal centrifuge-purified vaccines which contain intact virus (whole virus vaccines) are less reactogenic than the earlier vaccines and can be given at higher dosages, producing a high protective efficacy.[88,89] However, some reactogenicity is associated with these relatively purified viruses, particularly when given to young children,[90] and it seems likely that at least some of the reactogenicity associated with whole virus vaccines resides in the intact virus particle.[91]

From basic studies of immunity to influenza, it is concluded that protection is closely associated with antibody to the HA and NA antigens. None of the other antigens of the virus appears to be associated with immunity.[31,33] Thus, ideal influenza vaccine should contain the surface antigens of the virus but not other virus components (nucleoprotein, matrix protein, lipids, viral RNA) which are irrelevant to immunity.

The earliest attempts to produce subunit vaccines involved treating the virus with ether or mixtures of Tween®-ether,[92] with sodium deoxycholate,[93] or with tri(*n*-butyl) phosphate and Tween.[94] Such preparations appear to retain antigenicity in man; there is evidence of their reduced toxicity for man, and improved acceptability for use in children has been shown.[95] Nevertheless, with the above techniques the resulting vaccines contain other virus components in addition to the essential envelope antigens HA and NA, and such preparations should properly be termed "split"-product rather than "subunit" vaccines.

Recent studies have involved the use of elaborate chemical and physical separation procedures to produce vaccines which contain only HA and NA and are true subunit vaccines. Starting with purified inactivated virus preparations, Brady et al.[96] were able to disrupt the virus with the detergent Triton® N 100 and separate the HA and NA subunits by zonal centrifugation. This resulting subunit preparation, which was essentially free of virus components other than HA and NA, was adsorbed onto aluminium hydroxide before injection and was found to elicit antibody to HA and NA in man and to induce resistance to challenge with the homologous virus. Using a similar approach, Laver and Webster[97] disrupted virus with ammonium deoxycholate and purified the HA and NA "subunits" by a combination of ultracentrifugation and electrophoresis to yield a "subunit" vaccine which was antigenic in man. A somewhat different approach was used by Bachmeyer and co-workers.[98] Employing a selective solubility technique using a nonionic detergent, Neodol 23-6, these workers were able to release the HA and NA antigens from the virus particle leaving the remainder of the virus as an intact "core." The purified "subunit" preparations were effective antigens in animals and man when adsorbed onto an aluminium hydroxide carrier.

Recommendations for the antigenic composition of inactivated vaccines are made annually by the World Health Organization based on the international surveillance of influenza. Most national authorities follow the WHO recommendations, although occasionally they may, in some countries, depart from them as a result of local epidemiological circumstances. Vaccines for the 1975 to 1976 influenza season contained two influenza A strains, A/Port Chalmers/1/73 (H3N2) and A/Scotland/840/74 (H3N2), and one influenza B strain, B/Hong Kong/8/73. Current WHO recommendations for the composition of vaccines in the winter of 1976 to 1977 are A/Victoria/3/75 (H3N2) and B/Hong Kong/8/73. WHO has also recommended that national authorities may wish to consider the inclusion in current vaccines or stockpiling of swine Hsw1N1 influenza virus of the antigenic type recently isolated from man in New Jersey. In the past, WHO has not made general recommendations concerning the potency of inactivated vaccines, but may do so in the future.

In the U.K., requirements for the manufacture and testing of inactivated influenza vaccines are based on WHO recommendations formulated by an expert group in 1967.[99] In most countries, licensing requirements are broadly similar to those recommended by WHO. However, there are differences between countries, particularly with respect to vaccine potency (antigen content and the units in which this is measured) and acceptable passage histories of the seed virus for vaccine production. In the U.K., manufacturers submit samples of each batch of the final vaccine, together with detailed protocols of manufacture and testing procedures, for potency and safety testing at the National Institute of Biological Standards and Control. Release of vaccine depends on satisfactory results of tests run on each batch of the final product.

Licensing is based on the following requirements:

1. The strains of influenza A and B viruses must be from an acceptable source and must have an acceptable passage history. In general, viruses isolated in eggs are acceptable as seed. The antigenic character of the vaccine virus must be appropriate.

2. The premises for vaccine manufacture must be approved by the licensing authority.

3. The influenza virus for vaccine preparation must be cultivated in embryonated eggs. Inactivation of the virus (usually by formalin or β-propiolactone) must be complete and tested for routinely.

4. The virus must be purified by an approved process and reach an acceptable degree of purity as defined by the licensing authority.

5. Tests for HA content must be performed to assess the potency of the vaccine. Several variations of this test are recommended by different authorities. In the U.K. and some other countries, the units of potency are defined as international units (IU), in reference to an international HA preparation. In the U.S. the chick cell agglutination (CCA) unit is used.

Traditionally, tests for potency of influenza vaccines have been based on the degree of agglutination of erythrocyte suspensions by the vaccine, using standard methods and comparing the test vaccine with international unit[100] or CCA standards. However, tests for influenza HA content based on the agglutination of erythrocytes have a number of disadvantages, both practical and theoretical. First, such assays are subject to a high degree of intra- and interlaboratory variation. Second, the ability of a preparation of HA to agglutinate erythrocytes depends on the quantity of HA present as well as its physical state. HA may be present on the intact virus particle (which can be spherical or filamentous in shape) or may exist as subviral structures or as isolated HA in the form of polyvalent or monovalent structures. So-called "split" "subunit" influenza vaccines can contain a wide variety of subviral structures in various degress of aggregation and purity; therefore, they may differ considerably in their ability to agglutinate erythrocytes. Finally, the ability of HA to react with erythrocytes does not establish its antigenicity. Because of these problems, attempts have been made to devise alternative methods, independent of hemagglutination, for the standardization of influenza vaccines.

Therefore, a single-radial-diffusion (SRD) method for the assay of influenza HA antigen was proposed by Schild et al.[101] The SRD test gives a

APPENDIX

SEROLOGICAL TESTS FOR INFLUENZA DIAGNOSIS FOR CONTROLLED TRIALS OF VACCINES OR ANTI-INFLUENZA COMPOUNDS, SEROEPIDEMIOLOGICAL SURVEILLANCE, AND STRAIN CHARACTERIZATION

To be comprehensive, assays of the natural or artificially induced immune response to influenza for diagnosis or studies of vaccine efficacy or, indeed, to control trials of potential anti-influenza compounds must include assays of the antibody response to each of the major antigens of the virus HA, NA, NP, and M. A number of methods for antibody assay are routinely used, including virus neutralization tests (essentially detecting antibody to HA) and hemagglutination-inhibition (HI)[106] and neuraminidase-inhibition (NI)[107] tests. Essentially, these are multicomponent test systems which depend on testing serial dilutions of sera with known virus doses in the presence of an appropriate indicator system (susceptible host cells, erythrocytes, or substrate). In addition, complement fixation (CF) tests are routinely employed to detect antibody to "soluble" antigens, essentially NP, present in extracts from infected egg membrane. This test is only of diagnostic value, whereas the various inhibition tests described above also give information on immunity. Although widely used and generally acceptable, a disadvantage of these tests in general is that they are tedious and time consuming to perform; some are subject to nonspecific factors present in sera (in particular, the HI test). In our laboratory, we have recently developed and assessed the potential value of three new methods: immunodouble diffusion (IDD), single radial diffusion (SRD), and passive hemolysis. The first two of these methods have provided simple, two-component test systems involving viral antigen and antibody only. This is particularly advantageous in the case of the SRD test and leads to a high degree of accurary for this method. There is little doubt that the need to use an indicator system in the conventional tests leads to a higher degree of variability.

Analysis and Assay of Antibody

IDD tests

Using IDD techniques in gels, it has been shown that precipitin lines corresponding to each of the major antigens of the virus particle may be detected.[13,108,109] Figure 6 shows the production of precipitin lines with antisera to the HA, NA, NP, and MP in tests with influenza virus particles which have been disrupted by the addition of sodium sarcosyl (SS) detergent. Since the influenza virus particle is too large to diffuse in agar gels, such precipitin tests may only be carried out using purified diffusible antigens isolated from virus particles or by disrupting virus particles with appropriate detergents to release the diffusible subunits. IDD techniques were found to be valuable as qualitative tests detecting the spectrum of antibodies present in test sera, but they are not readily adaptable to quantitative studies.

Studies to compare the sensitivity of IDD and conventional diagnostic methods (CF and HI tests) with acute and convalescent human sera from persons exposed to A/Hong Kong/1/68 virus or horses exposed to equine influenza infection showed that the acquisition or reinforcement of

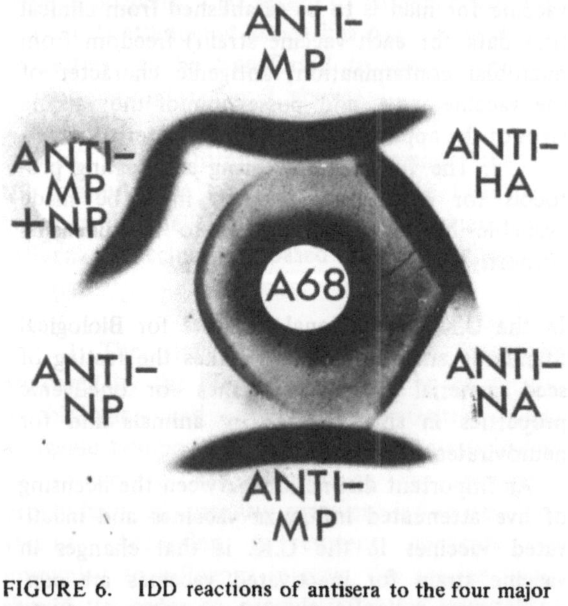

FIGURE 6. IDD reactions of antisera to the four major antigenic components of the influenza virus. The center well (A68) contains purified A/Hong Kong/1/68 (H3N2) (X-31 strain) disrupted by the addition of SS detergent to liberate diffusible, antigenic components. Wells marked ANTI-HA, ANTI-NA, ANTI-NP, and ANTI-M contain immune rabbit serum to the purified hemagglutinin and neuraminidase of A/Hong Kong/1/68 and to influenza A nucleoprotein and matrix protein antigens, respectively.

precipitin reactions to the ribonucleoprotein antigen was as sensitive as conventional techniques as an index of infection. In convalescent human sera, precipitin lines corresponding to ribonucleoprotein were most frequently detected, lines corresponding to neuraminidase less frequently, and lines corresponding to hemagglutinin least frequently. Precipitin methods were rapid and simple and were not susceptible to nonspecific reactions.

The value of IDD techniques in the analysis of antigenic variation of the HA and NA antigens of the influenza virus has been described elsewhere and will be briefly discussed later.

SRD Test for the Assay of Antibody to Influenza Antigens

Basis of the Test

The SRD test, first described for influenza antigens by Schild et al.,[110] provides a simple two-component immunological test system for the rapid, convenient, and accurate assay of antibody to each of the major antigens of the influenza virus. In these assays the antigen is evenly distributed throughout a gel. Antibody for assay is placed in circular wells in the gel and diffuses radially throughout the gel, reacting with the antigen and giving a visible, circular zone of reaction. The annulus area of the zone is a direct measure of the quantity of antibody added to the well. The various test systems for the assay of antibody to the known antigenic components of influenza A virus are summarized in Table 9.

Assay of Antibody to the Surface Antigens of the Virus HA and NA

For the assay of antibody to the surface antigens of the influenza virus, hemagglutinin and neuraminidase purified intact influenza virus at a standard final concentration of 0.2 mg virus protein per milliliter of gel is incorporated in the gel. Antibody to either of the two surface antigens is capable of reacting with the intact virus giving opalescent zones (Figure 7a). When viewed against a dark background, such zones are easily visible without staining. The linearity of a plot of antiserum dilution against annulus area is shown in Figure 7b. The opalescent zones produced by antibody to HA are more intense than those produced by anti-NA antibody. This is probably a reflection of the fact that there is a higher density

TABLE 9

SRD Systems for the Assay of Antibodies to Each of the Major Antigens of Influenza A/HK/1/68 (H3N2) Virus in Human Sera

	Detection of antibody to HA and NA intact virus particles of			Detection of antibody to NP and M-SS-disrupted virus particles of	Antibody to M only, SS-disrupted virus particles of
Antibody specificity	H3N2[a]	H3Neq1[b]	Heq1N2[c]	A/chicken/Germany "N"/49 (Hav2Neq1)	Germany "N" virus heated 100°C/2 min
Antibody to envelope proteins					
Anti-H3	+	+	–	–	–
Anti-N2	+	–	+	–	–
Antibody to internal proteins					
Anti-influenza A NP	–	–	–	+[d]	–
Anti-influenza A MP	–	–	–	–	+

Note: Antigens used at a standard concentration of 0.2 mg protein (Lowry) per milliliter of agarose gel of purified virus preparation with the exception of the test for M antibody only which employed heated "N" virus at a concentration of 0.02 to 0.05 mg virus per milliliter.

[a] A/HK/1/68 (H3N2) virus (X-31 strain).
[b] Recombinant A/HK/1/68 (H3)-A/equine/Prague/56 (Neq1).
[c] Recombinant A/equine/Prague/56 (Heq1)-A/HK/1/68 (N2); recombinants kindly provided by Drs. E. D. Kilbourne and J. Schulmann, Mt. Sinai Hospital, New York.
[d] Clear distinction could be made between antibody to NP and M since zones produced by the latter were of extremely high intensity.

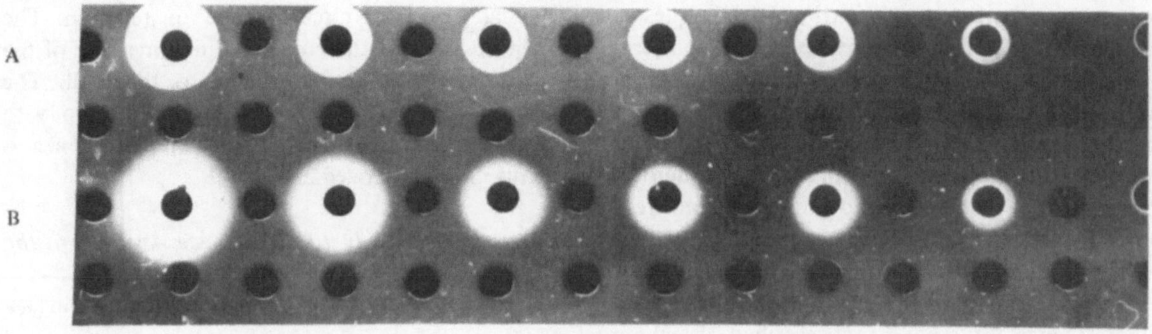

(a)

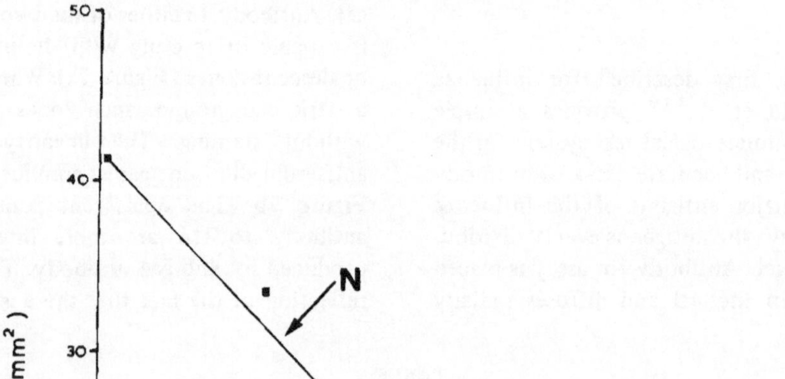

(b)

FIGURE 7. (a) The SRD reactions of anti-HA and anti-NA sera in agarose gels containing A/Hong Kong/68 virus (0.2 mg virus protein per milliliter agarose). (A) Reactions of dilutions of anti-H3 rabbit serum. From left to right the wells contain sera diluted 1:1, 1:15.5, 1:2, 1:3, 1:4, 1:8, and 1:16. (B) Reactions of dilutions of anti-N2 rabbit serum. From left to right the wells contain sera diluted as for anti-H3 above. With serum dilution, the intensity of the zones remains constant while their areas decrease. (b) Atypical SRD dose-response plots for antisera to hemagglutinin and neuraminidase. (HA) Plot for antihemagglutinin serum; (N) plot for antineuraminidase serum. Note that the dose-response relationship differs from that of passive hemolysis tests (Figure 9).

of HA subunits than of NA subunits on the virus surface.

In human sera antibody to both HA and NA antigens of the test strain of influenza virus may be present and both antibodies will contribute to the SRD reactions given by such sera. Where anti-NA and anti-HA coexist in a test serum, the zones produced by both antibodies overlap and are not additive. Thus, the overall diameter of the zone reflects the amount of the most abundant antibody, and the reaction is not easily attributed to either one or the other of these antibodies. However, using appropriate recombinant (antigenic hybrid) influenza virus, antibody to HA or NA may be assayed independently. Suitable test systems for the assay of antibody specific for the HA or NA antigens for the A/Hong Kong/1/68 (H3N2) virus are shown in Table 9. It is of interest that the SRD with intact virus particles does not depend on precipitation since monovalent (nonprecipitating) antibody molecules react efficiently in SRD tests but not in IDD tests.[111] Thus, the mechanism appears to depend on primary antibody-antigen binding which presumably modifies the light-scattering properties of the gel. Of considerable advantage is the fact that the SRD test does not detect responses due to nonspecific inhibitors of hemagglutinin; thus, it is not necessary to pretreat sera by receptor-destroying enzyme as is necessary for hemagglutination-inhibition tests. The test is particularly useful for serological control of chemoprophylactic trials for example, since a large number of sera may be tested on a single SRD plate, the plates can be sent by mail to a central laboratory.

Assay of Antibody to Internal Antigens NP and M

In SRD tests with intact virus particles, antibody to the two major internal antigens, the ribonucleoprotein (NP) and matrix protein (M) produces no reactions. However, if the virus particles are disrupted with a suitable detergent* before addition to the gel, these antigens are released and made available for reaction with antibody.[9,33,41,88] In SRD with SS-disrupted virus particles as antigen, antibody to both NP and M is detectable by the production of opalescent zones (Figure 8). In this modification of the SRD test, the antigens are rendered soluble and diffusible by the detergent, and the zones are formed as a result of precipitation of antigen by antibody. The precipitates are not readily visible but may be made so by the routine use of a protein stain, e.g., Coomassie blue. Zones produced by antibody to M are more intense than those produced by antibody to NP, and this feature may be used to distinguish the two antibodies when both are present. When overlapping zones are formed by a single serum specimen, the area of the intense inner zone represents M antibody and the area of the less intense outer zone represents antibody to NP. These reactions can usually be readily distinguished. In tests with human sera, antibody to influenza A virus M and NP may be assayed independently of antibody to surface antigens by using an influenza A virus with surface antigens unrelated to those of past or current human influenza viruses. A suitable virus is an avian influenza strain A/duck/Germany/N/49 (Hav2Neq1) (Table 9).

The test may be made entirely specific for M antibodies if SS-disrupted and heated "N" virus are used. Heating destroys the NP antigen but not the M antigen[33] (Table 9). The test may be made more sensitive if the "N" virus is used at a low concentration of 0.02 to 0.05 mg of virus protein per millileter of gel rather than at the concentration of 0.2 mg/ml used in the standard SRD tests for other antibodies.

Applications in Diagnosis, Serological Surveillance, and Studies of Immunity

SRD tests for the assay of antibody to the envelope and internal antigens have been evaluated by a number of laboratories for sensitivity as diagnostic and seroepidemiological tools. Serological evidence of infection is clearly demonstrated by the acquisition or increase in the diameter of reaction zones in one or both test systems. A summary of the results is given in Table 10, indicating that the SRD test with intact virus was slightly more sensitive than conventional HI tests in the diagnosis of infections and was the most sensitive of all tests included in the study. When these tests were carried out with recombinant influenza A viruses with irrelevant HA or NA antigens to detect antibody to each of these antigens independently, a proportion of infected individuals (Table 10) developed only HA or NA antibodies and not both. The detection by SRD test of antibody rises to the NP and M antigens

*Sodium sarkosyl (SS) NL90 is used routinely.

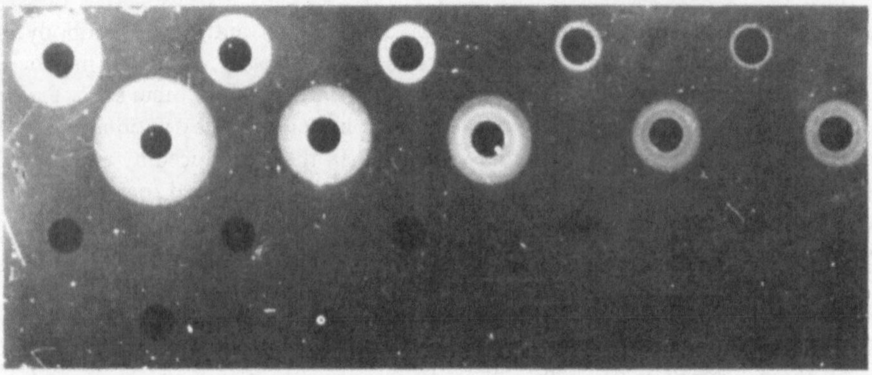

(a)

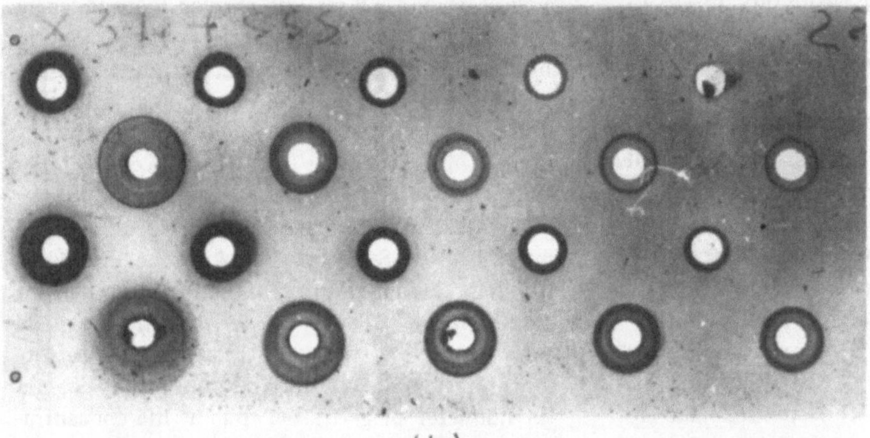

(b)

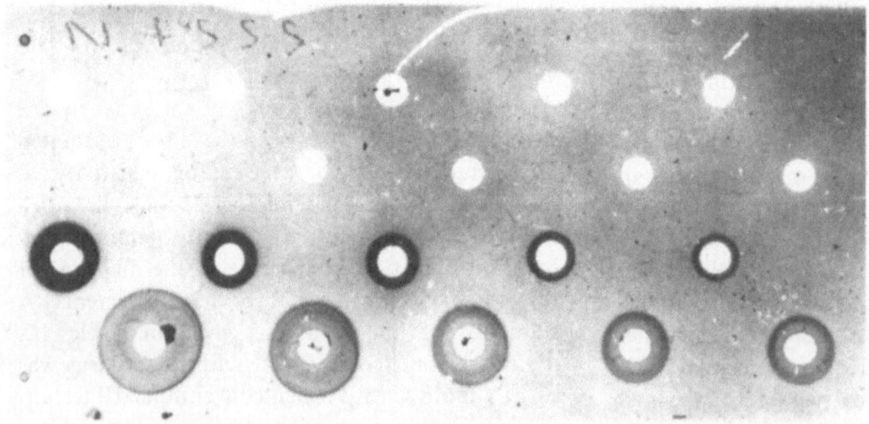

(c)

FIGURE 8. Reactions of antisera to influenza A surface and internal antigens in immunoplates containing intact and disrupted virus as antigen. On each of the immunoplates a, b, and c, the wells in Row 1 contain (left to right) serial twofold dilutions (1:1 to 1:16) of antiserum to A/HK/1/68 HA; the wells of Row 2 contain a similar dilution series of antiserum to A/HK/1/68 NA; Row 3 contains dilutions of antiserum to influenza A M; and Row 4 on each plate contains dilutions of antiserum to influenza A NP. (a) Intact A/HK/1/68 (X-31) virus. (b) A/HK/1/68 virus disrupted by SS detergent. (c) A disrupted avian influenza A "N" virus with surface antigens distinct from those of the A/HK/1/68 virus. Plate a, containing intact A/HK/1/68 virus, detects only antibody to the surface antigens, whereas Plate c, with disrupted avian influenza a virus, detects only antibody to the internal antigens, M and NP. Note that for serological studies involving chemoprophylactic agents or vaccines, 56 wells may be used on each immunoplate, thus allowing analysis of large numbers of sera. The plates are transportable by mail.

TABLE 10

Comparisons of Conventional Diagnostic Tests (HI and CF) and SRD Test Methods in the Serodiagnosis of Infection in Suspected Cases of (Clinical) Influenza Due to A/Hong Kong/1/68 Virus

	Individuals with diagnostic[a] antibody rises by conventional diagnostic tests		Individuals with significant[b] antibody in SRD test with stated antigens				
Serum pairs tested	CF tests[c]	HI tests[d]	NP[e]	HA, NA or both[f]	HA + NA	HA only[g]	NA only[h]
150	55 (37%)	80 (53%)	75 (50%)	86 (57%)	75 (50%)	6 (7%)[i]	5 (5.8%)[j]

[a]At least fourfold.
[b]20% or greater increase in zone area.
[c]Complement-fixation tests employing A/PR8/34 soluble NP antigen in extracts of infected egg membranes.
[d]Standard HI tests (29) employing eight HAU of influenza virus.
[e]SRD test with disrupted avian influenza A virus in immunoplates.
[f]SRD test with intact A/Hong Kong/1/68 (X-31) virus in immunoplates.
[g]SRD test recombinant containing A/Hong Kong/1/68 HA and an irrelevant NA.
[h]SRD test with recombinant containing A/Hong Kong/1/68 NA and an irrelevant HA.
[i]Only 22 of 75 individuals (30%) with antibody rises to NP had detectable anti-MP in the convalescent serum.
[j]Percentages expressed as proportion of sera showing SRD rises in tests with intact A/Hong Kong/1/68 virus (HA and/or NA).

was considerably more sensitive than conventional CF tests with influenza "soluble" NP. The SRD test indicated that only a small proportion (30%) of infected individuals who developed antibody to NP, also developed unequivocal evidence of antibody to M.

The SRD test was also evaluated for sensitivity in detecting antibody to A/Hong Kong/1/68 and A/England/42/72 virus in randomly collected human serum samples for seroepidemiological surveys.[112] It was shown that SRD tests with intact viruses in the gel were of comparable sensitivity to HI tests in detecting antibody. Sera with HI titers of 1:20 or greater generally produced opalescent zones of 2.5 mm in diameter or more. However, discrepancies were observed in a small proportion of sera in the results of HI and SRD tests; 7% of sera gave positive reactions on HI but not SRD tests and 4% gave responses in SRD but not HI tests. The age group distribution of antibody to the two test viruses in the population were closely similar by both methods. In other investigations[113] the sera of subjects who received various influenza vaccines and who were subsequently naturally exposed to A/Hong Kong/1/68 virus during an epidemic were investigated by conventional tests and by SRD tests for antibody to HA, NA, and NP. The SRD test was found to be more sensitive than conventional CF tests in detecting resting levels of NP antibody and considerably more sensitive in detecting significant rises in antibody to NP. The SRD test was slightly less sensitive than HI in detecting antibody to HA but more sensitive in detecting rises in anti-HA. The test has been used successfully to estimate the number of persons infected with influenza during large-scale chemotherapeutic trials of amantadine (Chapter 3, Volume II). For detecting anti-NA antibody, the SRD test was less sensitive than neuraminidase inhibition tests but equally sensitive for the detection of antibody rises to NA. Anti-M antibody was detected far more frequently in individuals who were naturally infected and became severely ill with influenza, relatively infrequently in persons with mild influenza, and rarely in persons receiving potent inactivated influenza A vaccines.

Passive Hemolysis Technique for Anti-HA Assay

More recently, Schild et al.[114] and Russell and colleagues[115] used slightly differing techniques to develop assay systems for antibody to hemagglutinin based on passive hemolysis of erythrocytes which are "sensitized" by coating with influenza virus particles or purified HA antigen. Sensitized erythrocytes are lysed at 37°C by specific anti-HA antibody in the presence of complement. For convenience, the method has been adopted as a modified SRD test (i.e., single radial hemolysis) in which erythrocytes coated with the virus (or isolated HA antigen) together with guinea pig complement are incorporated in agarose gels. Antisera for assay are added to wells in the gel. On overnight incubation at 37°C, zones of clear hemolysis develop surrounding antibody containing wells. Figure 9a shows the hemolysis reactions of a series of dilutions of three samples of convalescent human serum in gels containing chicken erythrocytes sensitized with A/Port Chalmers/1/73 (H3N2) virus. The sizes of the zones of hemolysis decrease regularly with serum dilution. A plot of log zone diameter against log serum dilution gives a straight line plot, and parallel plots (Figure 9b) are given by serum samples of differing potencies. Undiluted sera with titers of 1:5120 in conventional HI tests gave zones of hemolysis approximately 11.0 mm in diameter, those with HI titers of 1:256 gave zones approximately 6.0 mm in diameter, while sera with low HI titers (1:10 to 1:15) gave zones 2.5 to 3.5 mm in diameter.

In detecting levels of anti-HA, this test was found to be more sensitive than the conventional HI test and was considerably more sensitive than HI tests in detecting changes in antibody levels. The method was effective and convenient for use as a diagnostic technique for tests with acute and convalescent paired serum samples. One advantage of this method is that, for routine assessment of antibody levels, sera need be tested only at one dilution, usually undiluted, and pretreatment of sera for testing is limited to heating at 56°C for 30 min. It is not necessary to remove nonspecific inhibitors of hemagglutination as in HI tests, since they do not interfere with hemolysis. Another advantage is that up to two- to fourfold changes in the concentration of antigen on the erythrocytes do not have a significant effect on zone area, thus, the method is readily standardized, and minor discrepancies in antigen concentration are not critical. As a routine control for nonspecific lysis, sera are tested in parallel on immunoplates containing normal chick erythrocytes and complement, and the results of sera positive in this test are discounted.

Under "standard"[114] conditions using virus at a concentration of 500 hemagglutinating units per milliliter of 10% chick erythrocytes, the passive hemolysis system was found to be specific for antibody to HA. Antibody to NA did not give reactions. However, if considerably higher concentrations of virus were used, NA antibody produced "hazy" zones of incomplete hemolysis.

The method was effective in detecting antibody in human, ferret, rabbit, or goat sera, but thus far no hemolysis with mouse or chicken sera has been detected, even though these are potent in specific anti-HA antibodies. Hemolysis was found to be highly specific for the antigenic variety of hemagglutinin. No cross reactions were detected between antibodies to different subtypes of hemagglutinin (H0, H1, H2, and H3). Furthermore, within the H3 hemagglutinin subtype, failure of cross reactions was detected with postinfection ferret sera to the variants A/Hong Kong/1/68 (H3N2) and A/Port Chalmers/1/73 (H3N2) even though these viruses give definite cross reactions in HI tests. The apparent high degree of strain specificity of this test may be an important advantage of this technique as applied to the analysis of influenza variants. Preliminary investigations suggest that, even in recently convalescent human sera, it is essentially the IgG antibodies which produce the hemolysis reaction.

The method of passive hemolysis may have wide application for serological studies with virus antigens such as hemagglutinins which have a natural affinity for receptors on the erythrocyte surface. Antibody to rubella virus hemagglutinin is readily assayed by this method, and this may be a useful method for diagnosis of rubella infection. Additionally, antigens which do not normally attach may be coupled to the erythrocyte surface by reagents such as chromium chloride, making it potentially possible to use the method to detect antibody for a wide range of antigens.

Applications of IDD and SRD to Studies of Antigenic Variation in Influenza Viruses

Such studies are very relevant to any discussion of chemoprophylaxis since experience with amantadine has indicated that certain variants are less inhibited by the compound than others (see Chapter 5).

Table 11 gives the results of HI tests with postinfection ferret sera to compare the HA antigens of influenza A (H3N2) viruses prevalent from 1968 to 1975. The results provide clear

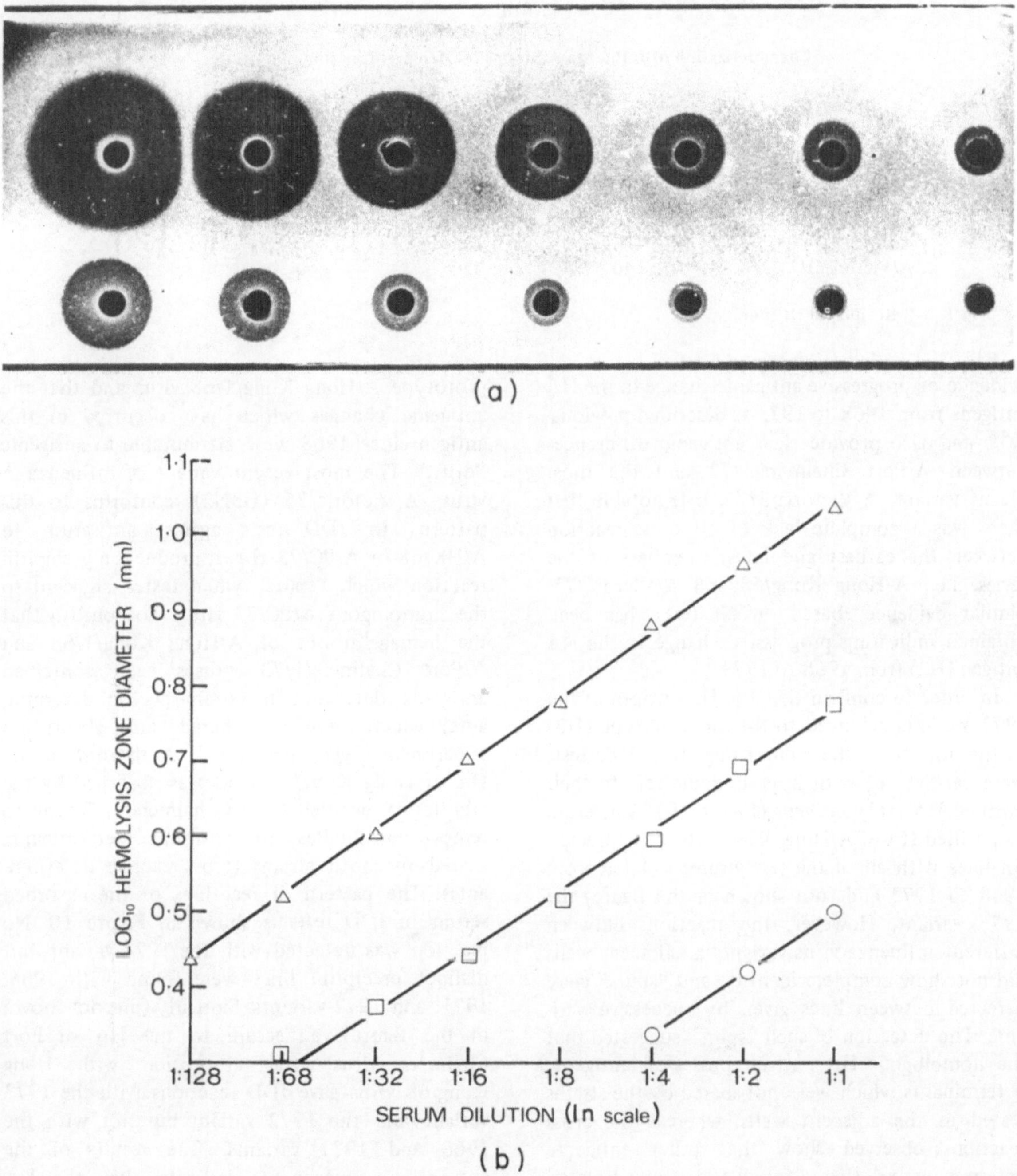

FIGURE 9. Single radial hemolysis (passive hemolysis) reactions of human sera in agarose gel immunoplates containing guinea pig complement (final concentration 1:30) and chick erythrocytes (1%) treated with A/Port Chalmers/1/73 virus. Clear areas, indicate zones of lysed erythrocytes produced by antibody to hemagglutinin. (a) Wells in upper row contain serial twofold dilutions (1:1 to 1:64) of a potent human serum having an HI titer of 1:2560 with A/PC/1/73 virus. Lower row contains similar dilutions of a serum with an HI titer of 1:256. (b) Dose-response graphs of serial dilutions of three samples of human sera of different HI antibody potencies. Serum with HI titer 1:2560 (Δ), 1:256 (□), and 1:20 (○). A plot of $\log_{10}$ zone diameter against serum dilution ($\log_2$ scale) gave parallel straight-line relationships for each of these sera.

TABLE 11

Characterization of Influenza A Strains by Hemagglutination-inhibition

Virus strains	Postinfection ferret serum to A/HK/1/68	A/Eng/42/72	A/PC/1/73[a]	A/Vict/3/75
A/HK/1/68 (H3N2)	7,680[a]	7,680	640	<40
A/Eng/42/72 (H3N2)	1,920	15,360[a]	3,840	640
A/PC/1/73	480	1,280	3,840	640
A/Victoria/3/75	<40	320	640	5,120[a]

[a] Reciprocal HI titer.

evidence of progressive antigenic change in the HA antigens from 1968 to 1973 as described previously[45] and also provide clear antigenic differences between A/Port Chalmers/1/73 and the most recent variant, A/Victoria/3/75. It is notable that there was a complete lack of HI cross reaction between the earliest and latest members of the series, i.e., A/Hong Kong/68 and A/Victora/75. Similar evidence, based on NI tests, has been obtained indicating progressive change in the NA antigen (H2) from 1968 to 1973.[45]

In order to confirm that the HA antigen of the 1973 variants belonged to the same subtype (H3) as the antigen of the Hong Kong virus, IDD tests were carried out with hyperimmune sera to their purified HA (H3) antigens (Figure 10). Antiserum to purified HA of A/Hong Kong/1/68 gave precipitin lines with all of the test viruses isolated from 1968 to 1973 and (not shown in the figure) the 1975 variant. However, the reactions between different influenza virus variants in adjacent wells did not show complete identity, and "spurs" were detected between lines given by successive variants. The detection of such "spurs" suggested that the homologous HA antigen possessed antigenic determinants which were not shared by the strains tested in the adjacent wells, whereas the cross reactions observed show that other antigenic determinants are shared by all test strains isolated from 1968 to 1973. Similarly (but not shown in the figures), antiserum to the purified HA of A/Port Chalmers/1/73 virus gave reactions with all the test viruses used (A/Hong Kong/1/68, A/England/42/72, and A/Port Chalmers/1/73), and "spur" formation was demonstrated between these strains tested in adjacent wells.

Thus, the pattern of precipitin reactions indicated that the HA of A/Port Chalmers/1/73 virus was of the same subtype (H3) as the prototype A/Hong Kong/1/68 virus and that the antigenic changes which have occurred in this antigen since 1968 were attributable to antigenic "drift." The most recent variant of influenza A virus, A/Victoria/75 (H3N2), conforms to this pattern. In IDD tests against antiserum to A/HK/68 or A/PC/73 HA, it produces a precipitin reaction which "spurs" when tested adjacent to the homologous A/PC/73 virus. To confirm that the hemagglutinins of A/Hong Kong/1/68 and A/Port Chalmers/1/73 viruses each contained antigenic determinants (strain-specific determinants) which were not shared, cross-absorption experiments were performed. Antiserum to the HA of Hong Kong/68 virus was absorbed by the addition of purified A/Port Chalmers/1/73 virus to remove antibodies to antigenic determinants shared by both strains (cross-reactive determinants). The pattern of reactions of the absorbed serum in IDD tests is shown in Figure 10. No reaction was detected with the 1973 variant, but distinct precipitin lines were found with 1968, 1971, and 1972 variants. Similarly (but not shown in the figure), antiserum to the HA of Port Chalmers virus after absorption with Hong Kong/68 virus gave IDD reactions with the 1973 variant and the 1972 variant but not with the 1968 and 1971 variants. The results of the absorption experiments indicate that the HA antigens of the 1968, 1973, and 1975 variants contained antigenic determinants which were unique and thus provided evidence that certain antigenic determinants of the HA antigen undergo complete antigenic change during antigenic "drift," whereas other determinants remain unaltered and provide the antigenic basis for the recognition of the antigen as H3. Since other studies[34,39,41] have indicated that it is these antigenically variable determinants of hemagglu-

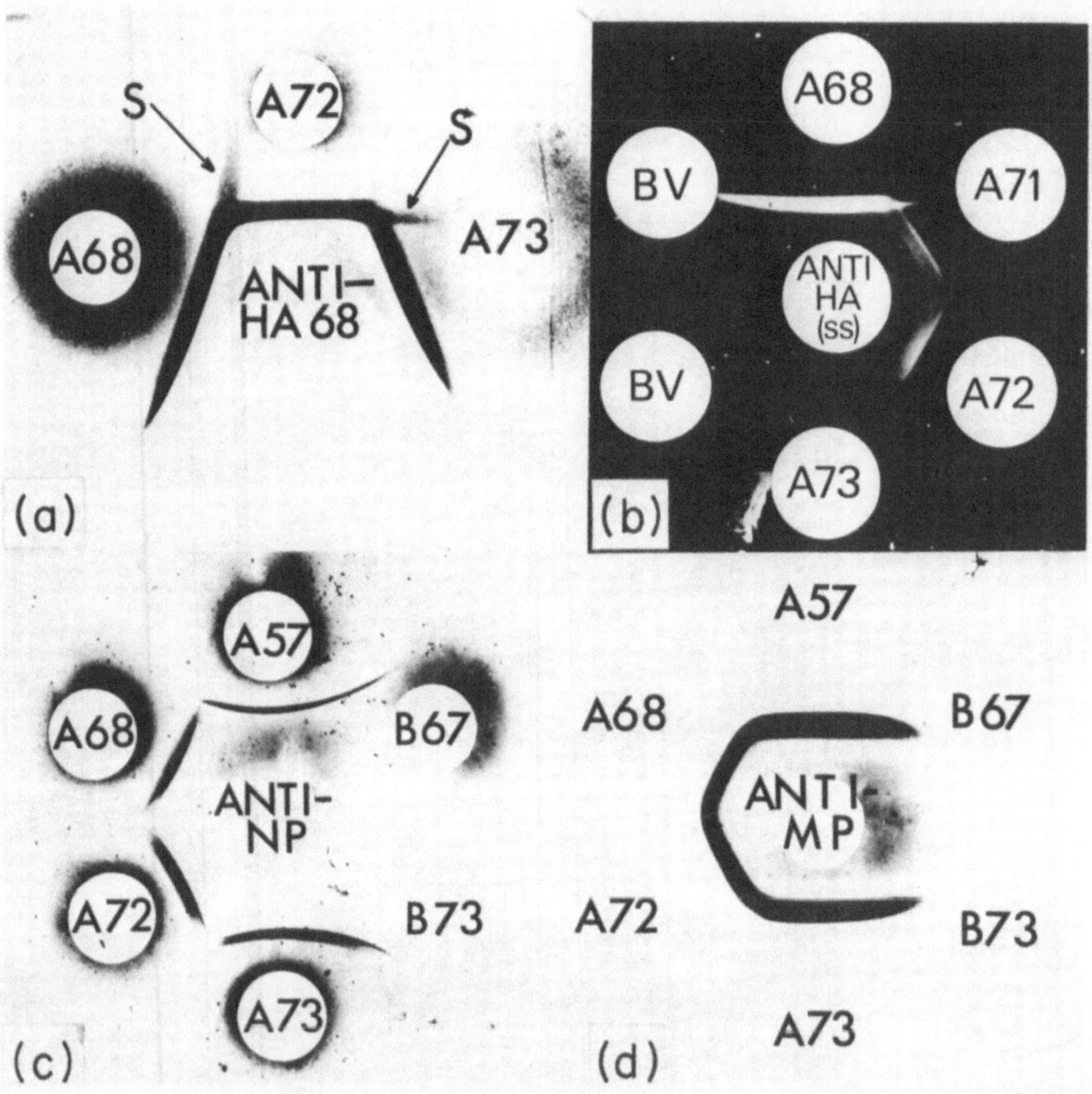

FIGURE 10. IDD studies to show antigenic variation in the HA antigen of the Hong Kong virus from 1968 to 1973 and lack of variation in NP and M antigens. Antigens: Wells marked A57, A68, A71, A72, and A73 contain purified influenza A/Singapore/1/57, A/Hong Kong/1/68, A/Hong Kong/107/71, A/England/42/72, and A/Port Chalmers/1/73 viruses disrupted with SS detergent; BV and B/73 wells contain influenza B/Hong Kong/8/73 virus; and B/67 wells contain B/Roma/1/67 virus. Antisera: Wells marked ANTI-HA 68 contain rabbit antisera to the purified HA of A/Hong Kong/1/68 virus. ANTI-HA 68 (SS) represents the same antiserum tested after adsorption with purified, concentrated A/PC/1/73 virus in order to obtain a strain-specific population of anti-HA antibody. ANTI-NP and ANTI-M represent wells containing rabbit antiserum to purified NP and M antigens. (a) Evidence of progressive antigenic variation in the HA antigen in strains isolated from 1968 to 1973, Spurs (s) showing antigenic determinants not shared by the HA antigens of the strains tested in adjacent wells are indicated. Such antigenic variants must be assessed with any virus-inhibitory compound to establish the comparative degree of inhibition of replication. (b) After adsorbing antiserum to A/HK/1/68 HA by the addition of purified, concentrated A/PC/1/73 virus, a relatively strain-specific population of antibodies remained and reacted with A/HK/1/68 virus and intermediate strains isolated in 1971 and 1972 but failed to react with A/PC/1/73. This population of antibodies is directed against antigenically variable determinants of HA which are not possessed by the 1973 variant. (c and d) Precipitin reactions given by antisera to internal components of influenza A (NP and M) which fail to show evidence of antigenic change in these antigens from 1957 to 1973. These antigens are distinct for influenza Type A and influenza Type B viruses and form the basis for classification of influenza viruses into types.

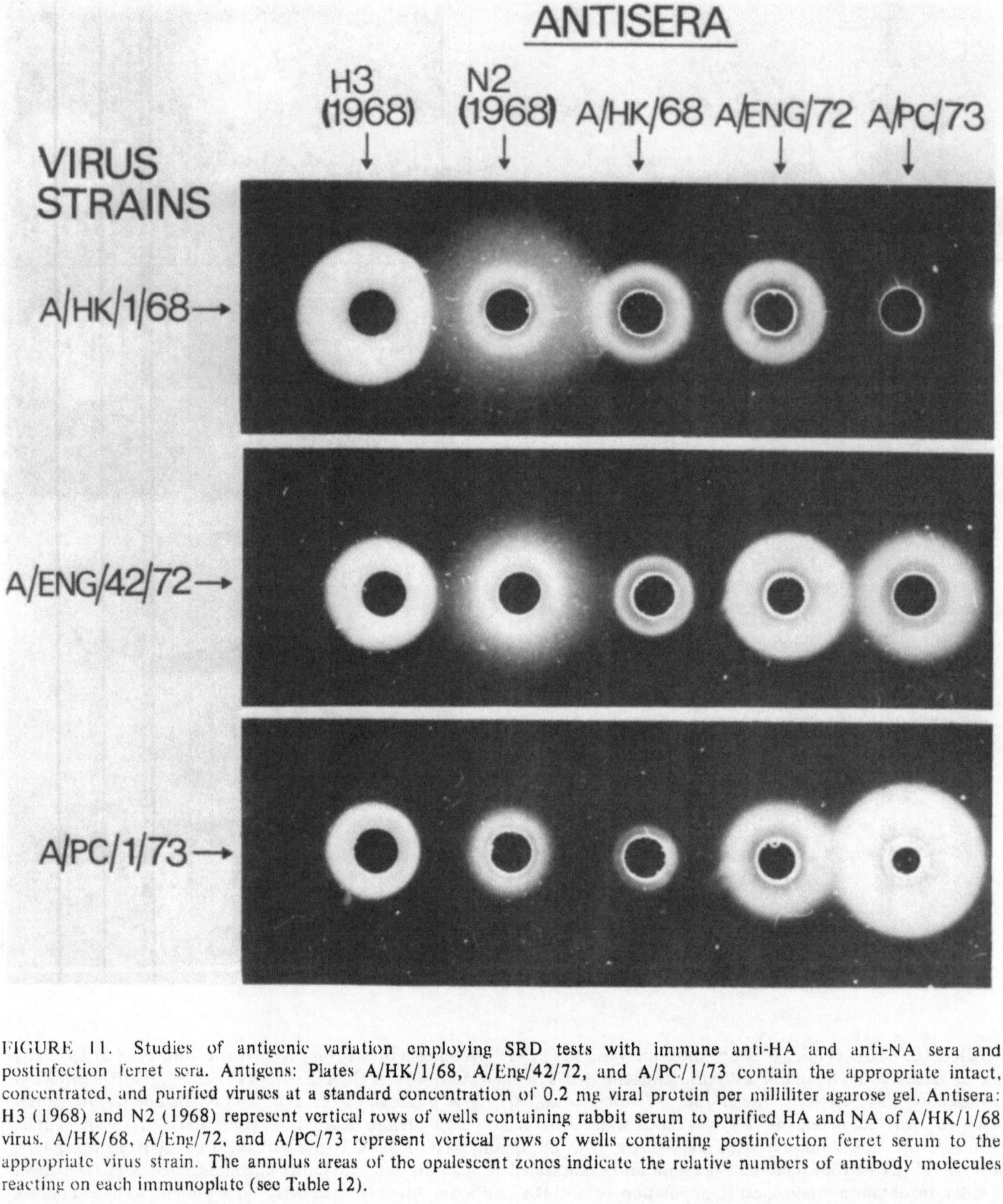

FIGURE 11. Studies of antigenic variation employing SRD tests with immune anti-HA and anti-NA sera and postinfection ferret sera. Antigens: Plates A/HK/1/68, A/Eng/42/72, and A/PC/1/73 contain the appropriate intact, concentrated, and purified viruses at a standard concentration of 0.2 mg viral protein per milliliter agarose gel. Antisera: H3 (1968) and N2 (1968) represent vertical rows of wells containing rabbit serum to purified HA and NA of A/HK/1/68 virus. A/HK/68, A/Eng/72, and A/PC/73 represent vertical rows of wells containing postinfection ferret serum to the appropriate virus strain. The annulus areas of the opalescent zones indicate the relative numbers of antibody molecules reacting on each immunoplate (see Table 12).

tinin which provide critical sites for virus neutralization, changes at these sites are probably of considerable importance to immunity.

The test strains were also compared by SRD. This technique enables a high degree of accuracy and reproducibility for antibody assays and, thus, was valuable in the quantitation of antigenic differences between influenza variants. The comparative SRD test reactions of sera on immunoplates containing A/Hong Kong/1/68, A/England/42/72, and A/Port Chalmers/1/73 are shown in Figure 11, and measurements of the relative areas of opalescent zones for the 1968 to 1973 strains are given in Table 12. In tests with the 1972 to 1975 viruses, antisera to the purified hemagglutinin of A/Hong Kong/68 virus gave opalescent zones which were smaller in area than those produced with the homologous virus and

TABLE 12

Serological Cross Reactions of A/Hong Kong/1/68 (H3N2) with its Variants in Single Radial Immunodiffusion Tests

	Antisera to isolated A/HK/1/68 antigens[a]				Postinfection ferret sera[b] to					
	Antipure H3		Antipure N2		A/HK/1/68		A/Eng/42/72		A/PC/1/73	
Purified virus incorporated in immunoplate	Area[c] (mm^2)	Homologous reaction (%)	Area (mm^2)	Homologous reaction (%)	Area (mm^2)	Homologous reaction (%)	Area (mm^2)	Homologous reaction (%)	Area (mm^2)	Homologous reaction (%)
A/HK/1/68 (H3N2)	63.8	100	130	100	43.2	100	30.3	66	5.5	8.6
A/Eng/42/72 (H3N2)	34.8	55	56.5	43	13.4	31	45.7	100	37.1	58
A/PC/1/73 (H3N2)	22.2	34	16.7	13	6.1	14	33.6	73	63.8	100

[a]Hemagglutinin of chick origin; neuraminidase.
[b]Zones given postinfection sera reflect essentially antibody to HA since such sera are found to contain high levels of antibody to HA and low amounts of antibody to NA.
[c]Areas of opalescence per 10 μl serum.

progressively decreased with the chronological order of isolation of the strains from 1972 to 1975. Similar findings were obtained with antiserum to the purified NA of A/Hong Kong/68 virus when tested against the 1972 and 1973 strains (Figure 11). This application of the SRD technique has been found to be an accurate and reproducible quantitative method of assessing antigenic "drift" in influenza viruses. In addition, the SRD method may be combined with antibody absorption tests. Such tests applied to the variants isolated in 1968 to 1973[4,5] have provided valuable confirmatory data on the antigenic differences between them.

It is important to analyze the inhibitory effects of potential chemoprophylactic compounds on new influenza variants (characterized as described above), and this aspect is further discussed in Chapter 5.

REFERENCES

1. **Dowdle, W. R.**, in *Influenza: Virus, Vaccines and Strategy*, Proc. Working Group on Pandemic Influenza, Selby, P., Ed., Academic Press, London, in press.
2. **Dowdle, W. R., Coleman, M. T., and Gregg, M. B.**, Natural history of influenza Type A in the United States, *Prog. Med. Virol.*, 17, 93, 1975.
3. **Stuart-Harris, C. H. and Schild, G. C., Eds.**, Influenza – the human disease, in *Influenza: The Viruses and the Disease*, Edward Arnold, London, 1976, 96.
4. **Kavet, J.**, Influenza and Public Policy, Ph.D. thesis, Harvard University, Cambridge, 1972.
5. **Kilbourne, E. D.**, in *Preventive Medicine and Public Health*, Smillie, W. G., Ed., Macmillan, New York, 1963, 192.
6. **Melnick, J.**, Classification and nomenclature of viruses, *Prog. Med. Virol.*, 15, 153, 1973.
7. **McGeach, D., Fellner, P., and Newton, C.**, Influenza virus genome consists of eight distinct RNA species, *Proc. Natl. Acad. Sci. U.S.A.*, in press.
8. **Skehel, J. J.**, Polypeptide synthesis in influenza virus infected cells, *Virology*, 49, 23, 1972.
9. **Schild, G. C. and Dowdle, W. R.**, Influenza virus characterization and diagnostic serology, in *The Influenza Viruses and Influenza*, Kilbourne, E. D., Eds., Academic Press, London, 1975, 316.
10. **Laver, W. G. and Valentine, R. C.**, Morphology of the isolated haemagglutinin and neuraminidase subunits of influenza virus, *Virology*, 38, 105, 1969.
11. **Pereira, H. G.**, Influenza viruses, antigenic spectrum and variation, *Prog. Med. Virol.*, 11, 46, 1969.
12. **Stuart-Harris, C. H. and Schild, G. C., Eds.**, The antigens of the influenza viruses, in *Influenza: The Viruses and the Disease*, Edward Arnold, London, 1976, 51.
13. **Schild, G. C.**, Evidence for a new type-specific structural antigen of the influenza virus particle, *J. Gen. Virol.*, 15, 99, 1972.
14. **Skehel, J. J. and Schild, G. C.**, Polypeptide composition of influenza A viruses, *Virology*, 44, 396, 1971.
15. **Choppin, P. W. and Compans, R. W.**, The structure of influenza virus, in *The Influenza Viruses and Influenza*, Kilbourne, E. D., Ed., Academic Press, London, 1975, 15.
16. **Barry, R. D. and Glass, S. E.**, The replication of influenza and parainfluenza viruses, in *Chemoprophylaxis and Virus Infections of the Respiratory Tract*, Oxford, J. S., Ed., CRC Press, Cleveland, in press.
17. **Laver, W. G.**, Separation of two polypeptide chains from the haemagglutinin subunit of influenza virus, *Virology*, 45, 275, 1971.
18. **Lazarowitz, S. G., Compans, R. W., and Choppin, P. W.**, Influenza virus structural and nonstructural problems in infected cells and their plasma membranes, *Virology*, 46, 803, 1971.
19. **Brand, C. M. and Skehel, J. J.**, Crystalline antigen from the influenza virus envelope, *Nat. New Biol.*, 234, 145, 1972.
20. **Wrigley, N. G., Skehel, J. J., Charlwood, P. A., and Brand, C. M.**, The size and shape of influenza virus neuraminidase, *Virology*, 51, 525, 1973.
21. **Oxford, J. S.**, Polypeptide composition of influenza B viruses and enzymes associated with the purified virus particles, *J. Virol.*, 12, 827, 1973.
22. **Compans, R. W., Klenk, H. D., Caliguri, L. A., and Choppin, P. W.**, Influenza virus proteins. I. Analysis of polypeptides of the virion and identification of spike glycoproteins, *Virology*, 42, 880, 1970.
23. **Compans, R. W. and Caliguri, L. A.**, Isolation and properties of an RNA polymerase from influenza virus-infected cells, *J. Virol.*, 11, 441, 1973.

24. **Schild, G. C.**, Studies with antibody prepared against the purified haemagglutinin of influenza A virus, *J. Gen. Virol.*, 9, 197, 1970.
25. **Webster, R. G.**, Antigenic hybrids of influenza A virus with surface antigens to order, *Virology*, 42, 633, 1970.
26. **Jahiel, R. I. and Kilbourne, E. D.**, Reduction in plaque size and reduction in plaque number as differing indices of influenza virus antibody reaction, *J. Bacteriol.*, 92, 152, 1966.
27. **Dowdle, W. R., Downie, J. S., and Laver, W. G.**, Inhibition of virus release by antibodies to surface antigens of influenza viruses, *J. Virol.*, 13, 269, 1974.
28. **Hobson, D., Curry, R. L., and Beare, A. S.**, The role of serum haemagglutination-inhibiting antibody in protection against challenge infection with influenza A2 and B viruses, *J. Hyg.*, 70, 767, 1972.
29. **Slepushkin, A. N., Schild, G. C., Beare, A. S.,Chinn, S., and Tyrrell, D. A. J.**, Neuraminidase and resistance to infection with live influenza A/Hong Kong vaccines, *J. Hyg.*, 69, 571, 1971.
30. **Schulman, J. L., Khakpour, M., and Kilbourne, E. D.**, Protective effects of specific immunity to viral neuraminidase on influenza virus infection of mice, *J. Virol.*, 2, 778, 1968.
31. **Virelizier, J. L., Oxford, J. S., and Schild, G. C.**, Mechanisms of immunity to influenza in experimental animals, *Postgrad. Med. J.*, 52, 332, 1976.
32. **Couch, R. B. and Kilbourne, E. D.**, Induction of partial immunity to influenza by neuraminidase-specific vaccine, *J. Infect. Dis.*, 129, 411, 1974.
33. **Oxford, J. S. and Schild, G. C.**, Biophysical and immunological studies with influenza M protein, *Virology*, in press.
34. **Virelizier, J. L.**, Host defences against influenza virus: the role of anti-haemagglutinin antibody, *J. Immunol.*, 115, 434, 1975.
35. **Schulman, J. L. and Kilbourne, E. D.**, Induction of partial specific heterotypic immunity in mice by a single infection of influenza A virus, *J. Bacteriol.*, 89, 170, 1965.
36. **Werner, G. H. and Wolff, E.**, Immunité croisée hétérotypique chez la souris entre virus grippaux humains de types A et A2, *C. R. Acad. Sci. Ser. D*, 263, 1913, 1966.
37. A revised nomenclature system for influenza viruses, *Bull. WHO*, 45, 119, 1971.
38. **Schild, G. C. and Pereira, H. G.**, Characterization of the ribonucleoprotein and neuraminidase of influenza A by immunodiffusion, *J. Gen. Virol.*, 4, 355, 1969.
39. **Virelizier, J. L., Postlethwaite, R., Schild, G. C., and Allison, A. C.**, Antibody responses to antigenic determinants of influenza virus haemagglutinin. I. Thymus dependence of antibody formation and independence of antigenic memory, *J. Exp. Med.*, 150, 1559, 1974.
40. **Laver, W. G., Downie, J., and Webster, R. G.**, Studies on the antigenic variation of influenza virus: Evidence for multiple antigenic determinants on the haemagglutinin subunits of A/Hong Kong/68 and A/England/72 strains, *Virology*, 59, 230, 1974.
41. **Schild, G. C., Oxford, J. S., and Virelizier, J. L.**, The immune response to influenza: methods of analysis of antibodies and antigens, in *The Role of Immunological Factors in Infectious Allergic and Autoimmune Processes*, Beers, R. T. and Bassett, E. G., Eds., Raven Press, New York, 1976, 481.
42. Weekly Epidemiological Report No. 14, World Health Organization, 1976, 108.
43. **Webster, R. G. and Laver, W. G.**, Antigenic variation of influenza viruses, in *The Influenza Viruses and Influenza*, Kilbourne, E. D., Ed., Academic Press, London, 1975, 270.
44. Pandemic of Influenza 1918–1919, Ministry of Health Report No. 4, HMSO, London, 1920.
45. **Schild, G. C., Oxford, J. S., Dowdle, W. R., Coleman, M. T., Pereira, M. S., and Chakraverty, P.**, Antigenic variation in current influenza A viruses: Evidence for a high frequency of antigenic drift for the Hong Kong virus, *Bull. WHO.*, 51, 1, 1974.
46. Epidemiological Report No. 43, World Health Organization Weekly, 1975, 366.
47. **Smith, W., Andrewes, C. H., and Laidlaw, P. P.**, A virus obtained from influenza patients, *Lancet*, 2, 66, 1933.
48. **Stuart-Harris, C. H. and Schild, G. C.**, The epidemiology of influenza, in *Influenza: The Viruses and the Disease*, Edward Arnold, London, 1976, 113.
49. **Hirsch, A.**, *Handbook of Geographical and Historical Pathology*, New Sydenham Society, London, 1, 7, 1883.
50. **Creighton, C.**, *A History of Epidemics in Britain*, Vol. 1, Cambridge University Press, London, 1891.
51. **Creighton, C.**, *A History of Epidemics in Britain*, Vol. 2, Cambridge University Press, London, 1894.
52. **Burnet, F. M. and White, D. O.**, *Natural History of Infectious Disease*, 4th ed., Cambridge University Press, London, 1972, 202.
53. **Collier, R.**, *The Plague of the Spanish Lady*, Macmillan, London, 1974.
54. **Burnet, F. M. and Clarke, E.**, Influenza Monogr. No. 4 Walter and Eliza Hall Institute, Melbourne, 1942.
55. **Koen, J. S.**, A practical method for field diagnosis of swine diseases, *Am. J. Vet. Med.*, 14, 468, 1919.
56. **Andrewes, C. H., Laidlaw, P. P., and Smith, W.**, Influenza observation on the recovery of virus from man and on the antibody content of human sera, *Br. J. Exp. Pathol.*, 16, 566, 1935.
57. **Schild, G. C. and Stuart-Harris, C. H.**, Serological epidemiological studies with influenza viruses, *J. Hyg.*, 63, 479, 1965.
58. **Shope, R. E.**, Swine influenza. III. Filtration experiments and aetiology, *J. Exp. Med.*, 54, 373, 1931.
59. **Jensen, N. E.**, A new set of type A influenza viruses, *JAMA*, 164, 2025, 1957.
60. **Dunn, F. L.**, Pandemic influenza in 1957 – a review of international spread of the new Asian strain, *JAMA*, 166, 1140, 1958.

61. Woodall, J., Rowson, K. E. K., and McDonald, J. C., Age and "Asian" influenza, *Br. Med. J.*, 2, 1316, 1958.
62. Langmuir, A. D., Epidemiology of "Asian" Influenza. International conference of asian influenza, *Am. Rev. Respir. Dis.*, 83, 2, 1968.
63. Chang, W. K., National influenza experience in Hong Kong, 1968, *Bull. WHO*, 41, 349, 1969.
64. Cockburn, W. C., Delon, P. J., and Fereira, W., Origin and process of the 1968–69 Hong Kong influenza epidemic, *Bull. WHO*, 41, 345, 1969.
65. Sharrer, R. G., National experience in the USA 1968–69, *Bull. WHO*, 41, 361, 1968.
66. Centre for Disease Control, Atlanta, and Public Health Laboratory Service, London, unpublished reports.
67. Schild, G. C. and McGregor, I., Weekly Epidemiological Record No. 6, World Health Organization, 1976, 46.
68. Hope-Simpson, R. E., Discussion on the common cold, *Proc. R. Soc. Med.*, 51, 267, 1958.
69. Davey, M. L. and Reid, D., *Br. J. Prev. Soc. Med.*, 26, 28, 1972.
70. Reid, D., personal communication.
71. Hemmes, J. H., Winkler, K. C., and Kool, S. M., Virus survival as a seasonal factor in influenza and poliomyelitis, *Nature*, 188, 430, 1960.
72. Kilbourne, E. D., Epidemiology of influenza, in *The Influenza Viruses and Influenza*, Kilbourne, E. D., Ed., Academic Press, London, 1975, 483.
73. Assaad, F., Cockburn, W. C., and Sundaresan, T. K., Use of excess mortality from respiratory diseases in the study of influenza, *Bull. WHO*, 49, 219, 1973.
74. Langmuir, A. D. and Housworth, J. A., A critical evaluation of influenza surveillance, *Bull. WHO*, 41, 393, 1969.
75. Smith, J. W. G., Surveillance of Influenza, Report of the Public Health Laboratories Service, Epidemiology Research Unit, in press.
76. Pereira, M. S., The role of epidemiological surveillance in the immunoprophylaxis of influenza, *Postgrad. Med. J.*, 52, 323, 1976.
77. Easterday, B. C., Animal influenza, in *The Influenza Viruses and Influenza*, Kilbourne, E. D., Ed., Academic Press, London, 1975, 449.
78. Stuart-Harris, C. H. and Schild, G. C., Eds., Influenza viruses of lower animals and birds, in *Influenza: The Viruses and the Disease*, Edward Arnold, London, 1976, 78.
79. Blakemore, F. and Gledhill, A. W., Discussion on swine influenza in the British Isles, *Proc. R. Soc. Med.*, 34, 611, 1939.
80. Harkness, J. W., Schild, G. C., Lamont, P. H., and Brand, C. M., Studies on relationships between human and porcine influenza, *Bull. WHO*, 46, 709, 1972.
81. Schild, G. C., Brand, C. M., Harkness, J. W., and Lamont, P. B., Studies on relationships between porcine and human influenza viruses, *Bull. WHO*, 46, 720, 1972.
82. Schäfer, W., Vergleichende sero-immunologische Untersuchungen veber die viren der influenza und Klassischen Gefluegelpest, *Z. Naturforsch. Teil*, 10, 81, 1955.
83. Masurel, N. and Marine, N., Recycling of Asian and Hong Kong influenza A virus hemagglutinins in man, *Am. J. Epidemiol.*, 97, 44, 1973.
84. Stuart-Harris, C. H. and Schild, G. C., Immunization with influenza virus vaccines, in *Influenza: The Viruses and the Disease*, Edward Arnold, London, 1976, 164.
85. Hilleman, M. R., Woodhour, A., Friedman, A., Wibel, R. E., and Stokes, J., The clinical application of adjuvant 65, *Ann. Allergy*, 30, 152, 1972.
86. Reimer, C. B., Baker, R. C., Van Frank, P. M., Newlin, T. E., Cline, G. B., and Anderson, N. G., Purification of large quantities of influenza virus by density gradient centrifugation, *J. Virol.*, 1, 1207, 1967.
87. Kilbourne, E. D., Future influenza vaccines and the use of the genetic recombinants, *Bull. WHO*, 41, 643, 1969.
88. Mostow, S. R. et al., Studies on inactivated influenza vaccines. II. Effect of increasing dosage on antibody response and adverse reactions in man, *Am. J. Epidemiol.*, 92, 248, 1970.
89. Schoenbaum, S. C. et al., Studies with inactivated influenza vaccines purified by zonal centrifugation. II. Efficacy, *Bull. WHO*, 41, 531, 1969.
90. Glezen, W. P., Loda, F. A., and Denny, F. W., A field evaluation of inactivated izonal-centrifuged influenza vaccines in children in Chapel Hill, North Carolina, 1968–1969, *Bull. WHO*, 41, 566, 1969.
91. Ennis, F., personal communication.
92. Davenport, F. M., Hennessy, A. V., Brandon, F. M., Webster, R. G., Barrett, C. D., and Lease, G. O., Comparisons of serologic and febrile responses in humans to vaccination with influenza A viruses or their haemagglutinins, *J. Lab. Clin. Med.*, 63, 5, 1964.
93. Webster, R. G. and Laver, W. G., Influenza virus subunit vaccines: immunogenicity and lack of toxicity for rabbits of ether and detergent disrupted virus, *J. Immunol.*, 96, 596, 1966.
94. Neurath, S. R. et al., The effect of nonaqueous solvents on a quaternary structure of virus properties of haemagglutinins obtained by disruption of influenza viruses with tri(*n*-butyl)phosphate, *Microbios*, 2, 209, 1970.
95. Philips, C. F. et al., Killed subunit influenza vaccine in children, *Pediatrics*, 52, 416, 1973.
96. Brady, M. I., Furminger, I. G. S., and Stones, P. B., An adsorbed surface antigen influenza vaccine and its serological activity in volunteers, *Postgrad. Med. J.*, 52, 368, 1976.
97. Laver, W. G. and Webster, R. G., Preparation and immunogenicity of an influenza virus HA and NA subunit vaccine, *Virology*, 69, 511, 1976.

98. **Bachmeyer, H., Liehl, E., and Schmidt, G.,** Preparation and properties of a novel influenza subunit vaccine, *Postgrad. Med. J.*, 52, 360, 1976.
99. *WHO Tech. Rep. Ser.*, No. 384, 1968.
100. **Westwood, M., Woodward, P. M., and Perkins, F. G.,** The British reference preparation for influenza virus haemagglutinin, *J. Hyg.*, 69, 729, 1971.
101. **Schild, G. C., Wood, J., and Newman, R. W.,** A single radial immunodiffusion technique for the assay of influenza haemagglutinin antigen, *Bull. WHO*, 52, 223, 1975.
102. **Beare, A. S., Schild, G. C., and Craig, J. W.,** Trials in man with live recombinants made from A/PR/8/34 (HON1) and wild H3N2 influenza viruses, *Lancet*, 2, 129, 1975.
103. **Murphy, B. R.,** in *Influenza: Virus, Vaccines and Strategy*, Working Group on Pandemic Influenza, Selby, P., Ed., Academic Press, London, 1976.
104. **Maassab, H. F.,** Adaptation and growth characteristics of influenza virus at 25°C, *Nature*, 213, 612, 1967.
105. Document No. IPAC 335, Medical Research Council, London Immunological Products Advisory Committee, 1976.
106. WHO Expert Committee on Influenza, *WHO Tech. Rep. Ser.*, No. 64, 1953.
107. **Aymard-Henry, M., Coleman, M. T., Dowdle, W. R., Schild, G. C., and Webster, R. W.,** Influenza virus, neuraminidase and neuraminidase inhibition test procedures, Bull. WHO, 48, 199, 1973.
108. **Schild, G. C. and Pereira, H. G.,** Characterization of the ribonucleoprotein and neurominidase of influenza A by immunodiffusion, *J. Gen. Virol.*, 4, 355, 1969.
109. **Schild, G. C., Winters, W. D., and Brand, C. M.,** Serological diagnosis of human influenza infections by immunoprecipitin techniques, *Bull. WHO*, 45, 469, 1971.
110. **Schild, G. C., Aymard-Henry, M., and Pereira, H. G.,** A quantitative single radial diffusion test for immunological studies with influenza virus, *J. Gen. Virol.*, 16, 231, 1972.
111. **Haaheim, L.,** *Acta Pathol. Microbiol. Scand.*, in press.
112. **Chakraverty, P., Pereira, M. S., and Schild, G. C.,** The use of single radial diffusion tests in sero-epidemiological studies of influenza, Bull. WHO, 49, 327, 1973.
113. **Mostow, S., Schild, G. C., Dowdle, W. R., Coleman, M. T., and Wood, R. J.,** Application of the single-radial-diffusion test for the assay of antibody to type A influenza viruses, *J. Clin. Microbiol.*, 2, 531, 1975.
114. **Schild, G. C., Pereira, M. S., and Chakraverty, P.,** Single-radial-haemolysis: a new method for the assay of antibody to influenza haemagglutinin, *Bull. WHO*, 52, 43, 1975.
115. **Russell, S. M., McCahan, D., and Beare, A. S.,** A single-radial-haemolysis technique for the measurement of influenza antibodies, *J. Gen. Virol.*, 27, 1, 1975.

Chapter 4

THE REPLICATION OF INFLUENZA AND PARAINFLUENZA VIRUSES

Richard D. Barry and Sylvia E. Glass

TABLE OF CONTENTS

I. INTRODUCTION

Recent developments in the biochemistry of virus replication have uncovered unusual and frequently unexpected phenomena, particularly in regard to the synthesis and function of nucleic acids. These advances should greatly improve our chances of finding compounds that act as specific inhibitors of virus replication. This chapter deals with the replication of influenza viruses, concentrating particularly on the functional activity of the genetic constituents of the virus (the genome) and the way in which the genome is reproduced. These viruses are still known popularly as myxoviruses, although more recently they have been described as "orthomyxoviruses," a name chosen to distinguish the influenza viruses from viruses of the parainfluenza type, now called "paramyxoviruses."[1] A description of the replication of parainfluenza viruses is provided at the end of the chapter. To date these viruses have been little studied from the point of view of chemotherapy, although they are important pathogens of children (Chapter 1).

Knowledge of the general principles of RNA virus replication has accumulated rapidly in recent years. The finding that small RNA viruses can replicate normally in the absence of DNA synthesis[2,3] or DNA-directed RNA synthesis[4] firmly established the concept that viral RNA replication is independent of the genetic material of the host cell. The detection of ribonuclease-resistant, virus-specific RNA in cells infected with EMC virus[5] gave the first indication that viral RNA synthesis proceeds by way of a double-stranded, base-paired intermediate. The existence of such replicative forms has been established for many animal and bacterial RNA viruses,[6] and it is now accepted that the replication of viral RNA is a two-stage process. The first step is the synthesis of a complementary polynucleotide strand that acts as a template in the second step for the synthesis of progeny strands of viral RNA; in the case of the RNA tumor viruses, the complementary polynucleotide template is DNA.[7] In those cases where the isolated viral RNA is infectious, the RNA serves as both genome and messenger (plus strand), while the complementary RNA (minus strand) is simply a template for plus-strand formation. Cells infected with RNA viruses produce virus-specific RNA-dependent RNA polymerases;[8,9] these enzymes are not found in uninfected cells and are responsible for the synthesis of both template and progeny RNA. The enzyme specifically concerned with progeny RNA synthesis is referred to as a replicase.

The RNA of influenza virus lacks infectivity, and the mRNAs that can be detected on the polyribosomes of infected cells are complementary to virion RNA.[10] Influenza viruses differ from many other RNA viruses in that the RNA in the virion is not a messenger RNA. Together with the paramyxoviruses and the rhabdoviruses, they have been described collectively as "negative strand viruses."[11] An essential requirement for this type of virus is the presence within the virion of an RNA polymerase that can transcribe the virion RNA into mRNA following infection.

This chapter begins with a brief summary of the structure of influenza virus (Section II and Chapter 5) and is followed by a more detailed consideration of the nature of the genome (Section III). The replication of influenza viruses, with emphasis on transcription of the genome into mRNA, is then considered (Section IV). The final section (Section V) considers briefly the similarities and differences between the replication of influenza and parainfluenza viruses. Wherever possible, the text is illustrated with data from our laboratory, where the main influenza virus under experimental investigation is the avian influenza A virus fowl plague (FPV), and the parainfluenza viruses are represented by parainfluenza 1 (Sendai) virus.

II. THE STRUCTURE OF INFLUENZA VIRUSES

A. Morphology

Influenza viruses are roughly spherical, pleomorphic particles with diameters of 75 to 100 nm.[12] The envelope of the virus consists of an outer lipid membrane bearing radiating, spike-like glycoprotein projections and an inner layer of protein (M protein). Within this envelope is a helical structure, the ribonucleoprotein (RNP). The principal properties of these two substructures will now be considered. Further details, with particular emphasis on relevance to the mode of action of inhibitors, are given in Chapter 5.

1. Virus Envelope

The envelope consists of a lipid bilayer surrounded by spike-like projections, 8 to 10 nm in length.[12,13] Two types of spikes have been identified: the more abundant rod-shaped hemagglutinin and the mushroom-shaped neuraminidase.[14-17] The lipid bilayer encloses a shell of membrane (M) protein.[18,19] Lipids constitute 20 to 24% of the mass of influenza virus particles,[20,21] and although their composition is largely determined by the host cell,[22] some variations are strain specific.[17] Such minor differences have been attributed to virus-induced alterations in the metabolic state of the host cell.[23]

Electron spin resonance studies indicate that the lipid is a bilayer consisting of cholesterol and phospholipid in equimolar amounts and that the glycoproteins do not penetrate deeply into the lipid.[24] The lipid protects and encloses all other virus proteins, because treatment of virions with proteolytic enzymes removes only the hemagglutinin (HA) and the neuraminidase (NA).[18,24,25]

The HA and NA glycoproteins are responsible for adsorption to the mucoprotein receptors of host cells and red blood cells and subsequent elution from them.[16,26,27] Intact HA spikes isolated from virus particles by SDS or Tween® treatment are 14 nm long and 4 nm wide when visualized by an electron microscope.[16] The NA peplomers, in the presence of SDS, are mushroom shaped and consist of an oblong 8.5-nm "head" connected by a thin "stalk," 10 nm long, to a small "foot" of 4-nm diameter.[16,28]

Until comparatively recently, it was not realized that the lipid envelope was lined internally by a shell of M protein. This particular polypeptide was "discovered" when the technique of SDS-polyacrylamide gel electrophoresis was applied to the analysis of virus structure,[29,30] and it was found that about one third of the total mass of the virion protein is provided by the M protein, the most plentiful molecule in the virion. Unequivocal proof that the M protein forms a layer enclosing the RNP was provided by Schulze[18] who prepared viral "cores" by removal of the envelope spikes with proteolytic enzymes. These cores, when stained with uranyl acetate, were found to contain a 6-nm layer of protein clearly visible beneath the 6-nm layer of lipid. When fixed with glutaraldehyde and treated with Nonidet P40 (NP40), smaller, more dense cores were obtained that still contained M protein; Schulze proposed that the term "membrane protein" should be used to define this polypeptide.

2. The Ribonucleoprotein

The RNP of influenza viruses (the so-called "soluble" or "S" antigen) is antigenically stable and type specific. It provides the basis for the subdivision of influenza viruses into the familiar types, A, B,[31] and C.[32] When released from intact virus, the RNPs are a heterogeneous collection of filaments[13,33-36] that can be separated by velocity sedimentation into three size classes of similar diameter but differing lengths – 100, 60 to 90, and 30 nm. Each class contains distinct sizes of vRNA molecules.[36-39]

One of the earliest electron microscopy studies suggested that the RNP might exist in the virion in the form of a tightly coiled helix with between 6 and 12 turns.[13] This supercoil had a width of 60 nm and consisted of strands 6 nm in diameter. However, more recent studies in which the structure of the RNP was examined by negative straining with uranyl acetate, both in complete or ultrathin sectioned particles, suggest that RNP is a relatively short, loosely coiled structure with a diameter of 10 to 15 nm.[19,35,40] When RNP released from detergent-disrupted virions was examined by either negative staining[19,35] or positive staining with uranyl acetate,[36] it was found in the form of a strand folded back on itself with a loop at one end. This coil consisted of a regular double-helical arrangement of 10 to 15 nm in diameter, except at the terminal loop which was a single strand of 5 to 7.5 nm in diameter. Present evidence suggests that the RNP occurs in the virion as double-helical coils of protein.

B. Structural Proteins

Proteins represent 60 to 70% of the weight of influenza virus.[22] Three functionally and antigenically distinct proteins, the nucleocapsid and the HA and NA glycoproteins, have been recognized for many years.[13,33] More recently the individual polypeptides of the virus have been resolved by polyacrylamide gel electrophoresis.[18,41-45] The results from different laboratories, using various strains of influenza virus grown in either eggs or tissue culture, show differences in the number and proportion of the various polypeptides, but there are sufficient similarities to justify the general summary shown in Table 1. A representative example, obtained in our laboratory, is shown in Figure 3a. Seven or eight structural polypeptides have been resolved, and they fall into three distinct molecular weight groups: 20,000 to 30,000, 45,000 to 65,000; and 80,000 to 90,000. Only the polypeptides associated with the HA and NA activities are glycosylated, and these show the greatest variability in molecular weight between strains.

The largest polypeptides are the P proteins of which there were thought to be at least two.[45-47] Recent work in this laboratory has established that P_2 is a mixture of two polypeptides of similar molecular weight so that there are actually three P proteins.[48] Since there are only about 50 molecules of the P proteins per virion, it is unlikely that they play a structural role. They probably constitute the transcriptase of the virus because they are associated with RNP structures that have polymerase activity and can be isolated either from virus particles[49] or infected cells.[50] The presence of a variety of P proteins is consistent with a subunit enzyme structure such as is found in mammalian and bacterial polymerases.[51] The HA

TABLE 1

The Structural Polypeptides of Influenza Viruses[a]

Designation	Function	Mol wt[b] × 10^{-3}	% in virion	No. molecules per virion
$P_{1,2,3}$	Proteins involved in RNA synthesis	90	3	50
HA	Hemagglutinin	75	30	1,000
HA_1[c]	Hemagglutinin	50	–	–
HA_2[c]	Hemagglutinin	25	–	–
NA	Neuraminidase	60	7	200–200
NP	Nucleoprotein	60	25	700
M	Membrane protein	25	35	3,000

[a]Adapted from the report on the N.I.H. Influenza Workshop I.[182]
[b] To the nearest 5,000.
[c] HA is cleaved to HA_1 and HA_2 in certain strains and/or host cells.

of influenza viruses is a single glycoprotein with a molecular weight of about 80,000. In both the virus and infected cells, it is usually found as two smaller glycopeptides (HA_1 and HA_2) that are joined by disulfide bonds.[43] This is because there is posttranslational proteolytic cleavage of HA to HA_1 and HA_2.[52,53] and this depends on whether a virus strain is cytopathic for a particular host cell. The HA peptide contains 17 to 20% carbohydrate by weight that is covalently linked to HA_1 and HA_2. The isolated HA spikes are thought to be dimers or trimers of the 80,000 mol wt polypeptide, held together by noncovalent bonds, and have a molecular weight of 150,000 to 210,000.[16,54] The mushroom-shaped NA spike is also composed of glycopolypeptide molecules, but there are uncertainties concerning the molecular weight of the subunit polypeptide. Estimates vary from 45,000[44] to 72,000[43] and may reflect differences in the extent of glycosylation in different virus-cell systems or undefined characteristics of the migratory properties of glycoproteins on polyacrylamide gels. Also, it is not known whether structural subunits of the NA consist of a single glycoprotein or a mixture of two species of closely similar molecular weight. Two different subunits are present in the X-7 recombinant of influenza A,[43,55-57] but certain strains of influenza viruses apparently contain only a single glycoprotein entity.[42,43,56,58] Estimates of the molecular weight of the active neuraminidase enzyme vary from 150,000 to 300,000. The active neuraminidases of influenza A/X-7 or B/Lee, for instance, average just over 200,000 and are probably tetramers of their subunits.[57,59,60] The monomers are thought to be linked into pairs by disulfide bonds which in turn associate with another pair, mainly by noncovalent bonds.[60]

The NP polypeptide is the protein component of the RNP and displays a marked affinity in vitro for both c and v influenza RNA as well as for a small AMP-rich RNA from normal cells.[61] Immunodiffusion tests indicate that the NP is type specific and distinct from the M protein.[62]

The membrane or matrix protein forms the layer beneath the phospholipid envelope of the virus. Schulze[18] calculated that 3,400 molecules of M protein would be required to occupy a shell of 6-nm thickness and 80-nm diameter, the approximate dimensions of the outer protein layer of the viral core.

III. THE INFLUENZA VIRUS GENOME

Influenza viruses contain between 0.7 and 1.0% RNA as estimated from the dry weight of the virus, and it was thought to have a molecular weight of around 2×10^6.[63,64] The RNA is single stranded;[65] it does not self-anneal.[39,61] The virion RNA is of the opposite sense to that of virus-specific mRNA found in infected cell polyribosomes.[10]

Within the intact virus particle, the RNA genome is associated with multiple copies of the NP, in the form of the helical RNP, and constitutes about 10 to 15% of the weight of these structures. In this state the RNA is sensitive to degradation by ribonuclease, although it is degraded at a slower rate than free RNA. On the other hand, the NP is resistant to mild pronase treatment,[39,66] suggesting that neither component is completely enclosed by the other.[67]

A. Biological Properties

1. Genetic Phenomena

Some of the unusual biological properties of influenza viruses can be accounted for by the fragmented structure of the virus genome. In 1956 Burnet observed a high rate of recombination in genetic experiments and proposed that several independent linkage groups exist in influenza virus.[68] The same conclusion was reached by Hirst[69] and Simpson[70] who encountered as much as 50% genetic recombination in marker rescue experiments.

The phenomenon of multiplicity reactivation, whereby infectious virus is produced by a process of cooperation between two or more genetically damaged virions infecting a single cell, occurs readily with influenza viruses. This was first demonstrated unequivocally by Barry[71] using influenza inactivated by ultraviolet irradiation, and a detailed analysis of the data indicated that each virus particle contains about six independent genetic units. Further indirect evidence that the influenza genome might be a collection of functionally separate entities is apparent from the sequential inactivation of virus-induced macromolecule synthesis. Certain virus activities survive after chemical inactivation,[72] ultraviolet light inactivation,[73] or viral infectivity. Loss of infectivity was followed sequentially by inability to synthesize HA and then NA, and finally RNP and RNA.

2. Infectivity of Influenza Virus RNA

There have been a number of reports claiming the isolation of infectious RNA from influenza virus,[74–78] but these claims could not be substantiated by other workers.[79–81] The most exhaustive of all these studies were those of Sokol and co-workers, who could find no evidence of infectivity by influenza virus RNA. In the light of recent evidence, influenza viruses are almost certainly "negative-strand" viruses which by definition do not contain infectious RNA. The evidence to support this statement is provided in Section IV.

B. Structural Properties

1. Physical Characterization

The unusual genetic behavior of influenza viruses suggests that the genome consists of independently replicating segments.[68,69,71] Biophysical analysis of isolated RNA has since confirmed that the viral genome consists of discrete pieces. Sucrose velocity density gradient analysis of the RNA gave the first hint that the genome may be physically fragmented. Davies and Barry[82] demonstrated that RNA extracted from A/Singapore/1/57 influenza virus sedimented as a sharp band at approximately 18S, corresponding roughly to a molecular weight of 500,000. At least four such molecules would be required to provide a genome assumed to be at least of a molecular weight of 2×10^6. Subsequent experiments with other virus strains established that the RNA was actually a heterogeneous population of molecules sedimenting in the range of 8 to 22S.[65,83,84]

Further evidence that the influenza RNA exists in pieces was derived from sedimentation analysis of viral RNP. When released by either ether or detergent treatment, RNPs could then be separated into three components on sucrose velocity gradients at 68S, 56S, and 48S[34] or 70S, 60S, and 50S.[37,38] Each class contained distinct vRNA molecules with the 70S RNPs containing mainly the largest component and the 60S and 50S RNPs containing the smaller vRNAs.[37–39] Much more detailed information concerning the nature of genome RNA has come from polyacrylamide gel electrophoresis. Early studies resolved at least five pieces of RNA providing a total genome size of 2.4 to 2.7×10^6, assuming that each virus particle contained one molecule of each species.[85,86] This estimate has been reevaluated in light of more recent studies. The RNA of two influenza strains, A/WSN and X-31, could be resolved into six or seven pieces on 3% polyacrylamide gels and fell into three size classes with a collective molecular weight of 5×10^6 ranging from 0.29 to 1.21×10^6 daltons.[46,87] The analysis of FVP RNA, using 8 *M* urea polyacrylamide gels,[88] has resolved eight segments of RNA with a molecular weight range of 2.9 to 12.1×10^5, giving a total molecular weight of 6×10^6, as all pieces seem to be present in equimolar amounts.[89]

The results obtained in this study are shown in Figure 1, which consists of three slab gel preparations of ^{32}P-labeled FPV RNA run for various lengths of time. It is noteworthy that there are no components of lower molecular weight than segment 8 and that complete separation requires extensive electrophoresis. Further evidence will be presented below (see Table 3) that each piece is unique and therefore represents individual elements of the virus genome. The estimated molecular weight of individual RNA fragments of FPV and the molecular weight of what are probably the corresponding virus-specific polypeptides synthesized in chick embryo fibroblasts infected with FPV are shown in Table 2.

2. Chemical Characterization

The biological and physical data summarized above provide strong circumstantial support for the idea that the influenza virus genome consists of segments, each of which represents a single and separate gene. Chemical analysis of influenza RNA provides further information on the subject. What is presently known of the chemical structure of influenza virus RNA will now be summarized briefly.

The total base composition of the RNA of various virus strains was determined by a number of workers during the 1950's.[63,90,91] Significant differences were found between A and B viruses, and both types had a high uracil content (32 to 37%). More recently the base composition of the human influenza A strains PR8[65] and WSN[92] was reported to be 23% cytosine, 23% adenine, 20% guanine, and 34% uracil. The base ratios of four separate size classes of RNA obtained by density gradient centrifugation from PR8 were similar to each other and also to the base composition of the unseparated RNA.[65] Separate size classes of RNA differ, however, in base sequence as demonstrated by hybridization using double-stranded virus-specific RNA from infected cells.[93] It was found

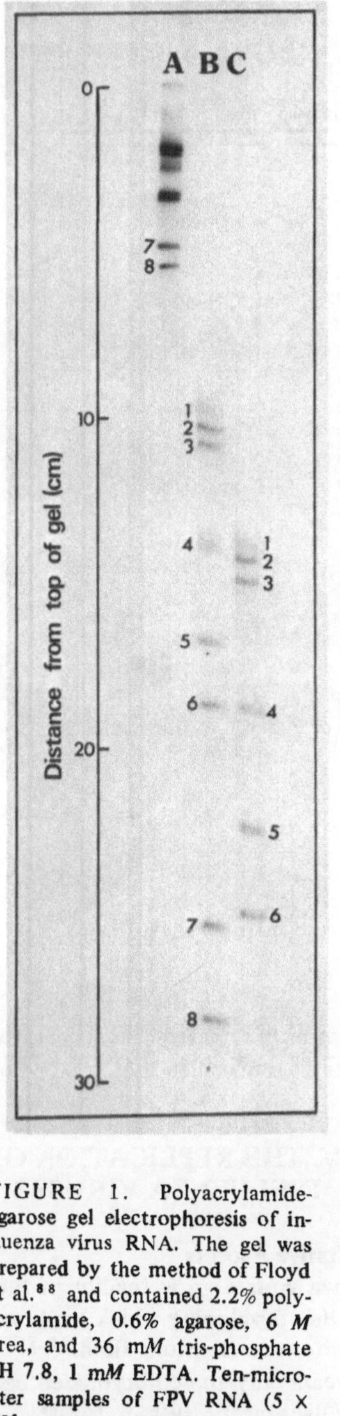

FIGURE 1. Polyacrylamide-agarose gel electrophoresis of influenza virus RNA. The gel was prepared by the method of Floyd et al.[88] and contained 2.2% polyacrylamide, 0.6% agarose, 6 *M* urea, and 36 m*M* tris-phosphate pH 7.8, 1 m*M* EDTA. Ten-microliter samples of FPV RNA (5 × 10^4 cpm) were loaded and electrophoresed at 180 V in 36 m*M* tris-phosphate pH 7.8, 1 m*M* EDTA, and 0.2% sodium dodecyl sulfate. After completion the gel was washed, dried, and autoradiographed. (A) Sample run for 6 hr. (B) Sample run for 24 hr. (C) Sample run for 32 hr.

that the single-stranded RNA with the highest molecular weight possessed unique sequences not present in the smallest class of RNA. The smallest RNA also contained unique sequences, as well as sequences from the larger classes. The latter were assumed to be breakdown products. When the three separate RNA classes were subsequently analyzed by oligonucleotide mapping,[94] each had a distinct fingerprint, which in turn suggested that different nucleotide sequences were present in each class of vRNA.

If individual segments of influenza RNA are generated by replication rather than by endonucleolytic cleavage, the 5′ termini should contain a polyphosphate group while the 3′ termini should contain a hydroxyl group. On the other hand, cleavage by any of the common endonucleases would result in a 5′ hydroxyl group and a 3′ monophosphate. When three RNA size classes derived from A/WSN were hydrolyzed with alkali and chromatographed on DEAE-cellulose, all RNA species were phosphorylated. The only polyphosphate-containing nucleotide was pppAp, suggesting that the 5′ terminal nucleotide of each of the RNA species was a polyphosphate-containing adenine.[95] The 3′ termini of A/WSN virus RNA were investigated by a labeling procedure that involved oxidation and reduction with ^{3}H-borohydride.[96] Each of three major size classes of RNA contained molecules with nonphosphorylated uridine at the 3′ terminus. Thus, there is evidence that, at least in the case of A/WSN, the nucleotide sequences of all segments begin at the 5′ terminus with ATP and end at their 3′ termini with uridine. The data suggest that RNA segments are independently replicated and do not arise by endonucleolytic cleavage. As mentioned previously, definitive evidence that the genome of influenza virus is not only fragmented, but consists of eight unique genetic elements, has been provided by McGeoch et al.[89] Using an improved method of gel electrophoresis, they separated the genome into eight sharp bands, each of which was represented in approximately equimolar amounts, although the lower molecular weight species were relatively more abundant. They recovered the bands and determined their oligonucleotide composition by two-dimensional electrophoresis. Each segment is unique and corresponds in molecular weight to a known viral protein. A summary of the chemical properties of each piece is shown in Table 3. In a similar study of the RNA of the human influenza

TABLE 2

The RNAs and Proteins of the Avian Influenza A, Fowl Plague Virus

	Genome RNAs		Virus-specified proteins	
RNA	Mol wt	Coding capacity (mol wt)	Mol wt	Name
1	1.21×10^6	118,000	96,000	P_1
2	1.02×10^6	100,000	87,000	P_2
3	1.00×10^6	98,000	85,000	P_3
4	0.83×10^6	82,000	75,000	Hemagglutinin HA
5	0.68×10^6	67,000	53,000	Nucleocapsid NP
6	0.58×10^6	57,000	45,000	Neuraminidase NA
7	0.34×10^6	33,000	25,000	Matrix M
8	0.29×10^6	27,000	23,000	Nonstructural NS_1
			11,000	Nonstructural NS_2

TABLE 3

RNA Species of Fowl Plague Virus

RNA species	Chain length from gel mobility	Chain length from fingerprint	^{32}P content % of total	Relative molar abundance
1	3,500	–	16.2	0.7
2	3,000	–	13.8	0.7
3	2,950	–	13.8	0.7
4	2,450	2,490	13.7	0.8
5	2,000	2,370	13.5	1.00
6	1,720	2,400	12.5	1.1
7	1,080	1,000	9.6	1.3
8	870	800	6.8	1.2

strains A/PR/8/34 (HON1), A/Hong Kong/8/68(H3N2), and recombinants derived from them, it was found that each contains eight RNA segments, and the fourth largest coded for the hemagglutinin.[97]

Experimental investigation of the nature of the genome of influenza virus can be summarized as follows. Early genetic studies of Burnet,[68] Barry,[71] and Hirst[69] all suggest that the genome consists of independently replicating pieces of RNA. Characterization of this RNA by sucrose velocity gradient sedimentation, polyacrylamide gel electrophoresis, and terminal group analysis indicates that the genome contains at least eight independently replicated segments of single-stranded RNA. The RNA segments are monocistronic and possess different nucleotide sequences.[89,93,94] The RNA of influenza virus is noninfectious[81] because vRNA is of opposite sense to that of mRNA.[10] It is anticipated that mRNAs of influenza virus in infected cells will consist of eight monocistronic molecules that are complementary to vRNA.

IV. THE REPLICATION OF INFLUENZA VIRUSES

A. Quantitative Aspects

Influenza viruses grow readily in the allantoic cells of the developing chick embryo and are released into the allantoic fluid.[98,99] For more than 20 years only the embryonated egg provided a suitable laboratory system for the analysis of influenza virus replication; even today, it remains the best system available for the production of high-titer virus. By comparison with their growth in eggs, influenza viruses (particularly the human type A strains) grow poorly in tissue culture;[100,101] abortive and/or defective virus production is also common in tissue culture.

Defective or "incomplete" virus production in embryonated eggs occurs only when cells are infected at high multiplicity.[102]

Quantitative studies of influenza virus growth in allantoic cells have established the principal features of the growth cycle. Adsorption takes place in two stages. Within seconds of inoculation, virus becomes attached to the surface of cells by means of glycoprotein receptors[103] (see Chapter 5). Uptake into cells, known as viropexis,[104] is a relatively slow process that can extend over several hours. The eclipse phase of intracellular multiplication occupies about 3 to 4 hr,[105] and release of virus from cells is also extended over several hours. Virus assembly and release take place at the cell surface.[106] Virus grown in allantoic cells has a plating efficiency of 10:1,[107] and single infectious particles initiate infection.[108]

Although tissue culture systems are far from ideal for the study of influenza virus replication, at least one strain of human influenza A (WSN) grows sufficiently well in cultured chick embryo cells (CEF) to produce cytopathic changes and can consequently be assayed by plaque titration. The WSN-CEF system has been used extensively in recent years for the study of the biochemistry of virus infection. However, by far the best experimental system for the study of influenza virus replication is chick embryo cells infected with the avian influenza A virus, fowl plague.[109] The efficiency of FPV growth in CEF is equivalent to that of other influenza virus in allantoic cells. Some characteristics of the multiplication of FPV are illustrated in Figure 2.

The principal biosynthetic events induced by virus infection will now be summarized briefly. Once virus has gained entry to the cell, the intact particle is dismantled and the genome released. The nature of this process or its whereabouts in the infected cell remains a mystery. During the first hour of infection there is a transient stimulation of cell RNA synthesis,[110] and transcription of the genome RNA by the virion-associated transcriptase begins.[111] Significant levels of transcription cannot be detected until approximately 40 min after infection, and transcription does not take place at all if the cells have been treated with actinomycin D. The effects of actinomycin D, and other inhibitors of DNA function (such as ultraviolet irradiation), on influenza virus replication were first reported by Barry and his colleagues.[112,113] They claimed that cell DNA transcription was required for influenza virus replication, and this idea received support from other inhibitor studies[114-117] and from the finding that replication cannot take place either in anucleate cell fragments[118] or in enucleated cells.[119] In addition to providing a host DNA

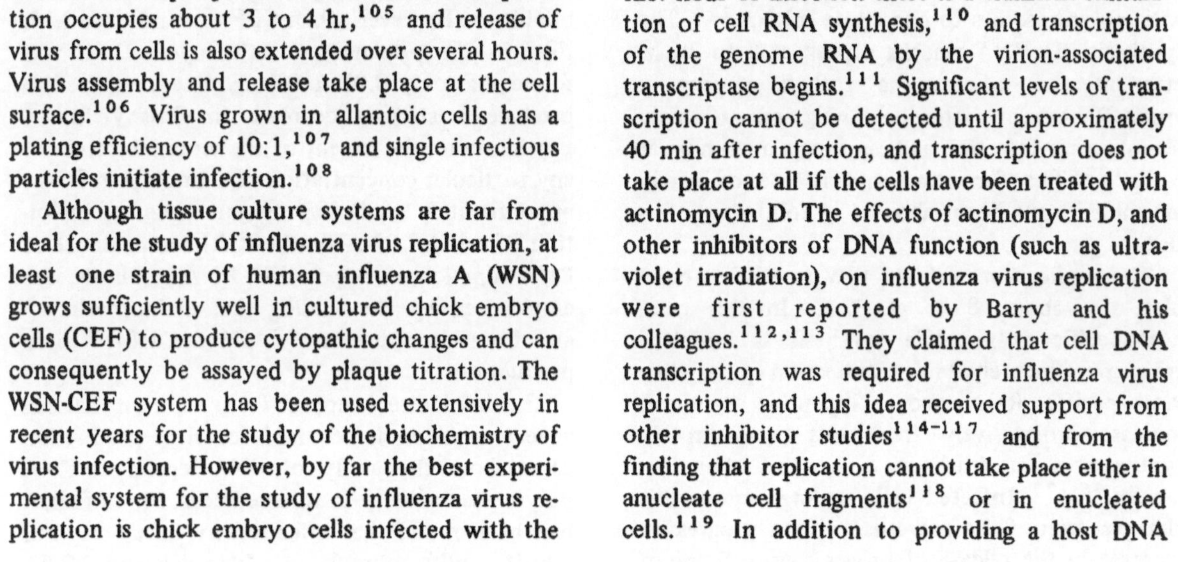

FIGURE 2. Characteristics of the multiplication of influenza virus (FPV) in cultured chick embryo cells. (A) Growth curve of FPV in CEF cells, as determined by hemagglutinin production. The multiplicity of infection was 120 PFU/cell. (B) Virus-induced RNA synthesis in CEF cells infected with FPV (•–•), in relation to the onset of virus yield, as determined by plaque assay.

function, nuclei may play an important topographical role in virus replication; Breitenfeld and Schäfer[120] found that newly synthesized RNP could be detected first in the nuclei of infected cells. A detailed consideration of the role of the nucleus in virus replication is provided in Section IV. B.

Once transcription begins, complementary (c)RNA is produced that not only serves as mRNA for virus-specific protein synthesis, but as template for the synthesis of genome RNA. A significant proportion of cRNA becomes associated rapidly with polyribosomes to constitute mRNA.[121] The synthesis of cRNA reaches a peak at 2 to 2½ hr after infection, whereas the synthesis of progeny virion RNA occurs later, beginning approximately 90 min after infection and reaching a maximum at 3 hr.[122] The formation and release of newly assembled virus begin about 4 hr following infection.

The replication cycle of FPV in CEF is complete by about 8 hr at 37°C. In this case, host-specific polypeptide synthesis is inhibited from approximately 2 hr postinfection (p.i.), while virus-specific RNA and polypeptide synthesis increase progressively from this time. Protein synthesis is maximal at 5 hr and then decreases.[45,123] Infected cells undergo cytopathic changes from 8 hr onwards, with a progressive increase in the release of progeny virus. Some of the principal features of the replication cycle of influenza viruses will now be considered in detail.

B. Role of the Nucleus in Virus Replication

The first indication that cell nuclei play a part in the replication of influenza viruses was the detection, by immunofluorescence, of virus-specific antigens in infected cell nuclei.[120,124,125] The likelihood that cellular genetic expression might be involved in virus replication was investigated by Barry et al.,[112] who studied the inhibitory effects of actinomycin D on virus growth. At that time, it was known that actinomycin D combined with DNA[126] with the result that DNA transcription was inhibited.[4] Recently, it has been shown that enucleated cells do not support virus growth, and these findings also support the idea that functional cell nuclei are needed for influenza virus replication.[119,127]

1. Inhibitor Studies

The effects of actinomycin D on the replication of influenza viruses were first studied in allantoic cells. Virus growth did not take place if the drug was added at any time during the first 2 hr of infection, but relatively high concentrations (1 to 10 μg/ml) were needed.[112] Nevertheless, actinomycin D at these levels had no effect on the growth of the paramyxovirus, Newcastle disease virus (NDV). Since much lower concentrations of actinomycin D were needed to inhibit the division of L cells (0.005 μg/ml) or vaccinia virus production (0.1 μg/ml) in these cells,[4] some doubts were raised concerning the specificity of influenza virus inhibition. However, when CEF were infected with FPV in the presence of actinomycin D, it was found that only 0.02 to 0.1 μg/ml was required to produce a striking decrease in virus yield.[113] Furthermore, the extent of the virus inhibition at any particular concentration of actinomycin D was approximately equivalent to the extent of inhibition of cellular RNA synthesis. The importance of DNA function, as opposed to DNA synthesis, was emphasized by the finding that aminopterin[127] and fluorodeoxyuridine[112] did not inhibit virus production.

Further indirect support for the notion that cell gene expression is required for virus production came from the study of DNA-dependent RNA polymerase activity in infected cells. First of all, it had been noted that infection by influenza virus significantly stimulated DNA-directed RNA synthesis.[110,128] Further developments had to await the understanding of cell DNA transcription. We now know that there are at least two major forms of DNA-dependent RNA polymerase in eukaryotic cells;[129] one is nucleolar (form I) and involved in the synthesis of ribosomal RNA (rRNA), while the other is nucleoplasmic (form II) and involved in the synthesis of heterogeneous nuclear RNA (hnRNA) and mRNA. These polymerases can be distinguished by their sensitivity to α-amanitin.[130] In vitro assays indicate that α-amanitin inhibits DNA transcription by binding specifically to the form II polymerase. The possibility that the polymerase II enzyme is involved in influenza virus replication arose when it was found that α-amanitin inhibits virus growth.[116] During the first 2 hr of infection, α-amanitin, at concentrations similar to those required to inhibit cellular DNA-dependent RNA polymerase II in vivo, inhibits virus growth.[117] Inhibitor studies thus suggest that cell gene expression, presumably mediated by DNA-dependent RNA polymerase II, is essential during the early stages of influenza virus replication.

2. Studies with Enucleated Cells

An alternative means of determining whether cell nuclei are needed for influenza virus replication is to enucleate cells before infection or use fragments of cytoplasm. Monolayer cultures of HeLa cells cut at random with a glass-cutting wheel contain fragments, many of which consist of cytoplasm alone.[118] Influenza viruses will grow only in fragments containing nuclei (where growth is indicated by the ability of the fragment to adsorb red cells), whereas paramyxoviruses grew in both types of fragments. Once the cytochalasin B enucleation technique was introduced, these studies could be extended.[132] Monolayer cultures of cells treated with cytochalasin B and then centrifuged yield an almost completely enucleated cell population. Follett and co-workers[119] investigated the effects of enucleation on the ability of influenza virus to grow in BSC-1 cells. Although this is an abortive system, in the sense that a complete cycle of replication is not obtained, synthesis of virus-specific proteins will normally take place. Enucleated fragments did not support virus-specific protein synthesis. These fragments were still capable of macromolecule synthesis, however, because the replication of vesicular stomatitis virus in them was unimpaired. In other studies it was reported that FPV could induce neither protein nor RNA synthesis in enucleated BHK cells, although the structural proteins of both NDV and vaccinia were synthesized normally in these fragments.[127] The implication of these studies is that functional cell nuclei are essential for influenza virus replication.

3. Accumulation of Virus-specific Components in the Nucleus

Certain virus-induced activities have been detected in the nuclei of infected cells. For example, at least two virus-specific proteins (the NP and a nonstructural protein of unknown function) are found there. Furthermore, virus RNA apparently migrates to the nucleus immediately after penetration and may accumulate there later in infection.

a. Proteins

Virus-specific proteins were first detected in infected cells by immunofluorescence. In an attempt to identify these proteins, Breitenfeld and Schäfer[120] prepared antibody specific for either the "S" (RNP component) or the "V" (envelope components) antigens of the virus. Using FPV-infected CEF, they found that the "S" antigen appeared initially in the nucleus at 3 hr following infection and then appeared in the cytoplasm from 4 hr onwards. The "V" antigen was first detected at 4 hr in the cytoplasm and was never found in the nucleus. Both "S" and "V" antigens accumulated later in the cell membrane. These findings were confirmed by Maeno and Kilbourne[133] and Kelly and Dimmock[134] who used antibody specific for the individual glycoproteins and found that the synthesis and accumulation of HA and NA were confined to the cytoplasm.

A nonstructural virus-induced protein has also been detected in infected cell nuclei using immunofluorescence.[135,136] Dimmock and Dimmock and Watson detected intense fluorescence in the nucleoli of infected cells at 4 hr p. i. when they used hyperimmune antiserum that had been adsorbed extensively with disrupted virus. The intracellular location of newly synthesized, virus-induced polypeptides was also investigated by radioactive tracer methods. Taylor et al.,[137] while reporting that all virus-induced polypeptide synthesis took place in the cytoplasm, also noted that two peptides, possibly the NP and the M protein, migrated rapidly into the nucleus.[30] In the case of the M protein, however, they were mistaken, and the polypeptide concerned turned out to be a nonstructural (NS) polypeptide with a molecular weight of 25,000.[52]

Later studies established definitely that the NS protein is specifically associated with the nucleoli of infected cells.[138,139] On the other hand, the NP protein occurs in the nucleoplasm in the form of newly synthesized RNP particles; these structures contain vRNA and have the same buoyant density as those extracted from virions.[92,140] There is a complication, however, that another nonstructural polypeptide of molecular weight 11,000 has been described by Skehel[45] in FPV-infected CEF. This so-called NS2 polypeptide has also been found in the nuclei of MDCK or HeLa cells infected with A/WSN virus.[138,139] The migration of newly synthesized nucleoprotein into the nucleus and its subsequent return to the cytoplasm may be an essential feature of permissive infections. Immunofluorescent studies of abortive infection of L cells by FPV[141] or HeLa cells by PR8[142] suggest that the "S" antigen migrates to the nucleus and remains there; it was assumed consequently that abortiveness was due to the inability of the "S" antigen to leave the nucleus. Reinvestigation of this phenomenon by both Fraser[143] and Kelly and Dimmock[134] revealed, however, that even in abortive infections RNP antigen moved out of the nucleus into the cytoplasm, but the process was much slower and less efficient than in productive infections. The possibility remains very likely that abortive infections, with their consequent failure to provide infectious virus, arise from a defect in the transport of mature RNP from nucleus to cytoplasm.

The data discussed above suggest that at least two virus-specific components, the RNP and the NS protein, are associated with the nucleus during infection. The transient nature of the nuclear association of the RNP seems to be an important feature of permissive infection, but the nature of specific events with which each polypeptide is concerned while in the nucleus is still a matter of speculation. There is probably a functional distinction between the two, as the RNP is found in the nucleoplasm while NS is associated with the nucleolus. One suggested role for the NS protein is that it may be involved in the inhibition of the processing of ribosomal precursor RNA[144] or the eventual disaggregation of nucleoli[40] that occur in cells infected with influenza viruses. In conclusion, it is apparent that several virus-induced proteins accumulate in the nucleus. Since the NP polypeptide is a building block for the RNP, it is possible that these structures are assembled in the nucleus. In attempting to ascribe a role to the functionally separate NS polypeptide, it is reasonable to suggest that it may be involved in vRNA synthesis (see below).

b. RNA

There is evidence to suggest that infecting viral RNA may be rapidly transported to the nucleus, based on the fate of radioactive input virus RNA. Both cell fractionation and cytological techniques have been used, but the associated technical problems are considerable. A consequence of purifying sufficient input virus of high specific activity so that it can be detected is that probably only $1/10^4$ to $1/10^6$ of the virus retains infectivity. Perhaps the results should be treated with some caution. The earliest study examined the intracellular distribution of the components of influenza virus labeled with ^{32}P, using a chemical fractionation procedure.[145,146] It was found that 30% of the input radioactivity associated with cellular deoxyribonucleoprotein at 1.5 hr p.i.

Two more recent studies have employed cell fractionation techniques to follow the movement of radioactive input RNA in infected chick embryo fibroblasts. In the first, ^{14}C-uridine-labeled FPV was used for infection, and 2 hr later the total RNA of nuclei was extracted.[147] More than 50% of the input ^{14}C was found in the nucleus, associated with structures that had sedimentation characteristics similar to those of vRNA. More recently, Stephenson and Dimmock[148] reported a transient association of infecting RNA with nuclei. Using a FPV/BEL recombinant labeled with ^{32}P, they infected cells at 4°C and looked for the appearance of RNA in the nucleus. Within 15 min at least 50% of the input RNA was found there, and by 3 hr the amount had increased to 85%. If infected cells were warmed to 37°C, 80% of the virion RNA disappeared from the nuclei within 30 min. Pretreatment of the cells for 1 hr with actinomycin D or cycloheximide did not affect migration of RNA into and out of the cell nuclei.

In an attempt to follow the intracellular movement of viral RNA following adsorption, Armstrong and Barry[149] infected cells at a very high multiplicity with ^{3}H-labeled FPV of high specific activity and examined the cells by autoradiography. They found that grains begin to accumulate in the nuclei of infected cell within 30 min p.i. and increase in number during the next 4 hr. In contrast, Sendai virus RNA of similar specific activity could only be detected in the cytoplasm of infected cells.

The experiments of Stephenson and Dimmock[148] and Armstrong and Barry[149] suggest that influenza virus RNA is transported to the nucleus early in infection. While such studies suggest an intimate involvement of influenza virus RNA with host cell nuclei, they do not allow any definite conclusions to be reached about its significance. At present there is no definite biochemical evidence that virus-induced RNA synthesis takes place in the cell nucleus. Cell fractionation studies revealed the presence of newly synthesized virion RNA in both the cytoplasm and the nucleoplasm.[92]

In an attempt to locate virus-induced RNA synthesis, Armstrong and Barry[150] exposed infected cells to 5-min pulses of ^3H-uridine and determined their incorporation by autoradiography. They detected two distinguishable phases of induced RNA synthesis, both of which occurred predominantly in the nucleoplasm. They attributed the first peak of RNA synthetic activity that occurred approximately 1 hr after infection to the α-amanitin-sensitive transcription of cell DNA.

4. *Significance*

Influenza viruses require the presence of a cell nucleus before any virus-directed RNA or protein synthesis can take place.[119,127] Since these viruses do not multiply in the presence of inhibitors of DNA function, such as actinomycin D or mitomycin C,[112-115] or in the presence of α-amanitin,[117,131] it is likely that cellular DNA-dependent RNA synthesis is required for the initiation of successful infection. The possibility that an individual segment of the virus genome is a cell gene product has been excluded by detailed hybridization studies.[151] The functional significance of cell transcription in influenza replication is not yet understood.

Apart from the need for cell gene transcription, it is likely that events associated with virus-induced RNA synthesis take place in the cell nucleus. It is known, for instance, that at least two virus-induced proteins, NP and NS, accumulate in infected cell nuclei.[120,136] An early report[122] suggested that virus particle-associated RNA was made in the nucleus. The topographical studies of Armstrong and Barry[150] certainly suggest that some major phases of virus-induced RNA synthesis occur in cell nuclei. In conclusion, the nucleus may be important in influenza virus replication for at least three reasons. In the first place, RNA synthesis directed by host DNA is needed in the early stages of replication. Following this, nuclei may provide the site of both viral RNA synthesis and assembly of the RNP.

C. Transcription

1. *The Virion Transcriptase*

RNA-dependent transcriptases have been identified in both influenza A viruses[152-154] and influenza B viruses.[155] Transcriptase activity can be detected by the incorporation of radioactively labeled ribonucleoside triphosphate into an acid-insoluble product when the reaction mixture contains virus particles, detergents (such as Triton N101 or Nonidet P40), divalent cations, and all four ribonucleoside triphosphates. The complete ingredients of a typical reaction mixture are listed in the footnote to Table 4, and further technical data are given in the Appendix of Chapter 5. The following features can be noted. The reaction is dependent on the presence of all four ribonucleoside triphosphates and only proceeds when the virus particles are disrupted by detergent. This indicates that the enzyme is located inside the virus particle. The template for the reaction is RNA, since incorporation is completely inhibited by the presence of ribonuclease, whereas deoxyribonuclease does not affect the reaction.[152] Inhibitors of DNA-dependent polymerase, such as actinomycin D, do not influence transcriptase activity. Maximum activity for the reaction depends on the presence of both Mn^{2+} and Mg^{2+} ions.[156] The activity of the enzymes associated with different strains of virus vary over an approximately 20-fold range.[152] However, these differences disappear in the presence of 0.3 m*M* GpG (see below). It seems likely, therefore, that the observed strain differences in enzyme activity are due to a failure of initiation of RNA synthesis in vitro, rather than quantitative differences in levels of enzyme contained by particular strains of virus.[156]

TABLE 4

Requirements for In Vitro RNA Polymerase Activity of Influenza Virus (FPV) Particles

Components	% incorporation of 3H UTP
Complete[a]	100
-ATP,GTP, and CTP	2
$-Mn^2$	34
$-Mg^{2+}$	42
$-Mn^{2+}$ and Mg^{2+}	0
-NP40	0
-Dithiothreitol	60
$+Ca^{2+}$ 1 m*M*	2
+RNase A 100 μg/ml	0
+Actinomycin D 200 μg/ml	100
+Rifampicin 200 μg/ml	100
+Streptolydigin 200 μg/ml	100
+Rifamcyin AF/013 150 μg/ml	100

[a] The complete assay mixture contained the following: 50 m*M* tris HCl buffer (pH 8.2); 2.0 m*M* each of ATP, CTP, and GTP; 0.4 m*M* 3H UTP (32 Ci/mol); 8 m*M* $MgCl_2$; 0.2 m*M* $MnCl_2$; 5 m*M* dithiothreitol; 150 m*M* NaCl; and 0.5% Nonidet-P40. The total reaction volume was 150 μl containing 50 to 150 μg of viral protein.

time. It seems likely that the major site of primary transcription is the cytoplasm. However, these experiments are subject to the limitations of cell fractionation, and it should be possible to obtain more reliable data when experiments involving autoradiography have been performed.

Certain in vitro properties of the virus-associated transcriptase enzyme do not correlate with its behavior in vivo. For example, the in vitro activity of the transcriptase is not sensitive to actinomycin D[152] or to α-amanitin,[131] and it also has a lower temperature optimum.[162] Primary transcription, in the sense employed by Bean and Simpson,[111] cannot be demonstrated in cells treated with actinomycin D, α-amanitin, or high concentrations of mitomycin C.[163] Therefore, it is possible that the initiation of cRNA synthesis in vivo is dependent on host factor(s) that are lacking in the in vitro reaction mixtures. The unexpectedly long delay in the initiation of cRNA synthesis could be explained as follows. A host function, mediated by newly synthesized factor(s), must be activated before RNA transcription can begin. These factors in turn could be the products of the increased cellular DNA-dependent polymerase activity observed during the first hour after infection.[110]

The inhibitor studies described previously suggest that transcription of host cell DNA is required during the first hour after infection, but that there is no need for the products of the transcription process to be translated in order for virus-directed RNA synthesis to occur. It seems likely that RNA molecule(s) provide an essential function in the initiation of virion RNA transcription; the exact nature of this function is still obscure.

3. Secondary Transcription

Secondary transcription can be defined, in the case of cells infected with influenza virus, as the amplified rate of cRNA synthesis mediated by new transcriptase molecules, synthesized under the direction of newly produced mRNA. The amplified rate of transcription begins 90 to 120 min p.i., reaches a peak at 3 hr, and then declines. It represents the major synthesis of cRNA during infection.[111] An alternative method for determining cRNA is to hybridize RNA extracted from infected cells with excess amounts of virion RNA.[164] Scholtissek and Rott found that maximum cRNA synthesis occurred at 2 hr p.i. in CEF infected with FPV. Similar results were reported by Krug,[92] who examined the synthesis of virus-specific RNPs in the cytoplasm and nucleoplasm of MDCK cells infected with A/WSN. For the first 3 hr after infection, RNPs contained only cRNA, but from 4 hr on only RNP with vRNA could be detected. As the rate of cRNA synthesis began to slow down from 3 hr p.i., vRNA began to accumulate, and finally the release of progeny commenced.

No matter when infected cells are treated with actinomycin D, the synthesis of cRNA ceases.[164] Apparently throughout the growth cycle the in vivo transcription of vRNA has a continued requirement for cell factors (see Section IV.C.2). For example, 0.8 μg/ml of actinomycin D when added at 2 hr p.i. preferentially inhibited the synthesis cRNA, without influencing the rate of vRNA synthesis. Some time later the synthesis of vRNA declined, but this may be a consequence of inhibition of cRNA synthesis. These findings were substantiated by Pons[165] who specifically investigated the nature of virus-induced RNA associated with polyribosomes and cytoplasmic RNPs when WSN-infected chick embryo fibroblast cells were treated with actinomycin D. It was found that the synthesis of cRNA, normally found on the polyribosomes, was inhibited to a much greater extent than the synthesis of RNA present in RNP structures. As mentioned previously, such findings imply that transcription of the viral genome requires a continued transcription of the host cell DNA.

D. Replication of vRNA

The early events involved in the synthesis of progeny vRNA are at present obscure because the enzyme(s) responsible for vRNA synthesis has not been identified. Furthermore, although it has been claimed that there are replicative intermediates in infected cells, the experimental conditions under which such structures were identified were unusual and may not represent the normal state of affairs in the infected cells.

The temporal aspects of vRNA synthesis have been determined by several groups of workers. Using cRNA as a probe, the appearance of vRNA in FPV-infected CEF was investigated by hybridization.[164] Virion RNA was detected first about 1.5 hr p.i., the rate of synthesis was maximal at 3 hr, and from this time onwards vRNA was the principal virus-specific RNA synthesized in infected cells. Krug[92] found that RNP containing vRNA could be detected at 2 hr p.i., and from 4 hr onwards only vRNPs were being synthesized. In these experiments, the synthesis of cRNA preceded that of vRNA, but it is not known what factors are responsible for the switch from predominantly cRNA to predominantly vRNA synthesis.

The synthesis of vRNA is rapidly and preferentially inhibited by cycloheximide. The effects of cycloheximide were demonstrated by hybridization of the total RNA from infected cells treated with the drug to excess amounts of cRNA synthesized in vitro.[164] An alternative approach was to determine the nature of the RNA of cytoplasmic virus-specific RNPs in influenza-infected cells treated with cycloheximide.[165] Both sets of data indicate that the continued synthesis of vRNA depends on continuous protein synthesis. Therefore, both secondary transcription and the enzymatic synthesis of vRNA require and depend on the continued synthesis of protein.

At present the nature of the replicase remains obscure, but at least three possibilities are worthy of consideration: (1) that replication and transcription are carried out by the same enzyme; (2) that a presently unrecognized, virus-induced protein acts as the replicase; or (3) that the transcription enzyme is modified by the addition of an extra protein, to form the replicase.

The search for the enzyme involved in RNA replication began in earnest with a report that the cytoplasm of cells infected with influenza virus contained a virus-specific RNA polymerase.[166] Detailed characterization of this activity indicated that it had all the in vitro properties of the virion transcriptase[167,168] and was associated with viral RNPs in the cell cytoplasm.[169,170] Peak levels of activity for this enzyme were not detected until 5 to 6 hr after infection, a time well past that at which other aspects of replication are maximal. Consequently, the cell-associated RNA polymerase activity may simply represent the accumulation of newly synthesized transcriptase complexes awaiting assembly into progeny virus.[150]

An RNA-dependent RNA polymerase activity has also been found in preparations of highly purified nuclei from FPV-infected cells.[171] The in vitro activity of this enzyme is identical to that of the virion transcriptase. However, despite this shortcoming the enzyme has peak activity at the time expected for an enzyme involved in either secondary transcription or replication. The induction of the nuclear enzyme activity in vivo was inhibited by actinomycin D or cycloheximide.[171]

The possibility exists that either in the nucleus or the cytoplasm there are enzymes that are only detectable as transcriptases in vitro, but which may have replicase activity in vivo. The conditions of assay available at present may be such that only transcriptase activity is favored; a rapidly turned over protein might be needed to convert transcription to replicase activity.

The replicating complex or RI would be expected to be a partially duplex molecule consisting of a strand of template cRNA to which newly synthesized vRNA polyribonucleotides in various stages of completion are attached. There have been a number of reports of RIs in cells infected with influenza viruses.[65,85,93,172,173] However, these double-stranded structures could only be detected at a time when the synthesis of vRNA and cRNA was declining or was inhibited by actinomycin D. Such structures could also arise merely as artifacts during phenol extraction.[174] Certain workers were unable to detect them in anything but negligible amounts,[164] and the nature of RFs and RIs in viral replication has yet to be determined. It is possible, however, that the template for vRNA synthesis is not a naked cRNA molecule, but an RNP structure containing cRNA. Such structures are common at relatively early stages of infection,[92] and cRNA assembled in an orderly fashion into RNP might constitute a stable template for the replicase.

E. Messenger RNA

Although influenza viruses have the properties of negative-strand viruses, it was reported some years ago that polyribosomes extracted from infected cells at 6.5 hr p.i. contained both vRNA and cRNA.[175] However, RNA isolated from infected-cell polyribosomes earlier in infection is almost totally cRNA.[10] These conflicting reports draw attention to the possibility that genome RNA may have direct messenger activity. With a fragmented genome, it is at least possible that some of the vRNA segments might be mRNA.

To add to the confusion Siegert et al.[176] reported that purified influenza vRNA, when added to an *Escherichia coli* cell-free protein-synthesizing system, was translated. A high percentage of the product was low molecular weight material, but certain polypeptides that correspond in size to certain viral structural proteins could be detected by gel electrophoresis, and the RNP antigen was identified further by immunodiffusion tests. The experience of other workers seems to be that some polypeptide synthesis can be induced by vRNA in the *E. coli* system.[275] However, in vitro systems derived from eukaryotic cells, such as reticulocytes or wheat germ, fail to respond to vRNA.[177,178] Kingsbury and Webster[177] reported that total RNA extracted from infected cells directed the synthesis of a polypeptide with the electrophoretic mobility and antigenicity of the virion M protein. Unfortunately, Seigert et al.[176] did not test their vRNA for the presence of cRNA, so that it is impossible to know whether the messenger activity they detected was due to vRNA or to a small amount of cRNA that had been accidently encapsidated.

It now seems certain, however, that cRNA constitutes the message. Glass et al.[121] prepared polyribosomes from cells infected with FPV with the intention of characterizing the newly synthesized, virus-induced RNA associated with these structures. They used the following three criteria to identify mRNA: (1) rapid incorporation of ^{3}H-uridine into RNA associated with polyribosomes, (2) continued association of this RNA with polyribosomes in CsCl density gradients, and (3) possession of a poly (A) sequence in the molecule. Significant amounts of rapidly labeled RNA with these properties were first detected at 45 min p.i. The mRNA was completely complementary to vRNA and sedimented at 8 to 20S, corresponding approximately to the sedimentation of genome RNA.[82] As it seems likely that certain virus polypeptides may be synthesized by particular classes of polyribosomes (see Section IV.F), it was anticipated that there might be some segregation of virus mRNA. The free and membrane-bound polyribosomes of infected cells were analyzed, but both contained the same classes of mRNA. There was no absolute segregation of mRNA sequences into either polyribosomes class, although each probably contained different amounts of the individual mRNAs.

Thus, it seems that virus-induced RNA with the properties of mRNA is complementary to virion RNA. Direct proof was obtained by Etkind and Krug,[178] who carefully purified cRNA from infected cells and induced protein synthesis with it in the wheat germ cell-free protein-synthesizing system. Four of the newly synthesized proteins comigrated on electrophoresis with four major nonglycosylated virus-specific proteins, and the product also contained candidates for the protein moieties of the two virion glycoproteins. The evidence now available indicates that influenza virus, despite its fragmented genome, is a conventional negative-stranded virus.

F. Virus-specific Protein in Infected Cells

As previously mentioned (Section IV.B.3), early studies of virus-induced protein synthesis were confined to immunofluorescent identification of the intracellular distribution of virus antigens. The introduction of SDS-polyacrylamide gel electrophoresis[179] has revolutionized our understanding of the virus. Together with cell fractionation procedures, gel electrophoresis during the past 6 years has provided a great deal of information concerning the sites of synthesis, maturation, and migration of individual influenza virus polypeptides.

Following infection, host-specific protein synthesis is replaced by virus-specific protein synthesis, but the extent and kinetics of this phenomenon appear to depend on the particular virus-cell system.[30,45,52] The synthesis of all viral polypeptides appears to occur in the cytoplasm.[137,180]

1. Identification

It now seems highly likely that influenza viruses contain at least eight structural components, six of which are primary gene products (P_1, P_2, P_3, NP, NA, and M), and the remaining two are the posttranslational cleavage products of the HA (see Tables 1 and 3). Once virus-induced protein synthesis has been established in infected cells, most of the structural components can be detected easily, and there also appears to be one, or possibly two, nonstructural polypeptides present as well. A typical example of the separation of ^{35}S-methionine-labeled polypeptides of FPV by means of gel electrophoresis, using the tris-glycine-buffer discontinuous system developed by Laemmli,[181] is shown in Figure 3A; it is compared to a similar gel of ^{35}S-methionine-labeled polypeptides from FPV-infected CEF obtained at 6 hr p.i. (Figure 3B). The major nonstructural polypeptide has a molecular weight of about 25,000 and was first detected in cells infected with A/WSN, where it was shown to migrate slightly faster than the M protein.[52] Using one-dimensional peptide maps of poor resolution, it was claimed that this peptide was distinct from the viral M protein.[52] This new polypeptide was considered to be a nonstructural component and

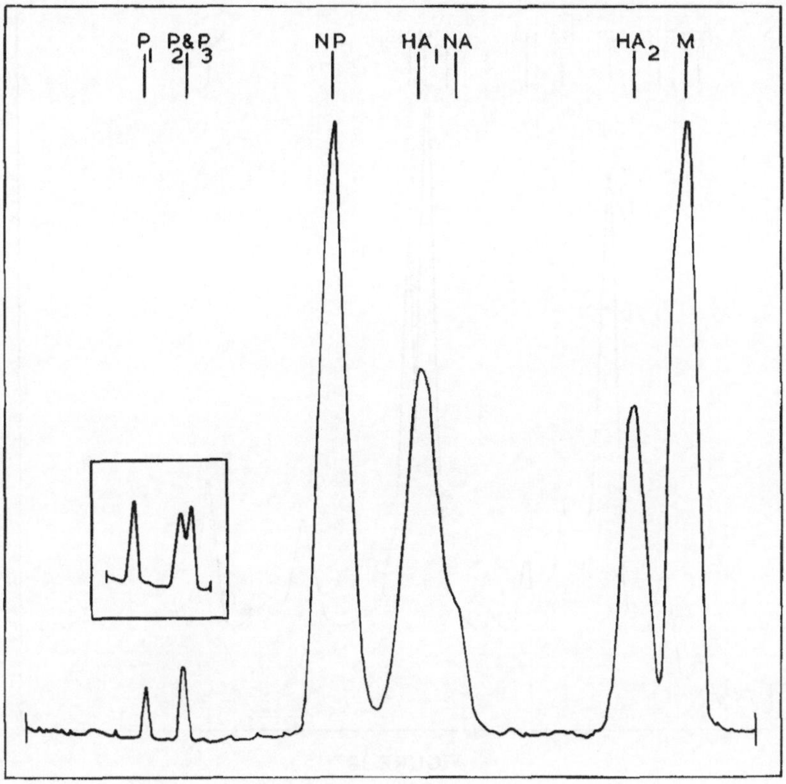

A.

FIGURE 3. Polyacrylamide gel electrophoresis of influenza structural polypeptides and virus-induced polypeptides in infected CEF cells. (A) Fowl plague virus, labeled with ^{35}S-methionine, subjected to discontinuous SDS-acrylamide gel electrophoresis in 0.1% SDS, tris-glycine buffer (pH 8.3) on an 11% acrylamide-bisacrylamide gel for 16 hr at 40 V. After completion the gel was washed, dried, and autoradiographed. A densitometer tracing of the autoradiogram is shown. Migration is from left to right. The insert demonstrates the separation of the P polypeptides into three components when polypeptides are run on a 7% gel. (B) Virus-induced polypeptides in FPV-infected CEF. Infected cells were labeled with ^{35}S-methionine between 4 and 6 hr p.i. and then subjected to SDS-polyacrylamide gel electrophoresis for 16 hr at 50 V. A densitometer tracing of the autoradiogram is shown, and the identity of the induced polypeptides is indicated at the top of the figure.

was designated NS.[182] The presence of NS was also reported for CEF infected with FPV,[45] where a second, smaller polypeptide (NS2) of 11,000 mol wt was also detected.

There are marked differences in the rates of synthesis of different viral polypeptides, suggesting that there is control of either or both the rate of transcription or translation of virus mRNA. This is illustrated by the observation that the P proteins and the NP are clearly not synthesized in equimolar proportions; in fact, the NP is synthesized at about 30 times the rate of the P proteins.[51] In general, the molar ratio of virus structural polypeptides in infected cells reflects the ratio seen in the virion.

2. Intracellular Localization

Virus-induced polypeptides have been detected in both the nucleus and the cytoplasm of infected cells. The occurrence and possible significance of virus-induced proteins in the nucleus have been considered previously (Section IV. B.3.a), so that this discussion will be confined to a consideration of the sites of synthesis and maturation in the cytoplasm. These studies are based on cell fractionation procedures, similar to that described by Caliguiri and Tamm.[182] By means of this technique, cytoplasmic extracts are separated into four components, namely, smooth membranes, an intermediate fraction, rough microsomal membranes, and free polyribosomes. The results suggest

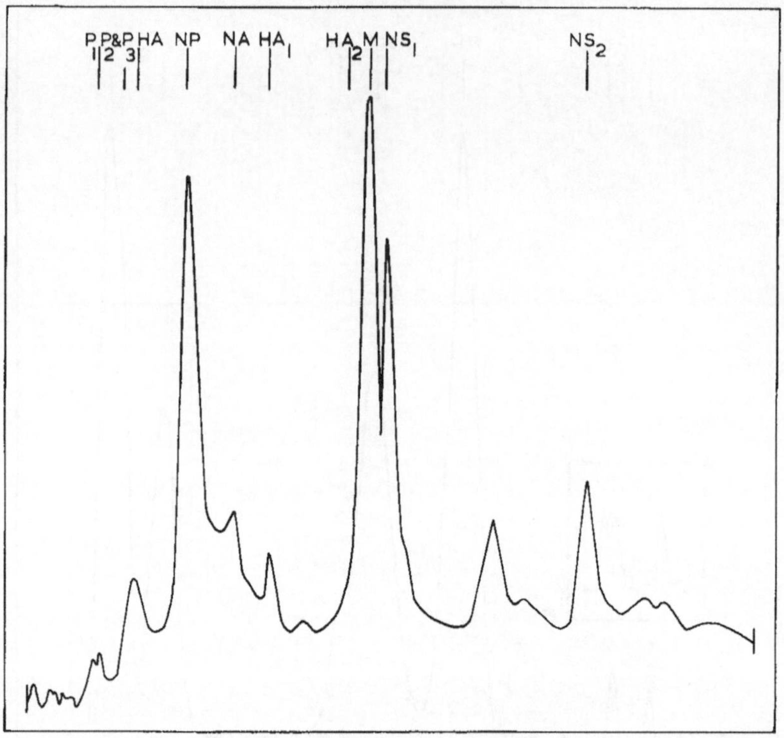

FIGURE 3B

that certain virus-induced proteins are specifically associated with certain components.[183-187]

The major component of the smooth membranes is the HA polypeptide, which in FPV-infected cells is cleaved to HA_1 and HA_2. The other virus envelope proteins, NA and M, were detected in small amounts in this fraction. The intermediate fraction contains the NP polypeptide as the major component, probably as viral RNP, since the cytoplasmic RNA polymerase was also found in this fraction.[170,183,186]

The rough microsomal fraction contained polypeptides HA, NP, and NS_1, and the predominant polypeptide found in the polysome fraction was NS_1.[183] The association of NS_1 with ribosomes appears to depend on ionic strength and does not behave as a structural component of the ribosomes, as was once considered a possibility.[138]

3. Plasma Membranes

Influenza viruses are assembled on preformed cell membranes (Section IV.G); assembly takes place by budding from the plasma membrane. The envelope constituents (HA, NA, and M) migrate to the plasma membrane from their sites of synthesis and maturation. The HA glycoprotein and probably other envelope proteins are synthesized on the rough endoplasmic reticulum.[183,186] A few moments later HA is present successively in the smooth membranes and finally the plasma membrane.[185] As has recently been pointed out,[188] throughout the migration process the HA and other envelope proteins are constituents of membranes along which they move; they are never detected as soluble proteins.

The M protein is found in relatively small amounts in infected cells, and its synthesis could be a rate-limiting step in virus production.[52,189] It is detected on smooth membrane and plasma membranes, but does not chase out of smooth membranes like the other envelope proteins.[186,187,190] The M protein is probably synthesized in close association with the plasma membrane and inserted into regions that contain virus HA.

4. Glycosylation

Carbohydrate is present in the influenza virus envelope as a constituent of both glycoproteins and glycolipids.[191] Glycosylation of the envelope polypeptides takes place in a stepwise manner during migration, and the glycoproteins are then inserted into the plasma membrane.[47,183-185] Glycosamine and mannose appear to be attached to HA polypeptides in the rough membranes very soon after or during polypeptide synthesis, whereas fucose appears to be attached later by transferases present in smooth membranes.[188] Glycosylation seems to be necessary for the integrity of the HA.[47]

The synthesis of HA and NA is inhibited by high concentrations of D-glucosamine or 2-deoxy-D-glucose[192,193] because they inhibit glycosylation.[191,194] Infected cells treated with these compounds produce a homogeneous unglycosylated HA precursor (HA_0) with a molecular weight of 64,000. This precursor is readily cleaved to yield heterogeneous fragments because of its increased susceptibility to proteolytic enzymes.

The glycosylated HA precursor undergoes post-translational cleavage to HA_1 and HA_2;[45,52,191] this occurs after insertion into the plasma membrane.[187] Cleavage is not required for virus assembly or for full expression of the biological properties of the virus.[195] In certain circumstances, cleavage of the assembled HA can be effected by serum plasminogen in the medium. The plasminogen is activated to the proteolytic enzyme plasmin by factors released from the host cells as a result of the cytopathic changes they undergo after infection, and the plasmin cleaves HA when it appears at the cell surface.[196]

5. Control of Protein Synthesis

The disproportionate levels of induced polypeptide synthesis throughout infection suggest some form of regulation. Individual polypeptides are not produced in anything like equimolar amounts, as can be seen from Figure 3. If equimolar amounts of each mRNA are present and translated with equal efficiency, the polypeptides should be produced in equal amounts. It is apparent that there is transcriptional and/or translational control. There is also evidence that some form of temporal control operates.

The major virus-specific polypeptides are first detected in infected cells at about 2 hr p.i. The rate of viral polypeptide synthesis increased steadily to reach a maximum at about 5 hr; this was maintained for a further 3 hr before gradually declining.[30,45] The individual viral proteins are not produced in equal amounts. The molar ratios of polypeptides synthesized in infected cells reflect the molar ratios of the polypeptides in the virion itself, with perhaps the NP in relative excess while the M protein is in relatively short supply.[30,52,197] There is some evidence to suggest that the rates of synthesis of viral polypeptides change relative to each other during the growth cycle. In particular, the synthesis of NS, NP, and P_2 in fowl plaque virus-infected cells can be detected before the synthesis of other virus polypeptides.[45,198] The use of cycloheximide to allow viral RNA synthesis in the absence of protein synthesis enabled Skehel[198] to examine the control mechanisms operating at the primary transcriptional level in FPV-infected cells. Removal of the drug at various times up to 12 hr p.i., followed by immediate pulse-labeling, showed that only the NS, NP, and P_2 polypeptides are synthesized. The other virus-specific polypeptides do not appear until about 30 min later. The reestablishment of the normal pattern of polypeptides synthesis after a cycloheximide block can be prevented if actinomycin D is added before removing the cycloheximide. It has been concluded that early in infection the virion transcriptase transcribes only those RNAs responsible for the NS, NP, and P_2 polypeptides.[198] Furthermore, transcription of the remaining RNA pieces does not require protein synthesis but is sensitive to actinomycin D.

The experiments described above assume that control operates at the level of transcription. Possible support is provided by Avery and Dimmock[199] who found that only 25% of the genome had been transcribed by 2 hr p.i. Such transcriptional control might be very transient because it has been found that by 1 hr p.i., even at 31°C, as much as 90% of the FPV genome was transcribed.[121] Using highly labeled virion RNA to increase the sensitivity of the technique, it was found that some mRNA sequences are represented in greater amounts than others; almost the entire genome is represented in cRNA on the polyribosomes early in infection, although some sequences are probably transcribed much less than others. The weight of the available evidence suggests that the synthesis of early proteins in influenza virus-infected cells is due to selective transcription of genome RNA.

A further aspect of the control of protein synthesis concerns the site of synthesis. As mentioned previously, the glycoproteins are closely associated with the membranes of the rough endoplasmic reticulum, whereas the P and NP proteins occur as soluble proteins in the postmicrosomal supernatant.[183,186,187,190] These findings imply that particular classes of mRNA may be segregated among the different types of polyribosomes. Such a segregation has been observed for VSV mRNAs.[200,201] However, in the case of influenza virus both the free and membrane-bound classes of polyribosome contain all types of virus mRNA.[121] In these circumstances it seems likely that some form of translational control is operating because the membrane-bound polyribosomes, for instance, synthesize predominantly the glycosylated polypeptides.

In conclusion, it can be stated that control processes operate in the synthesis of virus structural components, but little is yet known concerning the processes involved.

G. Maturation and Assembly of the Virus

The assembly of influenza viruses takes place by the process of budding from the cell membrane. This was first observed in electron micrographs of infected cells, where virus antigens could be detected on the cell surface where viruses were forming.[202] More recently it has been found that the envelope of the emerging virus has a membrane similar to that of the host cell, but contains a layer of spikes or projections on its outer surface. Within the virus membrane is an additional electron-dense layer not found below the normal cell surface, and this probably consists of M protein.[40,203,204] The process probably begins with the accumulation of newly synthesized glycoproteins in the plasma membrane because regions of the membrane that show no visible change acquire the ability to absorb red blood cells.[40] Budding usually occurs where there is an accumulation of RNP beneath the altered membrane,[205] but the presence of RNP may not be an essential trigger for maturation because empty particles are frequently found.

Radioactive tracer studies of virus-modified plasma membranes indicate that the HA is the first polypeptide to be incorporated, and once in position it seems to determine the point of insertion of the M protein.[187] The virus polypeptides do not replace existing cell polypeptides in the plasma membrane, but are incorporated in addition. The HA molecules are inserted at sites distinct from the points of virus assembly, but migrate subsequently in the plane of the membrane to form clusters of HA molecules. The M protein, on the other hand, is only found in areas of membrane containing HA polypeptides. This suggests that the site of insertion of M protein is determined by the presence of HA and not vice versa.

The rates of incorporation of individual polypeptides into virions differ from their rates of synthesis so that different polypeptides are probably assembled by distinct pathways.[187,190] Whereas the M protein is very quickly assembled, there is a delay of about 20 min before HA is incorporated, corresponding to the time needed for its migration to the plasma membrane and subsequent cleavage. The delay of NP incorporation is approximately 45 min.[206] It seems that once M is associated with specific sites in the plasma membrane, modified by the presence of HA, budding can occur.

Very little is known about the assembly of the RNP and the process whereby eight distinct genome fragments are successfully assembled into the majority of mature virions. Newly synthesized NP rapidly associates with viral RNA to form nucleocapsids,[170,186] and there is a possibility that this process takes place in the cell nucleus (Section IV. B.3.a). Because of the small size of the influenza nucleocapsids, they cannot be identified with certainty by electron microscopy in the cell nucleus or cytoplasm, but clusters of strands are apparent within the budding particle.

The problem of the release of influenza virus is closely concerned with the functions of the NA glycoprotein. That the enzyme plays an important role in virus release has been suggested by the ability of antibody specific for neuraminidase to inhibit release[26,207] and the finding that the enzyme is required to remove neuraminic acid from the virus envelope in order to avoid aggregation of progeny virus on the cell surface[208] (see Chapter 6 for further details).

V. REPLICATION OF PARAINFLUENZA VIRUSES

The viruses originally classified as myxoviruses were influenza virus, mumps virus, and Newcastle disease virus (NDV) as they were similar in size, were sensitive to ether, and had a strong affinity for cellular mucoproteins.[209] Other characteristics of these viruses that were originally considered to be important were their ability to agglutinate red cells[210] and the presence of the enzyme neuraminidase as a structural component of the virus.[211] Some years later the newly recognized parainfluenza viruses 1 to 3 were added to the myxovirus group.[212] On further investigation it became apparent that significant morphological and biological differences existed between the influenza viruses and the remaining myxoviruses.[213] The myxoviruses were subsequently divided into two major subgroups: orthomyxoviruses or influenza viruses and paramyxoviruses or parainfluenza viruses. The parainfluenza virus group consists of parainfluenza viruses 1 to 4, simian virus 5 (SV5), mumps, and Newcastle disease virus (NDV). More recently the viruses of measles, rinderpest, and dog distemper were added to this group[1] because of morphological similarities to the original viruses of the group, although they have little or no affinity for mucoprotein and lack neuraminidase. There is a fundamental difference between the parainfluenza viruses and influenza viruses with respect to the organization of their genetic material. Whereas the influenza virus genome consists of eight pieces of ribonucleic acid (RNA) (see above), each paramyxovirus genome is a single molecule of RNA.[214-217]

A. Structure of Parainfluenza Viruses

There is a general morphological similarity between the influenza and parainfluenza viruses, although the latter are much larger. Electron microscopy reveals that paramyxoviruses are enveloped, are extremely pleomorphic (varying in size from 120 to 180 nm), and have an internal ribonucleoprotein (RNP) component with helical symmetry[218] (see Chapter 1). The outer lipoprotein membrane-like envelope is covered with regularly spaced spikes consisting of two glycoproteins.[219] The outer membrane of paramyxoviruses as represented by SV5 is double layered[220] and consists of an outer phospholipid layer from which the spikes extend and an inner protein layer. By analogy to influenza virus, it has been suggested that the submembrane protein is a matrix or membrane protein which surrounds the helical RNP.[220]

Electron micrographs of RNP from ruptured virus particles suggest that the nucleocapsid may be a double helix.[218] Subsequent work has established, however, that the nucleocapsid consists of single-protein structural units arranged in a herringbone pattern around the viral RNA.[221] Using optical diffraction methods, Finch and Gibbs[221] calculated that there are 11 or 13 structural units per turn of the helix in Sendai virus.

1. Ribonucleoprotein

The nucleocapsid of NDV is representative of the group and has a sedimentation coefficient of 200 to 225S[222,223] as calculated from rate zonal centrifugation, a mean modal length of 1 μm,[224] and a RNA content of 4.5×10^6 daltons.[222] Ribonucleoproteins of paramyxoviruses can be digested with pronase,[223] but are resistant to the action of ribonuclease,[225] as would be expected if the nucleocapsid protein surrounds the RNA. Although Bukrinskaya et al.[226] presented data that suggested that nucleocapsids of Sendai virus are infectious, others could not repeat this work.[225]

2. RNA

The RNA of parainfluenza 1 virus (Sendai) had a sedimentation coefficient of 57S when analyzed on sucrose density gradients.[215,227] Similarly, RNA isolated from NDV by the cold phenol/SDS method had a sedimentation coefficient of 57S in 0.1 *M* NaCl as determined in the analytical ultracentrifuge.[214] Rapidly sedimenting RNA was obtained from other strains of NDV[228] and from SV5[216] and had sedimentation coefficients of 50S and 57S, respectively.

Like all large, single-stranded polyribonucleotides, parainfluenza RNAs are greatly influenced in their sedimentation rate by the ionic constituents of the solution. Therefore, it is not surprising that under similar conditions of ionic strength and pH sedimentation, values vary somewhat. All parainfluenza RNAs are now considered to have a sedimentation coefficient of 50S. Length measurements[229] of Sendai RNA in the electron microscope indicate that it has molecular weight of 5.2 to 5.6×10^6. Parainfluenza viruses also encapsidate small RNA molecules that could be of cell origin, for it has been found that up to 60% of the RNA associated with Sendai virus was transfer RNA.[230]

A peculiar property of 50S RNA isolated from Sendai virus is its ability to self-hybridize.[231] Self-annealing of high molecular weight Sendai RNA indicated that up to 60% could be converted to a ribonuclease-resistant form, but there is considerable variation from one RNA preparation to another. The ribonuclease-resistant material has a sharp melting point[231] and a lower buoyant density on $CsCl/Cs_2SO_4$ gradients,[232] characteristics of double-stranded RNA. Since the extent of self-annealing of the virion RNA was found to be dependent on the RNA concentration, it seems that intermolecular annealing rather than intramolecular annealing takes place.[231] It is not known whether the two complementary molecules occur within the same virion or in different virions. However, it has been found by electron microscopy that virus particles with more than one unit length of helical nucleocapsid occur commonly. Thus, virions with both kinds of RNA probably exist. Populations of Sendai virus 50S RNA molecules, with mutually complementary base compositions, could not be separated on the basis of buoyant density.[232]

3. Virus Proteins

As all the paramyxoviruses are morphologically identical and have RNA of similar if not identical size, it might be expected that the number, size, and relative proportions of their constituent polypeptides would be similar. However, polyacrylamide gel electrophoresis of Sendai,[236,237] NDV,[238] SV5,[239] and measles virus[240] indicates that each type has a distinct and characteristic gel pattern, consisting of between five to eight constituents. These studies were completed, however, before the high-resolution, discontinuous system of Laemmli[181] was available, and it is highly likely that further examination will reveal basic similarities. A detailed study of the polypeptide composition of parainfluenza 1 (Sendai) virus has been undertaken recently,[241] and the characteristic electrophoretic pattern of polypeptides is shown in Figure 4. Altogether there are nine major components present. Some polypeptides, such as the NP and M proteins, are functionally equivalent to their counterparts in influenza virus. The complete list is shown in Table 6.

All paramyxoviruses possess a high molecular weight (hmw) polypeptide that together with the

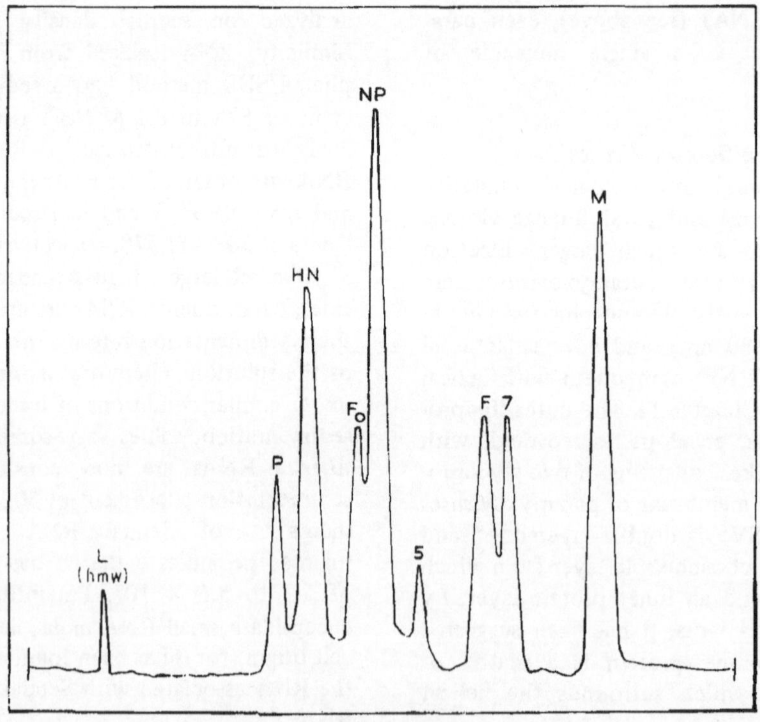

FIGURE 4. Polyacrylamide gel electrophoresis of parainfluenza 1 (Sendai) virus. Sendai virus, labeled with ^{35}S-methionine, subjected to discontinuous SDS-acrylamide gel electrophoresis. The gel contained 10% acrylamide and was run for 16 hr at 60 V. After completion the gel was washed, dried, and autoradiographed. Migration is from left to right.

TABLE 6

The Structural Polypeptides of Parainfluenza 1 (Sendai) Virus

Designation	Function	Mol wt × 10^{-3}	% in virion	No. molecules per virion
L	RNA transcription?	160	2	40
P	RNA transcription?	79	4	300
HN	Hemagglutinin/ neuraminidase	72	13	1,000
F_o	Precursor of F	65	13	1,000
NP	Nucleoprotein	60	32	2,600
5	?	54	3	300
F	Infectivity	48	5	600
7	Cellular polypeptide	42	9	1,100
M	Membrane protein	34	19	3,000

P polypeptide is probably a component of the transcription complex. It has been suggested[241] that the hmw polypeptide should henceforth be designated as the L protein. The virion contains three glycoproteins: HN, F_0 and F. The highest molecular weight glycoprotein (HN) contains both hemagglutining and neuraminidase characteristics.[242,243] A number of parainfluenza viruses have both cell-fusing and hemolytic activity. The polypeptide responsible for these activities is the F protein, and viruses must also possess F for infectivity. Each particle also contains a precursor of F (F_0) as a structural polypeptide that can be cleaved enzymatically to yield F.[244] Polypeptide 7 is a relatively abundant constituent of the virion, but is almost certainly a cell protein.[241] Evidence presently available suggests that polypeptide 7 is identical to muscle actin. The significance of actin polypeptides in the structure of parainfluenza virus particles is not known.

4. Transcriptase

The lack of infectivity of paramyxovirus genome RNA[228] and the production of large amounts of single-stranded molecules, smaller in size than virion RNA and complementary in base sequence,[215,245] suggest that genome transcripts might function as messenger RNA. RNA-dependent transcriptases were sought and found both for NDV[246] and Sendai virus,[247-249] but a feature common to all paramyxovirus transcriptase assays thus far reported is their relatively low specific activity. NDV has less than 3% of the activity of vesicular stomatitis virus[246] or influenza viruses. Nevertheless, as with influenza virus transcriptase (Chapter 5), the enzyme could be used to screen for antiviral compounds.

Relatively little is known of the properties of parainfluenza transcriptases, but it is likely that in many respects they resemble the transcriptases of rhabdoviruses (see Table 5). They require a single-stranded RNA template, and the product of transcription is cRNA that sediments on sucrose gradients as a heterogeneous 16S peak.[248] As Sendai virus-induced messenger RNA associated with polysomes is heterogeneous and much smaller than vRNA,[250] it seems likely that the transcriptase has multiple initiation sites and produces monocystronic messages.

Sendai virus transcriptase is associated with nucleocapsids.[251] When separated from other virion proteins, the nucleocapsid has ninefold increased transcriptase activity over that obtained with unfractionated, detergent-disrupted virus. Active nucleocapsids contain the P and NP polypeptides.[252]

B. Multiplication

1. Attachment, Penetration, and Uncoating

Like the influenza viruses, the neuraminidase-containing parainfluenza viruses grow to highest titer in chick allantoic cells and are released into the allantoic fluid. Most parainfluenza viruses also grow readily in continuously cultivated cells such as Hela, Vero, or Hep-2, although in some cell types (such as mouse L cells) nonproductive infections occur.[253] In the past it has been assumed that chick embryo cells (CEF) are completely permissive for the growth of either NDV or

Sendai viruses, and these cells have been used extensively for the experimental investigation of RNA synthesis. However, at least 50% of CEF are nonproductive for NDV, and only 10% produce amounts of virus equivalent to that obtained from allantoic cells.[254] However, an advantage of this system for the study of gene expression by parainfluenza viruses is that infected cells accumulate large amounts of cRNA.[255] Most parainfluenza viruses can be titrated readily by plaque assay.

The precise mechanism of attachment of paramyxoviruses to cell surface glycoproteins is still unknown. The roles of the HN and F polypeptides in this process still have to be elucidated. The entry of virus particles into cells may be effected in one of two ways. The phagocytosis of SV5 into cytoplasmic vesicles has been reported,[235] indicating that viropexis[104] may take place. On the other hand, it has been proposed that infection takes place by virus-cell fusion.[256] On available evidence there is no way to know whether viropexis, fusion, or both mediate infection.

Cell fusion as a mechanism of penetration has apparent advantages over viropexis because the nucleocapsid would be released directly into the cell cytoplasm. It is likely that transcription takes place on structurally intact nucleocapsids,[247] so that transcription and perhaps even genome replication may occur without need for any further uncoating of the incoming RNA. In such a process, infection by fusion would be an advantage.

2. RNA Synthesis

The molecular aspects of RNA transcription and replication by parainfluenza viruses have a great deal in common with those observed for influenza viruses. One very important point of difference, however, is that there is no evidence for nuclear involvement and/or DNA transcription in the replication of paramyxoviruses. These viruses multiply readily in the presence of DNA function inhibitors[112,113] and in enucleate cell fragments.[118] It now seems clear that all the molecular events involved in the replication and assembly of paramyxoviruses take place in the cytoplasm or at the plasma membrane and do not require cell gene expression. One of the advantages of being insensitive to actinomycin D is that virus-induced RNA synthesis can be investigated in the absence of cell RNA synthesis.

a. Early RNA Synthesis

In common with other negative-strand viruses, virus-induced RNA synthesis by Sendai virus begins with primary transcription that can be demonstrated in the presence of cycloheximide.[251] When compared to the influenza viruses, however (Section III.C.2), relatively little is known about the transcription process. With Sendai virus it was found that virus-induced RNA synthesis occurs in two stages.[215] Early in the eclipse phase, virus-induced RNA is much smaller in size than virion RNA and consists entirely of cRNA.[228,245] The second phase of virus RNA synthesis, here referred to as late RNA synthesis, occurs when replication of virion 50S RNA takes place, and this event occurs much later than cRNA synthesis. Thus, early RNA synthesis is probably concerned with the repetitive transcription of the infecting RNA genome.

Early RNA was found to be predominantly 18S in size,[215,227] although minor components sedimenting at 22 and 35S were detected.[227,245] Both 18 and 35S RNA are found on the polyribosomes,[257] and thus probably constitute newly synthesized mRNA. Polyacrylamide gel analysis of 18S RNA indicates that it consists of a number of components;[258] Vincent and Barry[250] estimated that polyribosome-associated Sendai mRNA consisted of at least six to seven components. Similarly, Collins and Bratt[259] separated the 18 to 22S putative mRNA of NDV into seven species, varying in molecular weight from 0.55 to 1.53 $\times 10^6$. Another property of 18S cRNA found in Sendai virus-infected cells, suggesting that it is probably mRNA, is the presence of poly-A.[260] It has also been found that Sendai 18S cRNA, when added to the reticulocyte in vitro protein-synthesizing system, is translated into at least three polypeptides that comigrate with the L, NP, and M proteins of the virus.[261] The mRNA of NDV, in common with a number of other virus mRNA's, is probably methylated in vivo by a virion-associated enzyme.[262] It is likely that the products of transcription are messengers for individual polypeptides. On the basis of this evidence it can be concluded that the large polyribonucleotide genome of parainfluenza viruses is transcribed into a series of monocystronic messengers that are equivalent functionally to the mRNA species of influenza virus.

Some doubt has existed concerning the extent of transcription of vRNA into message.[245,259,263] A detailed study by Roux and Kolakofsky[264] has shown, however, that the entire genome is transcribed; 60% of the genome is represented by 18S cRNA, while the remaining 40% is represented by cRNA that sediments at 33S.

b. Late RNA Synthesis

The replication of parainfluenza virus RNA is a mystery. Studies of the temporal aspects of virus-induced RNA synthesis suggest that vRNA synthesis begins several hours after that of cRNA[215] and that both types of RNA are then synthesized continuously throughout the remainder of the growth cycle. RNA replication presumably proceeds via a replicative intermediate, and it has been claimed that Sendai-infected cells contain double-stranded 24S RNAs, containing 50S RNA, that would serve this purpose.[265] It seems likely that early in infection virus-specific mRNA is needed for the synthesis of viral proteins, some of which are required to replicate the viral genome via a full-length antigenome intermediate.[266] However, nothing is yet known of the nature of the replicase.

3. Virus-specific Proteins

The pattern of polypeptide synthesis in infected cells is similar for NDV[267,268] and Sendai.[241] Once induced protein synthesis begins, all the structural polypeptides of the virion (see Table 6) can be detected, as well as two nonstructural proteins of molecular weight 36,000 and 22,000. Each structural and nonstructural polypeptide was produced at all times after the onset of synthesis, and they were synthesized in unequal amounts which corresponded to their representation in the virion, except that polypeptides P and NP were present in greater amounts in infected cells than in the virion. The relative amount of M protein found in infected cells was considerably less than that found in the virion, suggesting that, as with influenza, the synthesis of M is controlled and may be a rate-limiting step in virus assembly. The data suggest that large, polyprotein precursors are not involved in parainfluenza virus replication and that the polypeptides are synthesized from monocystronic mRNAs. The role of the nonstructural proteins is not known, but it is likely that they are involved in some aspect of RNA replication.

The glycoproteins HN and F are constituents of the virus envelope and are synthesized on the rough endoplasmic reticulum.[269] They are transferred from the rough endoplasmic reticulum via smooth intracellular membranes to the plasma membrane and into virions. In the course of migration, F_0 is converted to F by means of proteolytic cleavage.

4. Virus Assembly and Release

Proposed mechanisms of assembly and release of parainfluenza viruses have been extensively reviewed.[270] Nucleocapsids are formed in the cytoplasmic matrix and then align themselves beneath areas of the plasma membrane that appear to be modified by the incorporation of virus-specific proteins. The glycoprotein spikes are synthesized in the rough endoplasmic reticulum and migrate into the plasma membrane.[269] In contrast, the M protein is incorporated into the plasma membrane immediately after synthesis. Thus, the cell membrane contains modified patches that are recognized by the nucleocapsids. Once in position, the nucleocapsids are enveloped by the modified membrane, and mature virions are released in a budding process.

5. Defective, Interfering Virus

Incomplete or defective forms of Sendai virus can be generated by serial undiluted passage[271,272] and consist of particles that sediment at 400S rather than at 1,000S, the normal sedimentation coefficient of parainfluenza viruses. The defective particles have a greatly decreased infectivity and have genome RNA that sediments at 19 to 25S, as compared to 50S for standard virions. Hybridization studies show that the 19 to 25S RNA has nucleotide sequences in common with the 50S RNA of standard virus.[273] These defective particles probably resemble the defective or T particles of rhabdoviruses in that they encapsidate subgenomic fragments of the RNA genome.

Defective particles are probably generated by mixed infection of cells with standard and defective particles.[273] The defective particles interfere with the normal pattern of RNA replication because there is a decreased accumulation of 50S viral RNA. It has been suggested that the "subgenomic" RNAs defective virus compete with standard virus genomes for the replicases synthesized by standard virus.[274]

ACKNOWLEDGMENTS

We are endebted to many colleagues and friends for data and helpful comments. In particular we would like to thank Dr. D. McGeoch for the data provided in Figure 1 and Tables 2 and 3. Figure 3 and Table 4 are based on studies by Dr. A. R. Carroll and S. Inglis. Table 6 and Figure 4 are adapted from data obtained in our laboratory by Dr. R. A. Lamb. We gratefully acknowledge the skilled assistance of Mrs. R. Farrell in preparing the figures.

REFERENCES

1. **Andrewes, C. H. and Pereira, H. G.,** *Viruses of Vertebrates,* 3rd ed., Bailliere Tindall, London, 1972.
2. **Simon, E. H.,** Evidence for the non-participation of DNA in viral RNA synthesis, *Virology,* 13, 105, 1961.
3. **Cooper, S. and Zinder, N. D.,** The growth of RNA bacteriophage: role of DNA synthesis, *Virology,* 18, 405, 1962.
4. **Reich, E., Franklin, R. M., Shatkin, A. J., and Tatum, E. L.,** Effect of actinomycin D on cellular nucleic acid synthesis and virus production, *Science,* 134, 556, 1961.
5. **Montagnier, L. and Saunders, F. K.,** Replication forms of encephalomyocarditis virus ribonucleic acid, *Nature,* 199, 664, 1963.
6. **Bishop, J. M. and Levintow, L.,** Replicative forms of viral RNA, *Prog. Med. Virol.,* 13, 1, 1971.
7. **Temin, H. M.,** Formation and activation of the provirus of RNA sarcoma viruses, in *Biology of Large RNA Viruses,* Barry, R. D. and Mahy, B. W. J., Eds., Academic Press, London, 1970, 233.
8. **Baltimore, D. and Franklin, R. M.,** Preliminary data on a virus specific enzyme system responsible for the synthesis of viral RNA, *Biochem. Biophys. Res. Commun.,* 9, 388, 1961.
9. **Baltimore, D., Eggers, H. J., Franklin, R. M., and Tamm, I.,** Poliovirus induced RNA polymerase and the effects of virus specific inhibitors on its production, *Proc. Natl. Acad. Sci. U.S.A.,* 49, 843, 1963.
10. **Pons, M. W.,** Studies on the replication of influenza virus RNA, *Virology,* 47, 823, 1972.
11. **Mahy, B. W. J. and Barry, R. D.,** *Negative Strand Viruses,* Academic Press, London, 1975.
12. **Horne, R. W., Waterson, A. P., Wildy, P., and Farnham, A. E.,** The structure and composition of the myxoviruses. I. Electron microscopic studies of myxovirus particles by negative staining techniques, *Virology,* 11, 79, 1960.
13. **Hoyle, L., Horne, R. W., and Waterson, A. P.,** The structure and composition of the myxoviruses. II. Components released from the influenza virus particle by ether, *Virology,* 13, 448, 1961.
14. **Almeida, J. D. and Waterson, A. P.,** Two morphological aspects of influenza virus, in *The Biology of Large RNA Viruses,* Barry, R. D. and Mahy, B. W. J., Eds., Academic Press, London, 1970, 27.
15. **Drzeniek, R., Frank, H., and Rott, R.,** Electron microscopy of purified influenza virus neuraminidase, *Virology,* 36, 703, 1968.
16. **Laver, W. G. and Valentine, R. C.,** Morphology of the isolated haemagglutinin and neuraminidase subunits of influenza virus, *Virology,* 38, 105, 1969.
17. **Tiffany, J. M. and Blough, H. A.,** Estimation of the number of surface projections on myxo- and paramyxoviruses, *Virology,* 41, 392, 1970.
18. **Schulze, I. T.,** The structure of influenza I. The polypeptides of the virion, *Virology,* 42, 890, 1970.
19. **Schulze, I. T.,** The structure of influenza II. A model based on the morphology and composition of subviral particles, *Virology,* 47, 181, 1972.
20. **Uhler, M. and Gard, S.,** Lipid content of standard and incomplete influenza A virus, *Nature,* 173, 1041, 1954.
21. **Blough, H. A. and Merlie, J.,** The lipids of incomplete influenza virus, *Virology,* 40, 685, 1970.
22. **Frommhagen, L. H., Knight, C. A., and Freeman, N. K.,** The ribonucleic acid, lipid and polysaccharide constituents of influenza virus preparations, *Virology,* 8, 176, 1959.
23. **Choppin, P. W., Klenk, H.-D., Compans, R. W., and Caliguiri, L. A.,** The parainfluenza virus SV5 and its relationship to the cell membrane, *Perspect. Virol.,* 7, 127, 1971.
24. **Landsberger, F. R., Lenard, J., Paxton, J., and Compans, R. W.,** Spin-label electron spin resonance study of the lipid-containing membrane of influenza virus, *Proc. Natl. Acad. Sci. U.S.A.,* 68, 2579, 1971.
25. **Kendal, A. P., Apostolov, K., and Belyavin, G.,** The effect of protease treatment on the morphology of influenza A, B and C viruses, *J. Gen. Virol.,* 5, 141, 1969.
26. **Webster, R. G., Laver, W. G., and Kilbourne, E. D.,** Reactions of antibodies with the surface antigens of influenza virus, *J. Gen. Virol.,* 3, 315, 1968.
27. **Webster, R. G. and Laver, W. G.,** Antigenic variation in influenza virus, *Prog. Med. Virol.,* 13, 271, 1971.
28. **Webster, R. G. and Darlington, R. W.,** Disruption of myxoviruses with Tween 20 and isolation of biologically active haemagglutinin and neuraminidase subunits, *J. Virol.,* 4, 182, 1969.

29. **Haslam, E. A., Hampson, A. W., Egan, J. A., and White, D. O.,** The polypeptides of influenza virus II. Interpretation of polyacrylamide gel electrophoresis patterns, *Virology,* 42, 555, 1970.
30. **White, D. O., Taylor, J. M., Haslam, E. A., and Hampson, A. W.,** The polypeptides of influenza virus and their biosynthesis, in *The Biology of Large RNA Viruses,* Barry, R. D. and Mahy, B. W. J., Eds., Academic Press, London, 1970, 602.
31. **Hoyle, L. and Fairbrother, R. W.,** Antigenic structure of influenza viruses; the preparation of elementary body suspensions and the nature of the complement-fixing antigen, *J. Hyg.,* 37, 512, 1937.
32. **Francis, T., Quilligan, J. J., and Minuse, E.,** Identification of another epidemic respiratory disease, *Science,* 112, 495, 1950.
33. **Schafer, W. and Zillig, W.,** On the structure of the elementary bodies of fowl plague virus. Preparation, physico-chemical and biological properties of split products, *Z. Naturforsch.,* 96, 779, 1954.
34. **Paucker, K., Birch-Anderson, A., and von Magnus, P.,** Studies on the structure of influenza virus I. Components of incomplete and infectious particles, *Virology,* 8, 1, 1959.
35. **Schulze, I. T., Pons, M. W., Hirst, G. K., and Hauser, R. N.,** The RNA and proteins of influenza virus, in *The Biology of Large RNA Viruses,* Barry, R. D. and Mahy, B. W. J., Eds., Academic Press, London, 1970, 324.
36. **Compans, R. W., Content, J., and Duesberg, P. H.,** Structure of the ribonucleoprotein of influenza virus, *J. Virol.,* 10, 795, 1972.
37. **Duesberg, P. H.,** Distinct sub-units of the ribonucleoprotein of influenza virus, *J. Mol. Biol.,* 42, 485, 1969.
38. **Kingsbury, D. W. and Webster, R. G.,** Some properties of influenza nucleocapsids, *J. Virol.,* 4, 219, 1969.
39. **Pons, M. W.,** Isolation of influenza ribonucleoprotein from infected cells. Demonstration of the presence of negative-stranded RNA in viral RNP, *Virology,* 46, 149, 1971.
40. **Compans, R. W. and Dimmock, N. J.,** An electron microscopic study of single-cycle infection of chick embryo fibroblasts by influenza virus, *Virology,* 39, 499, 1969.
41. **Compans, R. W., Klenk, H.-D., Caliguiri, L. A., and Choppin, P. W.,** Influenza virus proteins. I. Analysis of polypeptides of the virion and identification of spike glycoproteins, *Virology,* 42, 880, 1970.
42. **Haslam, E. A., Hampson, A. W., Radiskevics, I., and White, D. O.,** The polypeptides of influenza virus III. Identification of the haemagglutinin, neuraminidase and nucleocapsid proteins, *Virology,* 42, 556, 1970.
43. **Skehel, J. J. and Schild, G. C.,** The polypeptide composition of influenza A viruses, *Virology,* 44, 396, 1971.
44. **Klenk, H.-D., Rott, R., and Becht, H.,** On the structure of the influenza virus envelope, *Virology,* 47, 579, 1972.
45. **Skehel, J. J.,** Polypeptide synthesis in influenza-virus-infected cells, *Virology,* 49, 23, 1972.
46. **Skehel, J. J.,** Characterization of the sub-viral particles derived from influenza virus, *Virology,* 44, 409, 1971.
47. **Schwarz, R. T. and Klenk, H.-D.,** Inhibition of glycosylation of the influenza virus haemagglutinin, *J. Virol.,* 14, 1023, 1974.
48. **Carroll, A. R.,** Influenza Virus Transcriptase, Ph.D. dissertation, University of Cambridge, 1976.
49. **Bishop, D. H. L., Roy, P., Bean, W. J., and Simpson, R. J.,** Transcription of the influenza ribonucleic acid genome by a virion polymerase. III. Completeness of the transcription process, *J. Virol.,* 10, 689, 1972.
50. **Caliguiri, L. A. and Compans, R. W.,** Analysis of the *in vitro* product of an RNA dependent RNA polymerase isolated from influenza virus-infected cells, *J. Virol.,* 14, 191, 1974.
51. **White, D. O.,** Influenza viral proteins: identification and synthesis, *Curr. Top. Microbiol. Immunol.,* 63, 1, 1974.
52. **Lazarowitz, S. G., Compans, R. W., and Choppin, P. W.,** Influenza virus structural and non-structural proteins in infected cells and their plasma membranes, *Virology,* 46, 830, 1971.
53. **Stanley, P. and Haslam, E. A.,** The polypeptides of influenza virus. V. Localization by iodination techniques, *Virology,* 46, 764, 1971.
54. **Skehel, J. J.,** Origin of pandemic influenza viruses, in *Evolution in the Microbiol. World,* SGM Symp. 24, Carlyle, M. J. and Skehel, J. J., Eds., Cambridge University Press, Cambridge, 1974.
55. **Webster, R. G.,** Estimation of the molecular weights of the polypeptide chains from the isolated haemagglutinin and neuraminidase sub-units of influenza virus, *Virology,* 40, 543, 1970.
56. **Laver, W. G. and Baker, N.,** Amino-acid composition of the polypeptides of influenza virus particles, *J. Gen. Virol.,* 17, 61, 1972.
57. **Bucher, D. J. and Kilbourne, E. D.,** A2 (N2) neuraminidase of the X-7 influenza virus recombinant: determination of molecular size and sub-unit composition of the active unit, *J. Virol.,* 10, 60, 1972.
58. **Gregoriades, A.,** Isolation of neuraminidase from the WSN strain of influenza virus, *Virology,* 49, 333, 1972.
59. **Kendal, A. P. and Eckert, W.,** The preparation and properties of ^{14}C-carboxyamido-methylated sub-units from A2/1957 influenza neuraminidase, *Biochim. Biophys. Acta,* 258, 484, 1972.
60. **Lazdins, I., Haslam, E. A., and White, D. O.,** The polypeptides of influenza virus VI. Composition of the neuraminidase, *Virology,* 49, 758, 1972.
61. **Scholtissek, C. and Becht, H.,** Binding of ribonucleic acids to the RNP antigen protein of influenza, *J. Gen. Virol.,* 10, 11, 1971.
62. **Schild, G. C.,** Evidence for a new type-specific structural antigen of influenza virus particles, *J. Gen. Virol.,* 15, 99, 1972.
63. **Ada, G. L. and Perry, B. T.,** The nucleic acid content of influenza virus, *Aust. J. Exp. Biol. Med. Sci.,* 32, 453, 1954.
64. **Frisch-Niggemeyer, W. and Hoyle, L.,** The nucleic acid and carbohydrate content of influenza A and virus fractions produced by ether disintegration, *J. Hyg.,* 54, 201, 1956.

65. **Duesberg, P. H. and Robinson, W. S.**, On the structure and replication of influenza virus, *J. Mol. Biol.*, 25, 383, 1967.
66. **Pons, M. W., Schulze, I. T., Hirst, G. K., and Hauser, R.**, Isolation and characterization of the ribonucleoprotein of influenza virus, *Virology*, 37, 250, 1969.
67. **Schulze, I. T.**, Structure of the influenza virion, *Adv. Virus Res.*, 18, 1, 1973.
68. **Burnet, F. M.**, The structure of influenza virus, *Science*, 123, 1101. 1956.
69. **Hirst, G. K.**, Genetic recombination with Newcastle disease virus, polioviruses and influenza, *Cold Spring Harbor Symp. Quant. Biol.*, 27, 303, 1962.
70. **Simpson, R. W.**, Genetic studies with influenza A virus, in *Cellular Biology of Myxovirus Infections*, Wolstenholme, G. E. W. and Knight, J., Eds., Little, Brown and Co., Boston, 1964, 187.
71. **Barry, R. D.**, The multiplication of influenza virus II. Multiplicity reactivation of UV irradiated virus, *Virology*, 14, 393, 1961.
72. **Scholtissek, C. and Rott, R.**, Behaviour of virus-specific activities in tissue cultures infected with myxoviruses after chemical changes of the viral ribonucleic acid, *Virology*, 22, 169, 1964.
73. **Gandhi, S. S. and Burke, D. C.**, Interferon production by myxoviruses in chick embryo cells, *J. Gen. Virol.*, 6, 95, 1970.
74. **Portocala, R., Boeru, V., and Samuel, J.**, Biosynthesis of influenza virus, starting from an ether-phenol virus extract, *Acta Virol.*, 3, 172, 1959.
75. **Maassab, H. F.**, An infectious ribonucleic acid derived from influenza infected cells, *Proc. Natl. Acad. Sci. U.S.A.*, 877, 1959.
76. **Maassab, H. F.**, Further studies on the infectivity of ribonucleic acid prepared from influenza-infected cells, *J. Immunol.*, 90, 265, 1963.
77. **Dubrovina, T. J.**, Conditions of isolation of infectious ribonucleic acid from influenza virus, *Acta Virol.*, 10, 117, 1966.
78. **Chetverikova, L. K., Polyak, R. J., Yukhnova, L. G., and Smorodentsev, A. A.**, Method for increasing the regularity of isolation of infectious ribonucleic acid from influenza virus, *J. Virol.*, 6, 183, 1970.
79. **Ada, G. L., Lind, P. E., Larkin, L., and Burret, F. M.**, Failure to recover infective ribonucleic acid from myxovirus preparations, *Nature*, 184, 360, 1959.
80. **Sokol, F. and Szurman, J.**, Isolation of ribonucleic acid from purified influenza virus by phenol extraction, *Acta Virol.*, 3, 175, 1959.
81. **Sokol, F. and Schramek, S.**, An improved method for the isolation of ribonucleic acid from myxoviruses, *Acta Virol.*, 8, 193, 1964.
82. **Davies, P. and Barry, R. D.**, Nucleic acid of influenza virus, *Nature*, 211, 384, 1966.
83. **Nayak, D. P. and Baluda, M. A.**, Isolation and partial characterization of the nucleic acid of influenza virus, *J. Virol.*, 1, 1217, 1967.
84. **Barry, R. D. and Davies, P.**, The sedimentation of influenza virus and its RNA in sucrose density gradients, *J. Gen. Virol.*, 2, 59, 1968.
85. **Duesberg, P. H.**, The RNA's of influenza virus, *Proc. Natl. Acad. Sci. U.S.A.*, 59, 930, 1968.
86. **Pons, M. W. and Hirst, G. K.**, Polyacrylamide gel electrophoresis of influenza RNA, *Virology*, 34, 385, 1968.
87. **Skehel, J. J.**, Estimations of the molecular weight of the influenza virus genome, *J. Gen. Virol.*, 11, 103, 1971.
88. **Floyd, R. W., Stone, M. R., and Joklik, W. K.**, Separation of single stranded ribonucleic acid by acrylamide-agarose-urea gel electrophoresis, *Anal. Biochem.*, 59, 599, 1974.
89. **McGeoch, D., Fellner, P., and Newton, C.**, The influenza virus genome consists of eight distinct pieces, *Proc. Natl. Acad. Sci. U.S.A.*, in press.
90. **Hoyle, L.**, Nucleotide composition of influenza virus ribonucleic acid, *Abst. 7th Int. Congr. Microbiol.*, p. 239, 1958.
91. **Hoyle, L.**, The nucleic acid of the influenza virus, *Symp. VII, 4th Int. Congr. Biochem.*, Vienna, 1958.
92. **Krug, R. M.**, Cytoplasmic and nucleoplasmic viral RNP's in influenza virus infected MDCK cells, *Virology*, 50, 103, 1972.
93. **Content, J. and Duesberg, P. H.**, Base sequence differences among the ribonucleic acids of influenza virus, *J. Mol. Biol.*, 62, 273, 1971.
94. **Horst, J., Content, J., Mandeles, S., Fraenkel-Conrat, M., and Duesberg, P. H.**, Distinct oligonucleotide patterns of distinct influenza virus RNA's, *J. Mol. Biol.*, 69, 210, 1972.
95. **Young, R. J. and Content, J.**, '5-Terminus of influenza virus RNA, *Nature*, 230, 140, 1971.
96. **Lewandowski, L., Content, J., and Leppla, S. H.**, Characterization of the sub-unit structure of the RNA genome of influenza virus, *J. Virol.*, 8, 701, 1971.
97. **Palese, P. and Schulman, J. L.**, Mapping of the influenza virus genome: identification of the hemagglutinin and neuraminidase genes, *Proc. Natl. Acad. Sci. U.S.A.*, in press.
98. **Burnet, F. M.**, Influenza virus infections of the chick embryo by the amniotic route. I. General character of the infections, *Aust. J. Exp. Biol. Med. Sci.*, 18, 353, 1940.
99. **Burnet, F. M.**, Growth of influenza virus in the allantois of the chick embryo, *Aust. J. Exp. Biol. Med. Sci.*, 19, 291, 1941.

100. **Henle, G., Girardi, A., and Henle, W.,** A non-transmissable cytopathogenic effect of influenza virus in tissue culture accompanied by formation of non-infectious hemagglutinin, *J. Exp. Med.*, 101, 25, 1955.
101. **Deibel, R. and Hotchin, J. E.,** Quantitative application of fluorescent antibody techniques to influenza virus infected cell cultures, *Virology*, 8, 367, 1959.
102. **von Magnus, P.,** Propagation of the PR8 strain of influenza A virus in chick embryos. II. The formation of incomplete virus following inoculation of large doses of seed virus, *Acta Pathol. Microbiol. Scand.*, 28, 278, 1951.
103. **Henle, W. and Henle, G.,** Interference between inactive and active viruses of influenza. I. Incidental occurrence and artificial induction of the phenomenon, *Am. J. Med. Sci.*, 207, 717, 1944.
104. **Fazekas de St. Groth, S.,** Viropexis, the mechanism of influenza virus infection, *Nature*, 1962, 294, 1948.
105. **Cairns, H. J. F.,** The asynchrony of infection by influenza virus, *Virology*, 3, 1, 1957.
106. **Cairns, H. J. F. and Mason, P. J.,** The production of influenza A virus in the cells of the allantois, *J. Immunol.*, 71, 38, 1953.
107. **Donald, H. B. and Isaacs, A.,** Counts of influenza virus particles, *J. Gen. Microbiol.*, 10, 457, 1954.
108. **Barry, R. D.,** Multiplication of influenza viruses. I. Formation of incomplete virus, *Virology*, 14, 389, 1961.
109. **Franklin, R. M. and Henry, C.,** The multiplication of fowl plague virus in tissue cultures of chick embryo cells, *Virology*, 10, 406, 1960.
110. **Borland, R. and Mahy, B. W. J.,** Deoxyribonucleic acid dependent RNA polymerase activity in cells infected with influenza virus, *J. Virol.*, 2, 33, 1968.
111. **Bean, W. J. and Simpson, R. W.,** Primary transcription of the influenza virus genome in permissive cells, *Virology*, 56, 646, 1973.
112. **Barry, R. D., Ives, D. R., and Cruickshank, J. G.,** Participation of DNA in the multiplication of influenza virus, *Nature*, 194, 1139, 1962.
113. **Barry, R. D.,** Effects of inhibitors of nucleic acid synthesis on the production of myxoviruses, in *The Cellular Biology of Myxoviruses*, Ciba Symposium, Wolstenholme, G. E. W. and Knight, J., Eds., Churchill Ltd., London, 1964, 51.
114. **Rott, R., Saber, S., and Scholtissek, C.,** Effect on myxoviruses of mitomycin C, actinomycin D and pretreatment of the host cell with ultraviolet light, *Nature*, 205, 1187, 1965.
115. **Nayak, D. P. and Rasmussen, A. F.,** Influence of mitomycin C on the replication of influenza viruses, *Virology*, 30, 673, 1966.
116. **Rott, R. and Scholtissek, C.,** Specific inhibition of influenza replication by α-amanitin, *Nature*, 228, 56, 1970.
117. **Mahy, B. W. J., Hastie, N. D., and Armstrong, S. J.,** Inhibition of influenza virus replication by α-amanitin: mode of action. *Proc. Natl. Acad. Sci. U.S.A.*, 69, 1421, 1972.
118. **Cheyne, I. M. and White, D. O.,** Growth of paramyxoviruses in anucleate cells, *Aust. J. Exp. Biol. Med. Sci.*, 47, 145, 1969.
119. **Follett, E. A. C., Pringle, C. R., Wunner, W. H., and Skehel, J. J.,** Virus replication in enucleate cells: VSV and influenza virus, *J. Virol.*, 13, 399, 1974.
120. **Breitenfeld, P. M. and Schafer, W.,** The formation of fowl plague virus antigens in infected cells as studied with fluorescent antibodies, *Virology*, 4, 328, 1957.
121. **Glass, S. E., McGeoch, D., and Barry, R. D.,** Characterization of the mRNA of influenza virus, *J. Virol.*, 16, 1435, 1975.
122. **Scholtissek, C., Rott, R., Hausen, P., Hausen, H., and Schafer, W.,** Comparative studies of RNA and protein synthesis with a myxovirus and a small polyhedral virus, *Cold Spring Harbor Symp. Quant. Biol.*, 27, 245, 1962.
123. **Lamb, R. A.,** Aspects of the Structure and Replication of Sendai Virus, Ph.D. dissertation, University of Cambridge, 1974.
124. **Watson, B. K. and Coons, A. H.,** Studies of influenza virus infection in the chick embryo using fluorescent antibody, *J. Exp. Med.*, 99, 419, 1954.
125. **Liu, C.,** Studies on influenza infection in ferrets by means of fluorescein-labelled antibody II. The role of soluble antigen in nuclear fluorescence and cross reactions, *J. Exp. Med.*, 101, 677, 1955.
126. **Kirk, J. M.,** The mode of action of actinomycin D, *Biochim. Biophys. Acta*, 42, 167, 1960.
127. **Kelly, D. C., Avery, R. J., and Dimmock, N. J.,** Failure of an influenza virus to initiate infection in enucleate BHK cells, *J. Virol.*, 13, 1155, 1974.
128. **Scholtissek, C. and Rott, R.,** Connection between the synthesis of ribonucleic acid and protein in the growth of the influenza virus group, *Z. Naturforsch.*, 168, 663, 1961.
129. **Roeder, R. G. and Rutter, W. J.,** Specific nucleolar and nucleoplasmic RNA polymerase, *Proc. Natl. Acad. Sci. U.S.A.*, 65, 675, 1970.
130. **Lindell, T. J., Weinberg, F., Morris, P. W., Roeder, R. G., and Rutter, W. J.,** Specific inhibition of RNA polymerase II by α-amanitin, *Science*, 170, 447, 1970.
131. **Mahy, B. W. J.,** The replication of fowl plague virus RNA, in *The Biology of Large RNA Viruses*, Barry, R. D. and Mahy, B. W. J., Eds., Academic Press, London, 1970.
132. **Prescott, D. M., Myerson, D., and Wallace, R. J.,** Enucleation of mammalian cells with cytochalasin B, *Exp. Cell Res.*, 71, 480, 1972.
133. **Maeno, K. and Kilbourne, E. D.,** Developmental sequence and intracellular sites of synthesis of three structural protein antigens of influenza A2 virus, *J. Virol.*, 5, 153, 1970.

134. **Kelly, D. C. and Dimmock, N. J.,** Fowl plague virus replication in mammalian cell-avian erythrocyte heterokaryons: studies concerning the actinomycin D and UV-light sensitive phase in influenza replication, *Virology,* 60, 210, 1974.
135. **Dimmock, N. J.,** New virus specific antigens in cells infected with influenza virus, *Virology,* 39, 224, 1969.
136. **Dimmock, N. J. and Watson, D. H.,** Non-virion antigens of influenza virus, in *Biology of Large RNA Viruses,* Barry, R. D. and Mahy, B. W. J., Eds., Academic Press, London, 1970, 619.
137. **Taylor, J. M., Hampson, A. W., and White, D. O.,** The polypeptides of influenza virus. I. Cytoplasmic synthesis and nuclear accumulation, *Virology,* 39, 419, 1969.
138. **Krug, R. M. and Etkind, P. R.,** Cytoplasmic and nuclear virus-specific proteins in influenza virus infected MDCK cells, *Virology,* 56, 334, 1973.
139. **Krug, R. M. and Soeiro, R.,** Studies on the intranuclear localization of influenza virus-specific proteins, *Virology,* 64, 378, 1975.
140. **Krug, R. M.,** Influenza viral RNP's newly synthesized during the latent period of viral growth in MDCK cells, *Virology,* 44, 125, 1971.
141. **Franklin, R. M. and Breitenfeld, P. M.,** The abortive infection of Earle's L cells by fowl plague virus, *Virology,* 8, 293, 1959.
142. **Loeffler, H., Henle, G., and Henle, W.,** Attempts to influence the incomplete reproduction cycle of influenza virus in Hela cells by antibodies, *J. Immunol.,* 88, 763, 1962.
143. **Fraser, K. B.,** Immunofluorescence of abortive and incomplete infections of influenza virus in hamster BHK 21 cells and mouse L cells, *J. Gen. Virol.,* 1, 1, 1967.
144. **Stephenson, J. R. and Dimmock, N. J.,** Inhibition of the processing of ribosomal RNA in avian cells infected with an influenza virus, *Biochim. Biophys. Acta,* 361, 198, 1974.
145. **Hoyle, L. and Frisch-Niggemeyer, W.,** The disintegration of influenza virus particles on entry into the host cell *J. Hyg.,* 53, 474, 1955.
146. **Hoyle, L.,** The use of radioactive influenza virus to determine the fate of the infecting particle on entry into the host cell, in *The Nature of Viruses,* Churchill Ltd., London, 1956.
147. **Ghendon, V., Ginsburg, V. P., Soloviev, G. Y., and Murkuskin, S. G.,** The fate of influenza virus RNA in cells treated with ultraviolet rays, *J. Gen. Virol.,* 6, 249, 1970.
148. **Stephenson, J. R. and Dimmock, N. J.,** Early events in influenza virus multiplication. I. Location and fate of the input RNA, *Virology,* 65, 77, 1975.
149. **Armstrong, S. J. and Barry, R. D.,** The role of the cell nucleus in the production of influenza viruses, in *Negative Strand Viruses,* Mahy, B. W. J. and Barry, R. D., Eds., Academic Press, London, 1975.
150. **Armstrong, S. J. and Barry, R. D.,** The topography of RNA synthesis in cells infected with fowl plague virus, *J. Gen. Virol.,* 24, 535, 1974.
151. **Cox, N. J. and Barry, R. D.,** Hybridization studies of the relationship between influenza virus RNA and cellular DNA, *Virology,* 69, 304, 1976.
152. **Chow, N. and Simpson, R. W.,** RNA dependent RNA polymerase activity associated with virions and sub-viral components of myxoviruses, *Proc. Natl. Acad. Sci. U.S.A.,* 68, 752, 1971.
153. **Penhoet, E., Miller, H., Doyle, M., and Blatti, S.,** RNA dependent RNA polymerase activity in influenza virions, *Proc. Natl. Acad. Sci. U.S.A.,* 68, 1369, 1971.
154. **Skehel, J. J.,** RNA-dependent RNA polymerase activity of the influenza virus *Virology,* 45, 793, 1971.
155. **Oxford, J. S.,** Polypeptide composition of influenza B viruses and enzymes associated with the purified virus particles, *J. Virol.,* 12, 827, 1973.
156. **Carroll, A. R., McGeoch, D., and Mahy, B. W. J.,** *In vitro* transcription by the influenza virion RNA transcriptase, *Collq. Inst. Nat. Santé Res. Med.,* 47, 95, 1975.
157. **Hefti, E., Roy, P., and Bishop, D. H. L.,** The initiation of transcription by influenza virion transcriptase in *Negative Strand Viruses,* Mahy, B. W. J. and Barry, R. D., Eds., Academic Press, London, 1975.
158. **McGeoch, D. and Kitron, N.,** Influenza virion RNA dependent RNA polymerase: stimulation by guanosine and related compounds, *J. Virol.,* 15, 686, 1975.
159. **Emerson, S. V. and Wagner, R. R.,** L Protein requirement for *in vitro* RNA synthesis by vesicular stomatitis virus, *J. Virol.,* 12, 1325, 1973.
160. **Flamand, A. and Bishop, D. H. L.,** Primary *in vivo* transcription of vesicular stomatitis virus complementation groups, *J. Virol.,* 12, 1238, 1973.
161. **Repik, P., Flamand, A., and Bishop, D. H. L.,** The effect of interferon upon the primary and secondary transcription of vesicular stomatitis and influenza viruses, *J. Virol.,* 14, 1169, 1974.
162. **Bishop, D. H. L., Obijeski, J. F., and Simpson, R. W.,** Transcription of the influenza ribonucleic acid genome by a virion polymerase. I. Optimal conditions for *in vitro* activity of the ribonucleic acid-dependent ribonucleic acid polymerase, *J. Virol.,* 8, 66, 1971.
163. **Choppin, P. W., Kilbourne, E. D., Dowdle, W., Hirst, G. K., Joklik, W. K., Simpson, R. W., and White, D. O.,** Genetics, replication and inhibition of replication of influenza virus – summary of influenza workshop VII. *J. Infect. Dis.,* 136, 713, 1975.
164. **Scholtissek, C. and Rott, R.,** Synthesis *in vivo* of influenza virus plus and minus strand RNA and its preferential inhibition by antibodies, *Virology,* 40, 989, 1970.

165. Pons, M. W., The inhibition of influenza RNA synthesis by actinomycin D and cycloheximide, *Virology*, 51, 120, 1973.
166. Ho, P. P. K. and Walters, P., Influenza virus induced ribonucleic acid nucleotidyl transferase and the effect of actinomycin D on its formation, *Biochemistry*, 5, 231, 1966.
167. Scholtissek, C. and Rott, R., Hybridization studies with influenza virus RNA, *Virology*, 39, 400, 1969.
168. Mahy, B. W. J. and Bromley, P. A., The *in vitro* product of an RNA dependent RNA polymerase induced by influenza virus, *J. Virol.*, 6, 259, 1970.
169. Skehel, J. J. and Burke, D. C., RNA synthesis in CEF cells infected with FPV, *J. Virol.*, 3, 429, 1969.
170. Compans, R. W. and Caliguiri, L. A., Isolation and properties of an RNA polymerase from influenza virus-infected cells, *J. Virol.*, 11, 441, 1973.
171. Hastie, N. D. and Mahy, B. W. J., RNA dependent RNA polymerase in the nuclei of cells infected with influenza virus, *J. Virol.*, 12, 951, 1973.
172. Pons, M. W., The effect of actinomycin D on the replication of influenza virus and influenza virus RNA, *Virology*, 33, 150, 1967.
173. Nayak, D. P. and Baluda, M. A., Ribonucleic acid synthesis in cells infected with influenza virus, *J. Virol.*, 2, 99, 1967.
174. Weissman, C., Feix, G., and Slor, N., *In vitro* synthesis of phage RNA: the nature of the intermediate, *Cold Spring Harbor Symp. Quant. Biol.*, 33, 83, 1968.
175. Nayak, D. P., The replication of influenza virus, in *Biology of Large RNA Viruses*, Barry, R. D. and Mahy, B. W. J., Eds., Academic Press, London, 1970, 371.
176. Siegert, W., Bauer, G., and Hofschneider, P. H., Direct evidence for messenger activity of influenza virion RNA, *Proc. Natl. Acad. Sci. U.S.A.*, 70, 2960, 1973.
177. Kingsbury, D. W. and Webster, R. G., Cell free translation of influenza virus messenger RNA, *Virology*, 56, 654, 1973.
178. Etkind, P. R. and Krug, R. M., Purification of influenza viral complementary RNA: its genetic content and activity in wheat germ cell-free extracts, *J. Virol.*, 16, 1464, 1974.
179. Maizel, J. V., Polyacrylamide gel electrophoresis of viral proteins, in *Methods in Virology* Vol. 5, Maramorosch, K. and Koprowski, H., Eds., Academic Press, New York, 1971.
180. Taylor, J. M., Hampson, A. W., Layton, J. E., and White, D. O., The polypeptides of influenza virus IV. An analysis of nuclear accumulation, *Virology*, 42, 744, 1970.
181. Laemmli, U. K., Cleavage of structural proteins during the assembly of the head of bacteriophage T4, *Nature*, 227, 680, 1970.
182. Caliguiri, L. A. and Tamm, I., The role of cytoplasmic membranes in poliovirus biosynthesis, *Virology*, 42, 100, 1970.
183. Compans, R. W., Influenza virus proteins II. Association with components of the cytoplasm, *Virology*, 51, 56, 1973.
184. Compans, R. W., Distinct carbohydrate components of influenza virus glycoproteins in smooth and rough cytoplasmic membranes, *Virology*, 55, 541, 1973.
185. Stanley, P., Gandhi, S. S., and White, D. O., The polypeptides of influenza virus VII. Synthesis of hemagglutinin, *Virology*, 53, 92, 1973.
186. Klenk, H.-D., Wollert, W., Rott, R., and Scholtissek, C., Association of influenza virus proteins with cytoplasmic fractions, *Virology*, 57, 28, 1974.
187. Hay, A. J., Studies on the formation of the influenza virus envelope, *Virology*, 60, 398, 1974.
188. Scholtissek, C. and Klenk, H.-D., Influenza virus replication, in *The Influenza Viruses and Influenza*, Kilbourne, E. D., Ed., Academic Press, New York, 1975, 215.
189. Klenk, H.-D. and Rott, R., Formation of influenza virus proteins, *J. Virol.*, 11, 823, 1973.
190. Meier-Ewert, M. and Compans, R. W., Time course of synthesis and assembly of influenza virus proteins, *J. Virol.*, 14, 1083, 1974.
191. Klenk, H.-D., Scholtissek, C., and Rott, R., Inhibition of glycoprotein biosynthesis of influenza virus by glucosamine and 2-deoxy-D-glucose, *Virology*, 49, 723, 1972.
192. Kilbourne, E. D., Inhibition of influenza virus multiplication with a glucose anti-metabolite-2-deoxy-D-glucose, *Nature*, 183, 271, 1959.
193. Kaluza, G., Scholtissek, C., and Rott, R., Inhibition of the multiplication of enveloped RNA-viruses by glucosamine and 2-deoxy-D-glucose, *J. Gen. Virol.*, 14, 251, 1972.
194. Gandhi, S. S., Stanley, P., Taylor, J. M., and White, D. O., Inhibition of influenza viral glycoprotein synthesis by sugars, *Microbios*, 5, 41, 1972.
195. Lazarowitz, S. G., Compans, R. W., and Choppin, P. W., Proteolytic cleavage of the haemagglutinin polypeptide of influenza virus. Function of the uncleaved polypeptide H.A., *Virology*, 52, 199, 1973.
196. Lazarowitz, S. G., Goldberg, A. R., and Choppin, P. W., Proteolytic cleavage by plasmin of the haemagglutinin polypeptide of influenza virus: host cell activation of serum plasminogen, *Virology*, 56, 172, 1973.
197. Etchison, J., Doyle, M., Penhoet, E., and Holland, J., Synthesis and cleavage of influenza virus proteins, *J. Virol.*, 7, 155, 1971.
198. Skehel, J. J., Early polypeptide synthesis in influenza virus-infected cells, *Virology*, 56, 394, 1973.

199. **Avery, R. J. and Dimmock, N. J.**, Temporal control of transcription of influenza virus RNA, *Virology*, 64, 409, 1975.
200. **Grubman, M. J., Moyer, S. A., Banerjee, A. K., and Ehrenfeld, E.**, Sub-cellular localization of vesicular stomatitis virus messenger RNA's, *Biochem. Biophys. Res. Commun.*, 62, 531, 1975.
201. **Both, G. W., Moyer, S. A., and Banerjee, A. K.**, Translation and identification of the viral mRNA species isolated from sub-cellular fractions of vesicular stomatitis virus-infected cells, *J. Virol.*, 15, 1012, 1975.
202. **Morgan, C., Hsu, K. C., Rifkind, R. A., Knox, A., and Rose, H. M.**, The application of ferratin-conjugated antibody to electron microscopic studies of influenza virus in infected cells, *J. Exp. Med.*, 114, 833, 1961.
203. **Bachi, T., Gerhard, W., Lindenmann, J., and Muhlethaler, K.**, Morphogenesis of influenza A virus in Erhlich axites tumor cells as revealed by thin sections and freeze-etching, *J. Virol.*, 4, 769, 1969.
204. **Apostolov, K., Flewett, T. H., and Kendal, A. P.**, Morphology of influenza A, B, C and infectious bronchitis virus (IBV) virions and their replication, in *Biology of Large RNA Viruses*, Barry, R. D. and Mahy, B. W. J., Eds., Academic Press, London, 1970, 3.
205. **Duc-Nguyen, H., Rose, H. M., and Morgan, C.**, An electron microscopic study of changes of the surface of influenza infected cells as revealed by ferratin-conjugated antibody, *Virology*, 28, 404, 1966.
206. **Hay, A. J. and Skehel, J. J.**, Studies on the synthesis of influenza virus protein, in *Negative Strand Viruses*, Vol. 2, Mahy, B. W. J. and Barry, R. D., Eds., Academic Press, London, 1975, 635.
207. **Seto, J. T. and Rott, R.**, Functional significance of sialidase during influenza virus multiplication, *Virology*, 30, 731, 1966.
208. **Palese, P., Tobita, K., Ueda, M., and Compans, R. J.**, Characterization of temperature sensitive influenza virus mutants defective in neuraminidase, *Virology*, 61, 397, 1974.
209. **Andrewes, C. H., Bang, F. B., and Burnet, F. M.**, A short description of the myxovirus group (influenza and related viruses), *Virology*, 1, 176, 1955.
210. **Hirst, G. K.**, Adsorption of influenza virus hemagglutinins and virus by red cells, *J. Exp. Med.*, 76, 195, 1942.
211. **Gottschalk, A.**, The influenza virus neuraminidase, *Nature*, 181, 377, 1958.
212. **Andrewes, C. H., Bang, F. B., Chanock, R. M., and Zhdanov, V. M.**, Parainfluenza viruses 1, 2, and 3: suggested names for recently described myxoviruses, *Virology*, 8, 129, 1959.
213. **Waterson, A. P.**, Two kinds of myxovirus, *Nature*, 193, 1163, 1962.
214. **Duesberg, P. H. and Robinson, W. S.**, Isolation of the nucleic acid of Newcastle disease virus (NDV), *Proc. Natl. Acad. Sci. U.S.A.*, 54, 794, 1965.
215. **Barry, R. D. and Bukrinskaya, A. G.**, The nucleic acid of Sendai virus and ribonucleic acid synthesis in cells infected with Sendai virus, *J. Gen. Virol.*, 2, 71, 1968.
216. **Compans, R. W. and Choppin, P. W.**, The nucleic acid of the parainfluenza virus SV5, *Virology*, 35, 289, 1968.
217. **Schluederberg, A.**, Measles virus RNA, *Biochem. Biophys. Res. Commun.*, 42, 1012, 1971.
218. **Horne, R. W. and Waterson, A. P.**, A helical structure in mumps, Newcastle disease and Sendai viruses, *J. Mol. Biol.*, 2, 75, 1960.
219. **Waterson, A. P., Jansen, K. E., Tyrrell, D. A. J., and Horne, R. W.**, The structure of parainfluenza 3 virus, *Virology*, 14, 374, 1961.
220. **Choppin, P. W., Compans, R. W., Scheid, A., McSharry, J. J., and Lazarowitz, S. G.**, Structure and assembly of viral membranes, in *Membrane Research*, Fox, C. F., Ed., Academic Press, New York, 1972, 163.
221. **Finch, J. T. and Gibbs, A. J.**, Observations on the structure of the nucleocapsids of some paramyxoviruses, *J. Gen. Virol.*, 6, 141, 1970.
222. **Kingsbury, D. W. and Darlington, R. W.**, Isolation and properties of Newcastle disease nucleocapsid, *J. Virol.*, 2, 248, 1968.
223. **Bikel, I. and Duesberg, P. H.**, Proteins of Newcastle disease virus and of the viral nucleocapsid, *J. Virol.*, 4, 388, 1969.
224. **Compans, R. W. and Choppin, P. W.**, The length of the helical nucleocapsid of Newcastle disease virus, *Virology*, 33, 344, 1967.
225. **Hosaka, Y.**, Isolation and structure of the nucleocapsid of HVJ, *Virology*, 35, 445, 1968.
226. **Bukrinskaya, A. G., Klimenko, S. M., Smirnov, Y. A., and Guschin, B. V.**, Infective substructures of Sendai virus from infected Ehrlich ascites tumor cells, *J. Virol.*, 2, 752, 1968.
227. **Blair, C. D. and Robinson, W. S.**, Replication of Sendai virus. II. Steps in virus assembly, *J. Virol.*, 5, 639, 1970.
228. **Kingsbury, D. W.**, Newcastle disease virus RNA. I. Isolation and preliminary characterization of RNA from virus particles, *J. Mol. Biol.*, 18, 195, 1966.
229. **Kolakofsky, D., Boy de la Tour, E., and Delius, H.**, Molecular weight determination of Sendai and Newcastle disease virus RNA, *J. Virol.*, 13, 261, 1974.
230. **Kolakofsky, D.**, Transfer ribonucleic acid nucleotidyl-transferase and transfer ribonucleic acid in Sendai virions, *J. Virol.*, 10, 555, 1972.
231. **Robinson, W. S.**, Self-annealing of subgroup 2 myxovirus RNAs, *Nature*, 225, 944, 1970.
232. **Robinson, W. S.**, Intracellular structures involved in Sendai virus replication, *Virology*, 43, 90, 1971.
233. **Waterson, A. P.**, The morphology and composition of Newcastle disease virus, in *Newcastle Disease: An Evolving Pathogen*, Hanson, R. P., Ed., University of Wisconsin Press, Madison, 1964, 119.

234. **Hosaka, Y., Kitano, H., and Ikeguchi, S.**, Studies on the pleomorphism of HVJ virions, *Virology*, 29, 205, 1966.
235. **Compans, R. W., Holmes, K. V., Dales, S., and Choppin, P. W.**, An electron microscopic study of moderate and virulent virus-cell interactions of the parainfluenza virus SV5, *Virology*, 30, 411, 1966.
236. **Content, J. and Duesberg, P. H.**, Electrophoretic distribution of the proteins and glycoproteins of influenza virus and Sendai virus, *J. Virol.*, 6, 707, 1970.
237. **Mountcastle, W. E., Compans, R. W., Caliguiri, L. A., and Choppin, P. W.**, Nucleocapsid protein subunits of simian virus 5, Newcastle disease virus, and Sendai virus, *J. Virol.*, 6, 677, 1970.
238. **Evans, M. J. and Kingsbury, D. W.**, Separation of Newcastle disease virus proteins by polyacrylamide gel electrophoresis, *Virology*, 37, 597, 1969.
239. **Mountcastle, W. E., Compans, R. W., and Choppin, P. W.**, Proteins and glycoprotein of paramyxoviruses: a comparison of simian virus 5, Newcastle disease virus and Sendai virus, *J. Virol.*, 7, 47, 1971.
240. **Hall, W. W. and Martin, S. J.**, Purification and characterization of measles virus, *J. Gen. Virol.*, 19, 175, 1973.
241. **Lamb, R. A., Mahy, B. W., and Choppin, P. W.**, The synthesis of Sendai virus polypeptides in infected cells, *Virology*, 69, 116, 1976.
242. **Scheid, A., Caliguiri, L. A., Compans, R. W., and Choppin, P. W.**, Isolation of paramyxovirus glycoproteins. Association of both hemagglutinating and neuraminidase activities with the larger SV5 glycoprotein, *Virology*, 50, 640, 1972.
243. **Scheid, A. and Choppin, P. W.**, Isolation and purification of the envelope proteins of Newcastle disease virus, *J. Virol.*, 11, 263, 1973.
244. **Homma, M. and Tamagawa, S.**, Restoration of the fusion activity of L-cell borne Sendai virus by trypsin, *J. Gen. Virol.*, 19, 423, 1973.
245. **Bratt, M. A. and Robinson, W. S.**, Ribonucleic acid synthesis in cells infected with Newcastle disease virus, *J. Mol. Biol.*, 23, 1, 1967.
246. **Huang, A. S., Baltimore, D., and Bratt, M.**, Ribonucleic acid polymerase in virions of Newcastle disease virus: comparison with the vesicular stomatitis virus polymerase, *J. Virol.*, 7, 389, 1971.
247. **Robinson, W. S.**, Ribonucleic acid polymerase activity in Sendai virions and nucleocapsid, *J. Virol.*, 8, 81, 1971.
248. **Stone, H. O., Portner, A., and Kingsbury, D. W.**, Ribonucleic acid transcriptase in Sendai virus and infected cells, *J. Virol.*, 8, 174, 1971.
249. **Hutchinson, J. E. and Mahy, B. W. J.**, RNA polymerase activity in virions of para-influenza 1 virus (Sendai), *Arch. Gesamte Virusforsch.*, 37, 203, 1972.
250. **Vincent, J. E. and Barry, R. D.**, Sendai virus-specific RNA associated with polysomes, in *Negative Strand Viruses*, Vol. 1, Mahy, B. W. J. and Barry, R. D., Eds., Academic Press, London, 1975, 409.
251. **Robinson, W. S.**, Intracellular structures involved in Sendai virus replication, *Virology*, 43, 90, 1971.
252. **Marx, P. A., Portner, A., and Kingsbury, D. W.**, Sendai virus transcriptase complex. Polypeptide composition and inhibition by virion envelope protein, *J. Virol.*, 13, 107, 1974.
253. **Wilcox, W. C.**, Quantitative aspects of an *in vitro* virus-induced toxic reaction. I. General aspects of the reaction of Newcastle disease virus with L cells, *Virology*, 9, 30, 1959.
254. **Hüppert, J., Gresland, L., and Lazar, P.**, Heterogeneity of chick embryo cells with regard to Newcastle disease virus multiplication, *J. Gen. Virol.*, 23, 281, 1974.
255. **Hüppert, J., Gresland, L., and Rosenbergova, M.**, Newcastle disease virus: virus particle and RNA synthesis in different host cells and at different temperatures, *J. Gen. Virol.*, 22, 129, 1974.
256. **Morgan, C. and Howe, C.**, Structure and development of viruses as observed in the electron microscope. IX. Entry of para-influenza 1 (Sendai) virus, *J. Virol.*, 2, 1122, 1968.
257. **Blair, C. and Robinson, W. S.**, Replication of Sendai virus. II. Steps in virus assembly, *J. Virol.*, 5, 639, 1970.
258. **Kaverin, N. V. and Varich, N. L.**, Newcastle disease virus specific RNA: polyacrylamide gel analysis of single stranded RNA and hybrid duplexes, *J. Virol.*, 13, 253, 1974.
259. **Collins, B. S. and Bratt, M. A.**, Separation of the messenger RNA of Newcastle disease virus by gel electrophoresis, *Proc. Natl. Acad. Sci. U.S.A.*, 70, 2544, 1973.
260. **Pridgen, C. and Kingsbury, D. W.**, Adenylate-rich sequences in Sendai virus transcripts from infected cells, *J. Virol.*, 10, 314, 1972.
261. **Kingsbury, D. W.**, Cell free translation of paramyxovirus messenger RNA, *J. Virol.*, 12, 1020, 1973.
262. **Colonno, R. J. and Stone, H. O.**, Methylation of messenger RNA of Newcastle disease virus *in vitro* by a virion associated enzyme, *Proc. Natl. Acad. Sci. U.S.A.*, 72, 2611, 1975.
263. **Portner, A. and Kingsbury, D. W.**, Complementary RNA's in paramyxovirus-infected cells, *Nature*, 228, 1196, 1970.
264. **Roux, L. and Kolakofsky, D.**, Isolation of RNA transcripts from the entire Sendai virus genome, *J. Virol.*, 16, 1426, 1975.
265. **Portner, A. and Kingsbury, D. W.**, Identification of the transcription and replicative intermediates in Sendai virus infected cells, *Virology*, 47, 711, 1972.
266. **Kingsbury, D. W.**, The molecular biology of the paramyxovirus, *Med. Microbiol. Immunol.* 160, 73, 1974.
267. **Samson, A. C. R. and Fox, C.**, Precursor protein for Newcastle disease virus, *J. Virol.*, 12, 579, 1973.
268. **Hightower, L. E. and Bratt, M. A.**, Protein synthesis in Newcastle disease virus-infected chick embryo cells, *J. Virol.*, 13, 788, 1974.

269. **Nagai, Y., Oguri, H., and Klenk, H.-D.**, Studies on the assembly of the envelope of Newcastle disease virus, *Virology*, 69, 523, 1976.
270. **Lenard, J. and Compans, R. W.**, The membrane structure of lipid-containing viruses *Biochim. Biophys. Acta*, 344, 51, 1974.
271. **Sokol, F., Neurath, A. R., and Vilcek, J.**, Formation of incomplete Sendai virus in embryonated eggs, *Acta Virol.*, 8, 59, 1964.
272. **Kingsbury, D. W., Portner, A., and Darlington, R. W.**, Properties of incomplete Sendai virions and subgenomic viral RNA's, *Virology*, 42, 857, 1970.
273. **Kingsbury, D. W. and Portner, A.**, On the genesis of incomplete Sendai virions, *Virology*, 42, 872, 1970.
274. **Portner, A. and Kingsbury, D. W.**, Homologous interference by incomplete Sendai virus particles: changes in virus-specific RNA synthesis, *J. Virol.*, 8, 388, 1971.
275. **Lamb, R. A.**, personal communication.

Chapter 5

SPECIFIC INHIBITORS OF INFLUENZA RELATED TO THE MOLECULAR BIOLOGY OF VIRUS REPLICATION

J. S. Oxford

TABLE OF CONTENTS

I. INTRODUCTION

Epidemics of influenza have not been controlled by vaccination or chemoprophylaxis, although influenza is one of the most serious virus diseases of man; in epidemic years it has caused considerable excess mortality and morbidity[1] (see Chapter 1). Particular features of the molecular biology and epidemiology (see Chapter 3) of influenza virus provide obstacles to the presently attempted methods of control which mainly involve inactivated vaccines; the problems of molecular biology will be discussed in this chapter. Hopefully, knowledge of the details of replication of the virus should lead to more logical approaches for the selection of virus-inhibitory compounds. It should be noted that until recently the search for inhibitors of influenza virus replication has been largely empirical. Following the outstanding success of a more or less random screening of fungus fermentation products and various organic chemicals in order to obtain specific inhibitors of bacteria and protozoa multiplication, it seemed possible that similar methods could be used to select inhibitors of virus replication. Unfortunately, this has proven *not* to be the case. Indeed, the most clinically used antiviral compounds, 5-iodo-2-deoxyuridine and methylisatin-3-thiosemicarbazone, were selected for antitumor and antituberculosis activity, respectively; only the latter was investigated for antiviral activity. After many years of testing many thousands of compounds of influenza virus replication, only one compound (amantadine) has been shown consistently by many laboratories to have in vitro and in vivo activity and to have some efficacy in man. This was discovered using the empirical approach. No influenza-inhibitory compound has yet been found using the more recent concepts of molecular biology, but this chapter will emphasize possible approaches. However, more laboratories involved in primary screening methods are now using a more selective procedure with influenza virus, e.g., use of virion-associated RNA polymerase enzyme and the neuraminidase enzyme, and it will be particularly interesting to see if more powerful and useful inhibitors of the virus are found. Buthala[2] has provided a useful survey of the earlier experiences in his laboratory using the empirical approach to screening (Tables 1 and 2).

The search for inhibitors of influenza virus began soon after the original isolation of human influenza virus in the Medical Research Council laboratories at Hampstead and the successful transmission of the virus to ferrets by Smith et al.[3] The subsequent use of embryonated hen's eggs and mice for the cultivation of the virus made such studies more practicable. Compounds of a wide range of chemical structures were shown early to inhibit influenza virus replication in laboratory

TABLE 1

Spectrum of Antiviral Activity of Selected Compounds

	Inhibitory activity					
Total tested	A/PR8/34	Coxsackie	*Herpes hominis*	Newcastle disease virus	Infective bronchitis virus	Vaccinia
384	8%	53%	31%	4.8%	4.4%	24%

From Buthala, D. A., *Ann. N.Y. Acad. Sci.*, 130, 17, 1965. With permission.

TABLE 2

Evaluation of Compounds for Antiviral Activity

In vitro		In vivo	
Total tested	Total active	Total tested	Total active
6405	6.0%	4230	0.33%

From Buthala, D. A., *Ann. N.Y. Acad. Sci.*, 130, 17, 1965. With permission.

studies; these compounds include general inhibitors of protein synthesis (such as *p*-phenylserine, canavanine, flavianate, and *p*-fluorophenyl alanine) and inhibitors of nucleic acid synthesis (such as thiouracil, 5-aryl pyrimidines, and proflavine) (reviewed by Hoyle;[4] see Table 3). In the summarized in vitro data, 2 of 51 purines and 23 of 165 pyrimidines were active against RNA viruses. None of 40 purines had activity in vivo, while 9 of 115 pyrimidines were active against certain RNA viruses. Most of the compounds listed failed to inhibit influenza A virus replication in mice, including such diverse molecules as substituted 6-aminopurines, cytosine and isocytosine derivation, nucleosides of adenine, ring analogs of adenine, 6-thiopurines, and 5-halouracils. Some inhibition of influenza A infection in mice was detected with 5-(2,4-dichlorophenoxy)-2-thiouracil, 6-azacytidine, and 5-allyl-5-pentadecyl-2-thiobarbituric acid.

Other substances, reviewed by Hoyle,[4] have been shown to marginally inhibit influenza virus replication in laboratory studies; these include derivatives of natural products including eucalyptus leaf extracts, streptothricins, and ecmoline from fish milk or medicinal plant extracts (Figure 1). Also of interest are the studies of Tamm and co-workers on the mode of action of benzimidazole derivatives.[10] Approximately 5 μmol of 5,6-dichloro-1-β-ribofuranosylbenzimidazole (DRB) (Figure 2) inhibited the growth of B/Lee virus in hens' eggs and mice. However, few of these early compounds showed great activity in experimental animals. In 1964 amantadine[23] was reported as an influenza inhibitor, thus providing a compound with well-defined inhibitory activity in vitro in laboratory animals and in man.

TABLE 3

Selective Review of Inhibitors of Influenza Virus Replication

Chemical	Virus	Antiviral activity: Mice	Antiviral activity: T.C. or eggs	Reference
Inhibitors of protein synthesis				
Methoxinine	A(H0N1)	ND	+	5
Canavanine	B/Lee	0	+	6
Tyrosine derivative	A(H0N1)	+	ND	7
Inhibitors of nucleic acid synthesis				
2-Mercapto-4-amino-5-(2-chlorophenyl)pyrimidime	A(H0N1)	+	ND	8
Isonicotinic acid 3,3-di-(*p*-chlorophenyl)-propylamide	A(H0N1) A(2N2) B/Lee	+	ND	9
5,6-Dichloro-1-β-D-ribofuranosyl-benzimidazole	B/Lee A(H0N1) A(H1N1)	+	+	10
1-β-D-Ribofuranosyl-1-2,4-triazole-3-carboxamide	A(H0N1) A(H2N2) A(H3N2) B/HK	+	+	11 12

TABLE 3 (continued)

Selective Review of Inhibitors of Influenza Virus Replication

Chemical	Virus	Antiviral activity: Mice	Antiviral activity: T.C. or eggs	Reference
Selenocystine	A(H0N1)	+	ND	13
Methylisatin-3-thiosemicarbazone	A(H2N2)	ND	+	14
Phenanthrolines	A(H2N2) B/Lee	ND	ND	15
1,10-Phenanthroline chelates	A(H1N1)	ND	+	16
Thiosemicarbazones and diphenyls	A(H0N1) B/Lee	+	+	17
Actinomycin D	A and B	ND	+	18
α-Amanitin	A and B	ND	+	19
Inhibitors of penetration				
α-Amino-*p*-methoxyphenylmethane sulfonic acid		+	+	20
Ammonium ions		0 ±	+ +	21 22
Amantadine derivatives	A(H2N2) A(H3N2)	+	+	23
Decalins	A(H0N1) A(H1N1)	0	+	24
Neuraminidase inhibitors				
N-substituted oxamic acid	A(H0N1)	ND	+	25
2-deoxy-2,3-dehydro-N-fluoroacetyl-neuraminic acid	A(H0N1) A(H3N2) B	+ 0	 +	 26
Dihydroisoquinolines	A and B	+	+	27
Miscellaneous with unknown point of action				
Urethane	A(H1N1)	+	+	28
1-(β-decylphenyl)-3-dimethylamino-1-propanol Cl	A(H1N1) A(H2N2)	ND	+	29
DL-Noformicin	A(H0N1) A(H2N2) B/Lee	 +	ND	30

Iodine. (*Lancet*, 1919, **1**, 423.) Baillie has used with success intravenous injections of iodine in ten bad cases with definite signs of broncho-pneumonia. The B.P. tincture of iodine was used, the dose being from 20 to 30 minims diluted with 9 c.c. of saline solution (0·85 per cent. in freshly distilled water). The drug was tolerated well, and the therapeutic effect was shown generally within 24 hours.

Mercuric Chloride. (*Riforma Medica*, 1918, **34**, 893.) Ferrarini advocates in grave cases a daily intravenous injection of 0·01 Gm. mercuric chloride dissolved in 1 c.c. of normal saline solution. This gave such good results in his experience with hundreds of cases that he has abandoned all other measures except heart tonics when needed. He has never had any untoward by-effects with these large doses.

Milk. (*L'Union Pharm.*, 1919, **60**, 9.) Domec recommends intramuscular injections of milk. Such injections act effectively against the toxaemia, and in the absence of specific serum he has found the treatment effective. The milk is carefully collected from non-tuberculous cows, is preserved in ampoules, and is sterilized. Injections are made in the gluteal region, care being taken to avoid the sciatic nerve.

Cinnamon Bark in the Treatment of Influenza. J. C. Ross. (*Brit. Med. Journ.*, March 18th, 1899, 660.) The author has obtained excellent results in treating cases of influenza with a strong decoction of cinnamon, prepared by boiling a pound of the bark *in vacuo*, with a sufficiency of water, till the fluid is reduced to a pint and a quarter. Of this a tablespoonful is given every half-hour for the first two hours, then every hour till the temperature falls to the normal, and afterwards four times a day for four days. The decoction may also be concentrated by evaporation, and made into tabloids, of which two represent half an ounce of the decoction of the strength referred to. This treatment has been particularly successful in nearly all cases in which it has been commenced within four or five hours from the onset of the attack, a rapid cure being invariably effected in such cases. The patient is to remain indoors from the commencement until at least twenty-four hours after the feverishness has subsided.

The efficacy of this treatment is fully confirmed by C. G. Grant (*Brit. Med. Journ.*, March 25th, 1899, 763).

Superiority proved by Analysis.

WINOX
WINE FOOD

is a high-class Pharmaceutical Wine, containing all the food elements of Wine, English Beef, Malt, and unfermented Grape Juice.

Complete analysis printed on every bottle.

The greatest repellent for Influenza. Invaluable for Anæmia, Neuralgia, Prostration, Nervous Break-downs, Bronchitis, Tuberculosis, and other Nervous and Bodily Weaknesses: these having been vouched for again and again by Doctors, Nurses, and all classes of Invalids.

WINOX is 40% richer than ordinary Tonic Wines in flesh forming properties, making it **the best Tonic during Convalescence.**

Chemists who do not already stock this 41% *profit proposition* should write for particulars to-day.

Just published "Facts about Winox versus ordinary Tonic Wines." Copy sent free.

WINOX LTD.
RICHMOND, SURREY.

FIGURE 1. Early attempts at chemotherapy of influenza virus. These remedies were used in the influenza pandemic of 1918. Even today investigators commonly report uncontrolled trials of new compounds and anecdotal information about efficacy.

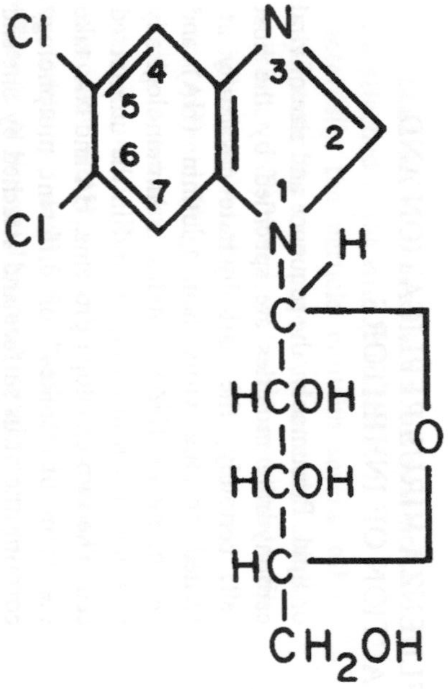

FIGURE 2. Molecular structure of 5,6-dichloro-1-β-D-ribofuranosylbenzimidazole (DRB). One of the earlier detailed studies of an inhibitor of influenza virus replication. (Redrawn from Tamm, I., Folkers, K., Shunk, C. H., and Horsfall, F. L., *J. Exp. Med.*, 99, 227, 1954. With permission.)

invaded cell. This translates the virus message into virus structural proteins and nonstructural proteins including enzymes, such as the RNA-dependent RNA polymerase enzyme, necessary for the replication of the virus RNA (Figure 5). An interesting feature of the influenza genome is that it exists in fragments of varying size, and each acts as a monocistronic message and codes for a particular virus-specified protein. Eight different species of RNA have been resolved by polyacrylamide gel electrophoresis, ranging in size from 3.4 to 9.8 × 10^5 daltons. Additional support for the existence of a segmented genome within the virion, which has very important implications, has come from nucleotide analysis of the fragments. Uridine was found at the 3′ terminus of RNA molecules in six of the eight classes; these termini would not have been found if cleavage of RNA by RNAase, for example, had occurred and resulted in production of segments.

However, recent electron microscope studies using detergent-disrupted virus and immediate processing have indicated that the RNP surrounding the virion nucleic acid appears as a double helical arrangement with an average diameter of 50 to 60 nm[31] (Figure 6). The coils are extruded from the disrupted virion as a single entity and not in small segments. It is possible that the continuous RNP may be a means of packaging discontinuous genetic segments and may aid the correct transcription of the RNA pieces in the package.

B. The Influenza Virus Replication Cycle

The single-cycle replication kinetics of influenza viruses vary considerably. In efficient and productive systems, however, the latent period is around 2 to 3 hr, and virus production reaches a plateau in 6 to 8 hr. As a first stage in the replication cycle, the influenza virion binds by its hemagglutinin projections to sialic acid residue containing receptors on the plasma membrane of the cell. The major sialoglycoprotein of the red cell membrane, called glycophorin,[32] has been isolated and purified as a homogeneous water-soluble form. Glycophorin represents approximately 10% of the total protein of the red cell ghost. It has blood group antigens and most of the membrane-bound sialic acid and is the major receptor for influenza virus. Future detailed analysis of the exact virus-binding sites could enable the synthesis of analogs capable of specifically inhibiting influenza virus adsorption. Present knowledge indicates that glycophorin is composed of a single polypeptide chain to which multiple oligosaccharides of varying composition and length are at-

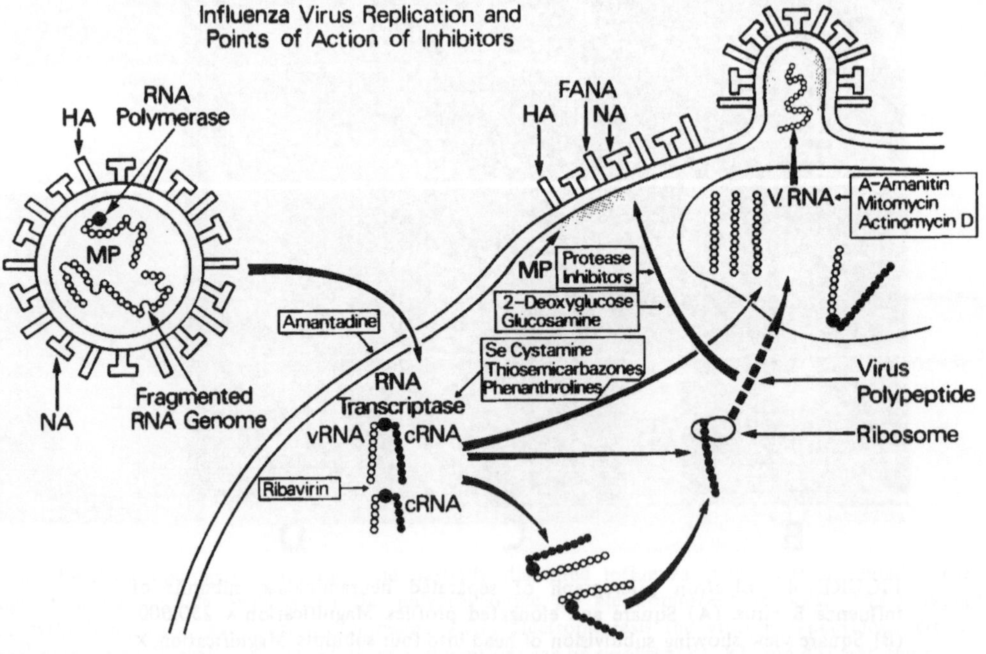

FIGURE 5. Summary of the molecular biology of the influenza virus replication cycle: (NA) neuraminidase; (HA) hemagglutinin; (cRNA) complementary, negative strand or messenger RNA; (vRNA) virion or positive strand RNA.

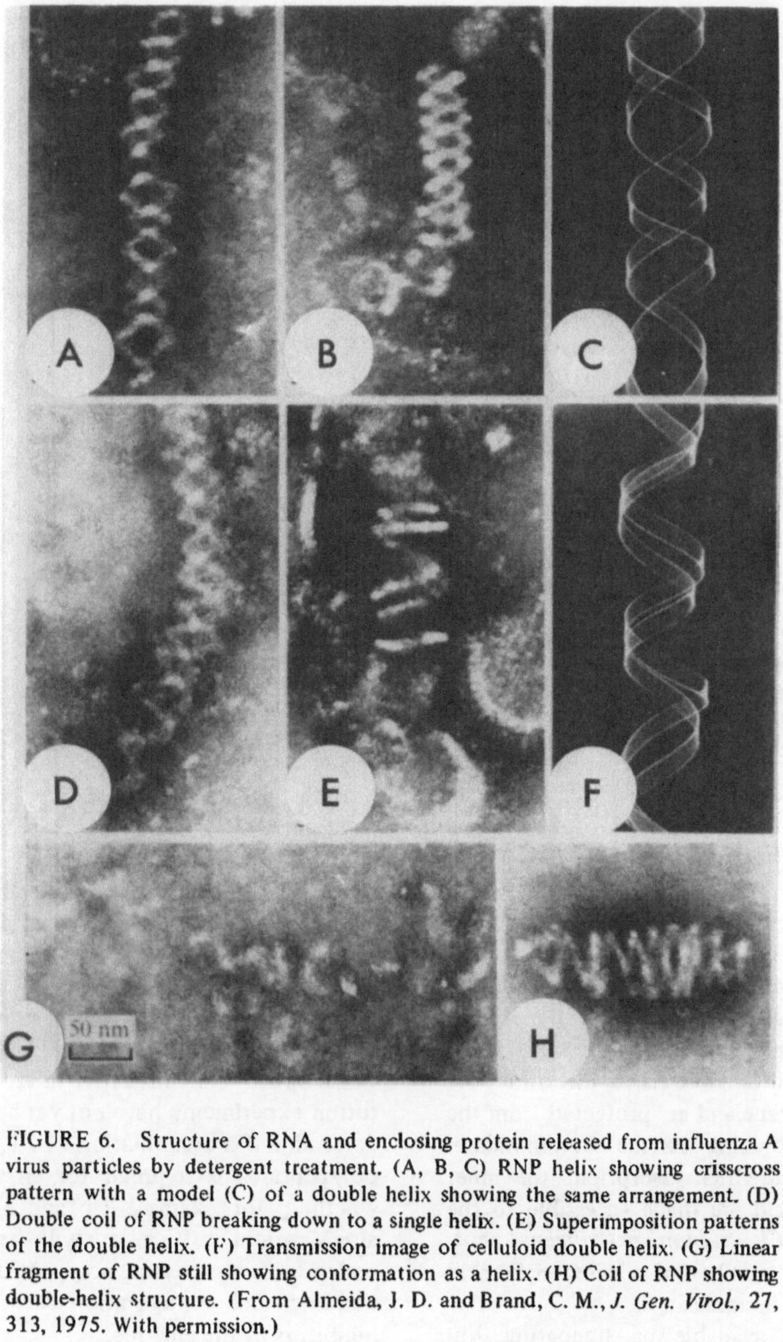

FIGURE 6. Structure of RNA and enclosing protein released from influenza A virus particles by detergent treatment. (A, B, C) RNP helix showing crisscross pattern with a model (C) of a double helix showing the same arrangement. (D) Double coil of RNP breaking down to a single helix. (E) Superimposition patterns of the double helix. (F) Transmission image of celluloid double helix. (G) Linear fragment of RNP still showing conformation as a helix. (H) Coil of RNP showing double-helix structure. (From Almeida, J. D. and Brand, C. M., *J. Gen. Virol.*, 27, 313, 1975. With permission.)

tached, mainly to the N-terminal half of the molecule. More detailed analysis of the glycoprotein is concerned with tryptic digestion studies, which have resulted in the production of at least four unique glycopeptides (Figure 7). One of the glycopeptides is partially associated with the lipid bilayer. The α-1 (16,000 daltons) and the β polypeptides (11,000 daltons) have receptor sites for influenza virus. Amino acid sequencing studies are in progress at present and should give a more comprehensive analysis of the precise virus-binding site. The virus probably enters the cell by pinocytosis.[33] In the latter study,[33] virus particles were observed in direct contact with the cell surface 10 min after inoculation; at 20 min particles were observed within cytoplasmic vacuoles (Figure 8). Particles within vacuoles often appeared ruptured, suggesting that partial uncoating might occur

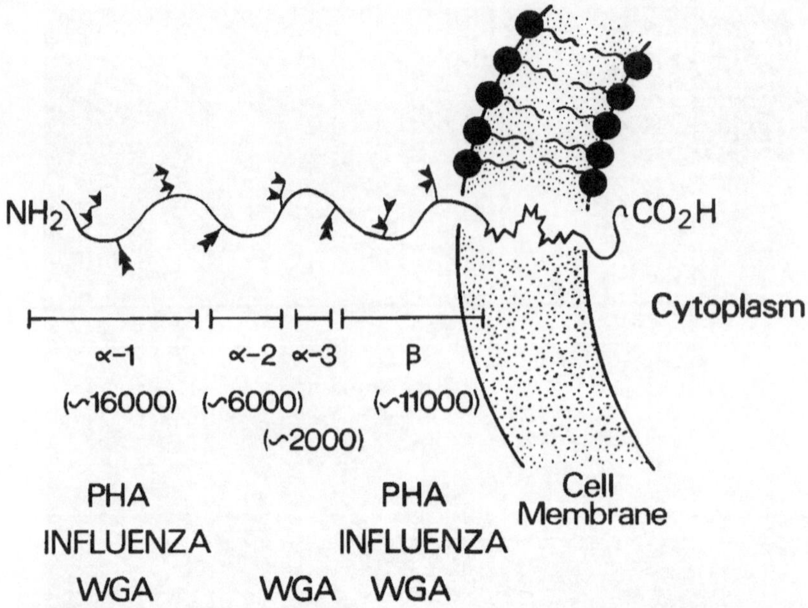

FIGURE 7. Receptor for influenza virus on the cell membrane. Glycophorin is fragmented by cyanogen bromide treatment into four polypeptides. The fragments can act as receptors for influenza virus hemagglutinin and/or wheat germ agglutinin (WGA) and phytohemagglutinin (PHA).

within the vesicle. Influenza virus adsorption may result in a spatial rearrangement of membrane components such as glycoproteins or glycolipids.[34] Following the adsorption of virus to sialic acid residues, it is reasonable to assume that some structural change in the cellular membrane or some spatial rearrangement of the membrane components, such as glycoproteins or glycolipids, may occur. In normal cells some of the sialic acid residues are exposed to the outside, but others are located intramembrane and are protected from the action of the hydrolyzing enzyme neuraminidase. Following influenza virus adsorption, the inner residues of sialic acid are made accessible to the action of added *Clostridium perfringens* neuraminidase, and the number of released sialic acid residues apparently increases.

Little data are available on uncoating, but cytoplasmic proteases and lipases presumably disrupt the virus envelope and activate the virion transcriptase enzyme. The latter enzyme is known to transcribe a copy (complementary strand RNA) of the virus RNA which is then used as a virus message[35] (Chapter 4). Product analyses of the RNA transcriptase reaction in vitro have shown that at least 95% of the product is complementary to the viral RNA, but it is not yet known if all of the eight viral RNA segments are copied.[36] It is possible, of course, that the in vitro reaction mixture lacks a factor which is important for in vivo transcription. Separation of detergent-disrupted virus by polyethylene glycol-dextran leads to a dextran phase containing all of the virion-associated transcriptase activity together with all of the viral RNA-NP P1 and P2 proteins. Further resolution of these subviral complexes into RNA-NP structures only results in loss of transcriptase enzyme activity. However, reconstitution experiments have not yet been published to determine if a combination of P1, P2, P3, and NP polypeptides is required for enzyme activity or whether, for example P1 and P2 polypeptides alone would suffice. It will be important to understand the exact polypeptide composition, bonding, and configuration for the synthesis of inhibitors of the enzyme.

It is still possible that some virion RNA species are complementary to their mRNA and that others are their own mRNA; transcription may be selective for those RNA species which are complementary to the mRNA. At present it is not certain where primary transcription (i.e., transcription not inhibited by cycloheximide or puromycin and carried out using infecting virus RNA as template) or secondary transcription (i.e., transcription from progeny viral-like RNA templates) occurs in the

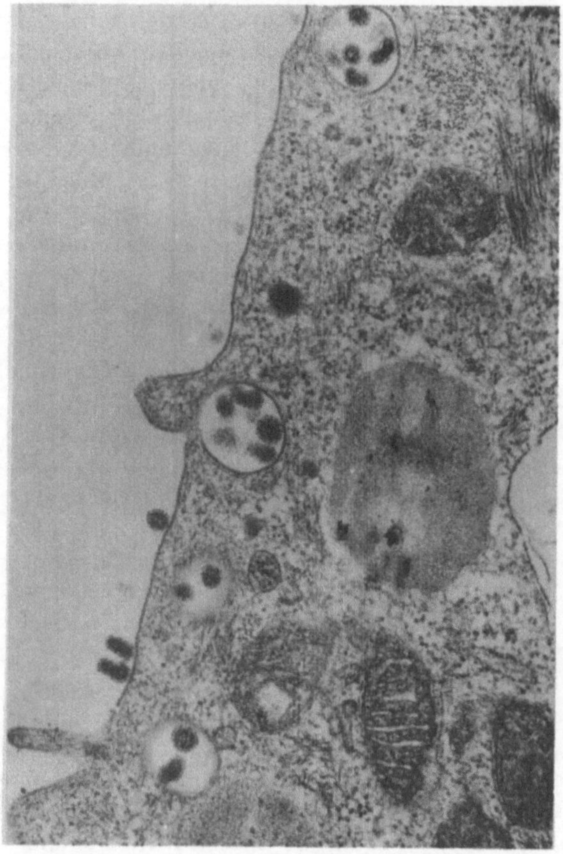

FIGURE 8. Electron micrograph of the early events of influenza virus adsorption and penetration to cells. Influenza virus particles are seen at the cell surface and enclosed in cytoplasmic vacuoles. Magnification × 52,000. The chorioallantoic membrane cells have been incubated for 5 min at 37°C after infection. (Courtesy of Dr. R. Dourmashkin, Clinical Research Centre, Harrow, England.)

cell. Certainly secondary transcription starts before 1-hr postinfection, which means that the earliest rounds of replication also start before 1-hr postinfection. Studies are in progress in several laboratories to determine if primary and secondary transcription occurs in the cytoplasm or nucleus and to determine the role of the cell nucleus in the process. If specific inhibitors of these processes are to be developed, it is of obvious importance to determine the exact location of enzyme activity in the virus-infected cell. Certain inhibitory compounds may more easily reach the cell cytoplasm than the cell nucleus for example. This could have great selectivity if the cell polymerase enzymes are restricted mainly to the nucleus. Obviously, specific inhibition of the transcriptase enzyme would act at a theoretically important stage in the life cycle of the virus, and this will be considered in greater detail below. At some stage the complementary RNA also enters the nucleus to become the template for the synthesis of virion RNA. It is not clear if a further "replicase" enzyme is required or whether the existing transcriptase can be modified. Again, inhibition of this stage would presumably halt the replication of the virus in the cell. Chemicals binding with the enzyme itself or the enzyme and template together, rather than the virus RNA alone, would appear to be more useful because of the previously demonstrated lack of specificity of antibiotics binding to nucleic acid. Detailed knowledge of the binding or initiation site of the polymerase to the RNA template would be of great interest.

Studies with bacteriophage have established the chemical basis for the attachment of mRNA to ribosomes. The mRNA attaches to a specific site on the ribosome, and specific protein factors are required in the binding process. As to the messenger binding site, divalent cations are required, and approximately 30 nucleotides of mRNA are involved. Ribosome-associated mRNA discriminating factors have been found in *Escherichia coli* which act at the level of initiation by affecting recognition by ribosomes of individual mRNA initiation sites. These are termed IF-3 interference factors.[37] Nudel et al.[38] (1973) have described similar factors in various mammalian cells. A factor was purified from reticulocyte ribosomes which discriminated between mRNAs of reticulocyte 9S hemoglobin and Mengo virus mRNA stimulating the translation of the former message. Also, Wigle and Smith[39] purified an initiation factor from Krebs cell supernatant for EMC virus messenger RNA which was inactive with hemoglobin mRNA. Thus, although the discrimination by initiation factors is poorly understood, the data suggest that these proteins confer to the ribosomes a higher affinity for several templates. A given factor is not specific for a particular mRNA but promotes translation of a class of mRNAs. The possibility of altering the affinity of ribosomes for influenza virus mRNA would have great potential for the development of virus-inhibitory compounds.

If a particular sequence of nucleotides or a particular RNA configuration were required, it might be possible to synthesize competitive inhibitors of the enzyme binding site which would thus block virus transcription. Such an approach is being investigated at present using EMC virus where the primary nucleotide sequence of the virus RNA is known.[95] A reservation of this approach would be immediately apparent if an RNA-dependent RNA polymerase were to be found in normal uninfected cells, since such a cell enzyme might be inhibited by chemicals blocking the influenza virus enzyme. Indeed, an RNA-dependent RNA polymerase has been isolated from rabbit reticulocytes,[40] but this remains to be confirmed in other laboratories. The enzyme preferentially used mRNA as a template for RNA synthesis, and the importance of such an enzyme is that it would provide a new site for the control of gene expression in the cell and a new pathway for synthesis of RNA without concomitant DNA transcription. An additional complication is that the replicase of phage QB has several subunits and the host bacterium contributes to these.[41] Studies with purified QB replicase have indicated that it contains four different polypeptide chains, only one of which is QB specific – the others preexist in the host cell. The replicase enzyme consists of four polypeptides (α, β, γ, and δ) with molecular weights of 74, 69, 47, and 33 $\times$ 10^3 daltons, respectively. A similar cell dependence of the influenza RNA polymerase would lessen the chance of developing virus enzyme-specific inhibitors. Recent evidence shows that no influenza viral complementary RNA can be detected in enucleated BHK cells,[42] possibly implying that primary transcription requires a nuclear-derived function. Also, there is accumulating evidence indicating that following influenza infection there is a stimulation of cellular DNA-dependent RNA polymerase activity, again suggesting a "turn on" of particular host function. Inhibition of the host cell DNA-dependent RNA polymerase by α-amanitin results in concomitant inhibition of influenza virus replication. This is all circumstantial evidence that a host factor may be required in influenza virus RNA transcription.

An interesting point of action of an inhibitor would be to stop the binding of virus messenger RNA to the cell ribosomes, preventing translation of virus message but allowing translation of cell mRNAs. It was speculated earlier that the virus-induced nonstructural polypeptide NS 1 (or polypeptide 8) might act as a ribosome initiator factor, thus ensuring recognition and translation of virus messenger RNA. Blockage of such a factor would inhibit virus replication. However, although the polypeptide associates with ribosomes, the attachment is loose and the function of NS 1 is at present unknown.[43] Of course, interferon may act by inhibiting viral mRNA attachment to ribosomes (Chapter 7, Volume II); in addition, chemical agents have been found by screening organic compounds for their capacity to inhibit protein synthesis in cell-free systems that use bacteriophage RNA as mRNA. The effects of inhibitors on protein synthesis directed by endogenous mRNA can be compared with any effect directed by phage mRNA. Aurintricarboxylic acid (ATA) and related triphenylmethane dyes have been shown to be inhibitors of initiation of protein synthesis. The compound binds to the 30S subunit of the bacterial ribosome and prevents attachment of phage RNA. However, an immediate problem with this type of molecule is lack of ready penetration of cell membranes, although this may be overcome by acetylation of the polar groups or by the use of liposome carriers (Chapter 8, Volume II). ATA has recently been shown to inhibit influenza RNA polymerase enzyme.[96]

If virus-specific tRNAs containing, for example, unusual bases are required at the stage of translation, this would be another point of action for inhibitors. Inhibitors of virus replication control factors, such as the postulated "equestron" of polio virus,[44] would be expected to interfere specifically with virus multiplication. Cooper proposed the equestron regulator because such a multifunctional complex with affinities for 45S ribosomal subunits and 5′ ends of viral RNA would explain the suppression of host protein synthesis, synchromous synthesis of viral RNA and protein, and the timing of maturation (see Chapter 4, Volume II).

C. Transport of Virus-induced Proteins Intracellularly

An important and vital problem for the successful completion of the influenza virus life cycle is the transport of newly synthesized virus polypeptides through the cell to the plasma membrane, where virus eventually buds off. Using recently developed cell fractionation techniques (which enable separation of plasma membranes, nuclei, ribosomes, etc. with little contamination of other cell components) together with pulse labeling with radioactively labeled amino acids, a number of important observations have been made. It will be of particular interest to see whether the accumulated detailed results can be used to select or synthesize inhibitors of the process.

All virion polypeptides were readily detected in extracts of CEF cells infected with A/fowl plague/27 virus and labeled with S^{35}-methionine.[45] Experiments to define the temporal sequence of polypeptide synthesis indicated that prior to 3-hr postinfection at 31°C the polypeptides P2, NP, and 8 (a nonstructural polypeptide with a molecular weight of approximately 23,000 daltons) are synthesized (Figure 9). The results from experiments using cycloheximide to inhibit protein synthesis during the early stages of infection suggest that selective transcription of three of the segments of the virus genome occurred, resulting in the synthesis of the above three polypeptides. Subsequently, in the later stages of virus replication, these three polypeptides continued to be produced together with the other five virus-coded polypeptides. In subsequent experiments virus-infected cells were fractionated, and the different cellular components thus isolated were examined for the presence of virus-coded polypeptides.

First it was established that the distribution of the various polypeptides between the subcellular components was similar at all times throughout the replicative cycle (Figure 10). The nuclei contained polypeptides NP, MP, and 8. Pulse chase experiments showed that these three polypeptides were in fact synthesized in the cytoplasm and only afterwards entered the nucleus.

The virus-specific glycopolypeptides (neuraminidase and hemagglutinin) together with MP were exclusively associated with membrane fractions of the cell. Proteolytic cleavage of polypeptide 3 (uncleaved HA) into HA1 and HA2 occurred after it was inserted into the plasma membrane. Polypeptide 3 was initially incorporated into the rough endoplasmic reticulum and was subsequently transferred to the plasma membrane via the smooth endoplasmic reticulum.

In summary, the above data suggest that the synthesis of the ten influenza virus-specific polypeptides in CEF cells is under strict control with respect to the amount of each polypeptide synthesized and the time of their synthesis during infection. Thus, polypeptides P1 and P2 are synthesized in much smaller amounts than the NP polypeptide at all times during infection. Little data are presently available about P3. The predominant localization of polypeptide 8 in the polysome fraction of infected cells suggests that this polypeptide might be involved in translation or transcription of virus mRNA. Obviously, at this stage the functions of polypeptides 8 and 9 must be speculative.

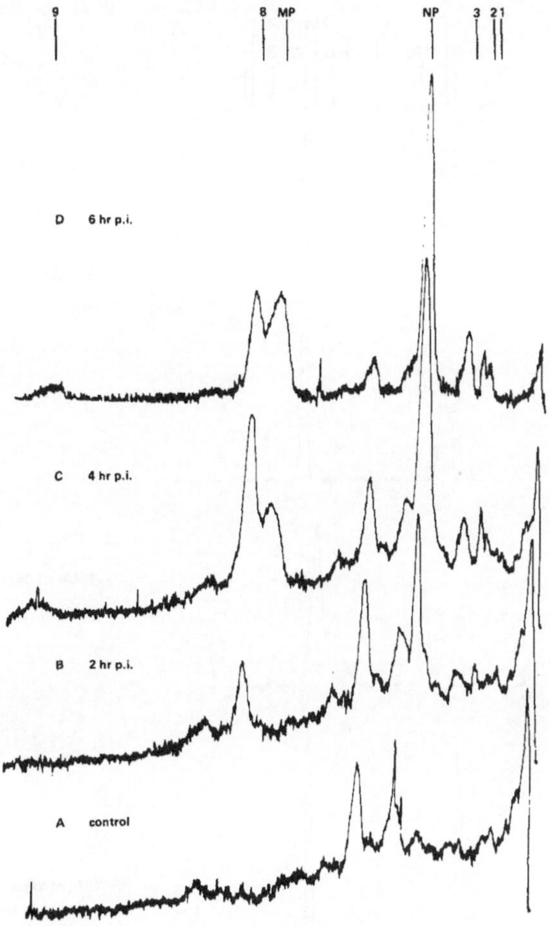

FIGURE 9. Densitometer tracing of autoradiograph showing production of influenza virus polypeptides in infected cells (SDS-PAGE). CEF cells were infected with influenza A/FPV and pulse labeled with S^{35}-methionine at varying times postinfection. SDS-lysed cell extracts are boiled for 2 min in the presence of β-mercaptoethanol, and the polypeptides are examined by polyacrylamide gel electrophoresis. As early as 2-hr postinfection virus-coded polypeptides 8 and NP are detectable. At 4-hr postinfection all virus-coded polypeptides are detected. (1 and 2) The migration position of the virus structural polypeptides P1 and P2. Virus-induced nonstructural polypeptides (8 and 9) is indicated. P3 polypeptide has not been resolved on these gels. (From Hay, A. J. and Skehel, J. J., in *Negative Strand Viruses*, Vol. 2, Mahy, B. W. J. and Barry, R. D., Eds., Academic Press, London, 1975, 635. With permission.

Influenza virus is released by budding from the plasma membrane of the cell. It is conceivable that compounds could interrupt and inhibit the complex process. Electron microscopy[46] has

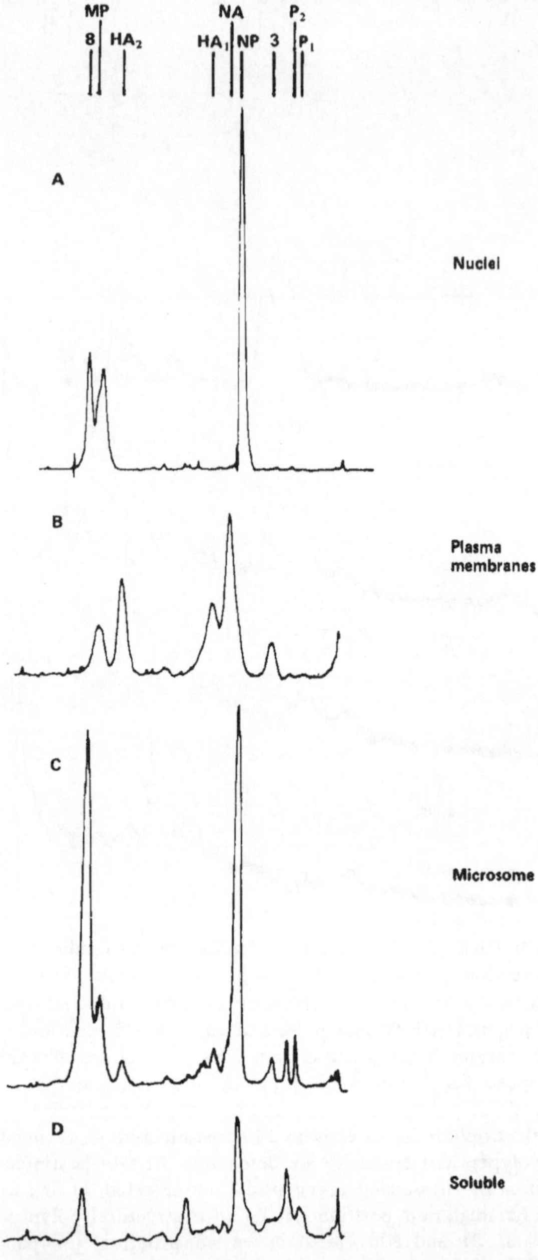

FIGURE 10. Association of influenza virus-induced polypeptides with different fractions of the infected cell. Infection of cells and separation of virus induced to polypeptides are described in Figure 9. (From Hay, A. J. and Skehel, J. J., in *Negative Strand Viruses,* Vol. 2, Mahy, B. W. J. and Barry, R. D., Eds., Academic Press, London, 1975, 635. With permission.)

provided some evidence that viral envelope proteins are incorporated first into areas of cell plasma membrane. Presumably the matrix (M) protein then associates with the inner surfaces of such regions (Figure 11). The HA and NA spikes may even penetrate through the lipid bilayer. The RNP containing RNA and P1, P2, and P3 polypeptides then binds to the membrane in the M protein-containing region, and budding takes place by an out-folding and pinching off of a segment of the membrane enclosing the RNP. The RNPs may be incorporated randomly into virions from an intracellular pool. Of course, various mechanisms are possible for the formation of such patches of membrane, and these should be considered if one is interested in inhibiting the process. Mechanisms (Figure 12) include insertion of soluble polypeptides directly into the plasma membrane or fusion with the cell surface membrane of preformed cytoplasmic membranes containing viral polypeptides.

It is known that influenza virus release is inhibited by antibody directed against the neuraminidase of the virus, and antibody molecules may act stereochemically by cross-linking neuraminidase projections exposed on the outside of the plasma membrane.[47] Nevertheless, the precise function of neuraminidase in the replication of influenza is at present unclear, although it may be concerned with virus release from cells. Cells infected with mutants defective in neuraminidase produce virus particles containing neuraminic acid; these form large aggregates, reducing the yield of released virus[26] (Chapter 6). The viral neuraminidase prevents such aggregation. Synthetic analogs of neuraminic acid (such as 2-deoxy-2,3-dehydro-N-trifluoroacetyl neuraminic acid, FANA), acting by competition with the substrate, have been synthesized and reduce the yield of influenza virus in a multicycle system of replication.[26] Also, compounds which inhibit influenza neuraminidase enzyme in vitro have been investigated for their potential as influenza virus-inhibitory compounds in laboratory animals and man. Earlier studies with a series of isoquinoline compounds showed 1-phenoxymethyl-3,4-dihydroisoquinoline to be a weak, noncompetitive inhibitor of bacterial and viral neuraminidases.[27] However, it became apparent early in these studies that the neuraminidase-inhibiting activity of the compounds was not the sole explanation of their virus-inhibitory effect because the compounds were effective against other viruses (such as rubella) which do not possess neuraminidase. In addition, and more seriously, further studies indicated that these compounds did not in fact inhibit neuraminidase enzyme but only blocked

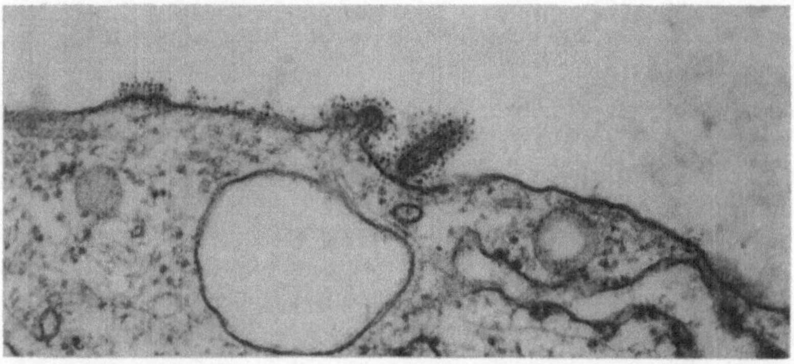

FIGURE 11. Release of influenza virus from infected cells by budding. Budding influenza A/FPV particles are labeled by the indirect immunoferritin-labeling technique. Antiserum in the sandwich is monospecific for A/FPV hemagglutinin. Some ferritin is also attached to an area of the plasma membrane with no visible evidence of virus structures. (Courtesy of Dr. Th. Bachi, Department of Medical Microbiology, University of Zurich.)

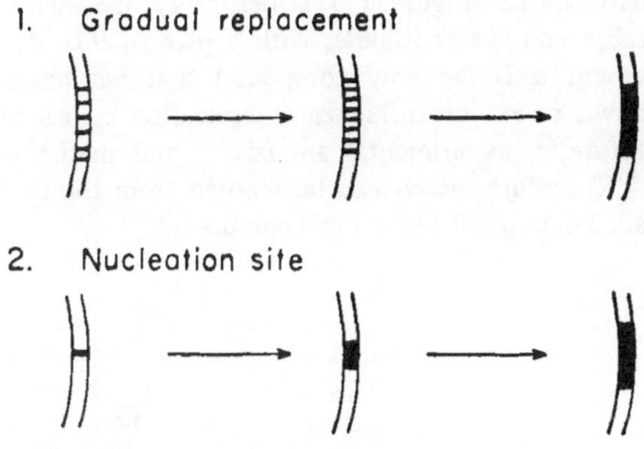

FIGURE 12. Alternative theories of the method of assembly of patches of influenza virus envelope proteins. (Courtesy of Dr. R. W. Compans.)

the color development of N-acetylneuraminic acid.[48]

III. ANTIGENIC DRIFT IN INFLUENZA AND REQUIREMENTS FOR AN IDEAL ANTI-INFLUENZA COMPOUND

Of great practical importance from the point of view of control of influenza by immunization or chemoprophylaxis is the fragmented genome of influenza virus, which results in a high frequency of recombination or reassortment between different influenza virus particles. Recombination may explain the periodic appearance of completely new influenza virus subtypes, some able to cause pandemics such as the 1918, 1948, 1957, and 1968 viruses (antigenic "shift"). In addition, influenza viruses undergo more gradual changes in the antigenicity of their external proteins, hemagglutinin and neuraminidase (antigenic "drift"). Control of influenza by vaccination is made extremely difficult by the changing HA and NA antigens (Chapter 3). At present, a further important consideration is the possible spread of the virus A/New Jersey/76 ($H_{swine}1N1$), which is almost identical to the virus which caused the influenza pandemic in 1918. Although the U.S. is producing large quantities of inactivated vaccine to immunize a considerable section of the community, much of this effort will be wasted if the expected epidemic does not materialize. Presumably, then, in future years individuals will have to be revaccinated because the protective effect, which in any case would not be expected to be greater than 60 to 70% (and may be considerably less), would not last for more than 1 year. An alternative plan would be to stockpile vaccine until the epidemiological situation becomes clearer and to place some reliance on chemoprophylaxis with amantadine (which inhibits the virus in vitro) to protect susceptible persons from the first wave of any epidemic while they are being immunized. The in vitro inhibition of A/New Jersey/76 virus by amantadine is discussed further below. Therefore, studies of possible virus-inhibitory compounds as an alternative approach to immunization are important. It should be realized, however, that the same property of mutability, and possibly recombination, of influenza virus which makes control by vaccination so difficult may present problems in the use of chemoprophylactic agents. Thus, it might be expected that drug-resistant virus mutants might occur and be selected; this has been demonstrated under laboratory conditions with amantadine and other viruses and inhibitory agents.

The most effective inhibitor of influenza virus would be independent for its mode of action on the rapidly changing surface HA and NA antigens of the virus; would inhibit both type A and B viruses; and would allow development, but without overt clinical signs, of influenza. Protective antibody would thus develop, preventing later reinfection with the same subtype. At present no single inhibitory compound with all these properties has been described in the literature. Figure 5 illustrates the possible points of action of influenza-inhibitory compounds, and it is also apparent from the figure that few inhibitors have presently been selected and investigated in detail.[52]

The remainder of this chapter will discuss mainly three groups of compounds.

1. Amantadine, is a symmetrical C-10 primary amine (Figure 13), structure 1), thermally stable and water soluble, with a pKa of 9.0. At present it is the only compound that has been shown to exhibit influenza A replication in tissue culture,[23] experimental animals,[53] and man.[54,55] Therefore, much can be learned from the detailed data published on this compound.

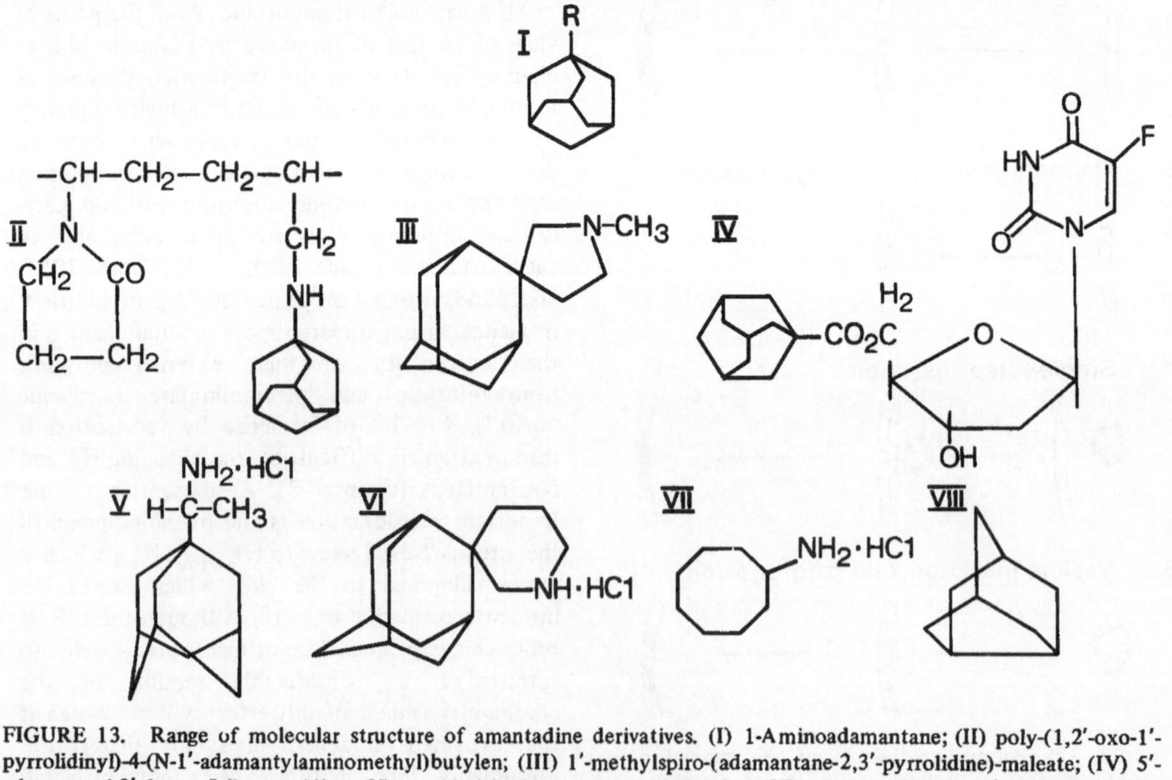

FIGURE 13. Range of molecular structure of amantadine derivatives. (I) 1-Aminoadamantane; (II) poly-(1,2′-oxo-1′-pyrrolidinyl)-4-(N-1′-adamantylaminomethyl)butylen; (III) 1′-methylspiro-(adamantane-2,3′-pyrrolidine)-maleate; (IV) 5′-adamantoyl-2′-deoxy-5-fluoruoridine; (V) α-methyl-1-adamantane methylamine; (VI) adamantane spiro-3′-piperidine; (VII) cyclooctylamine; (VIII) noradamantane.

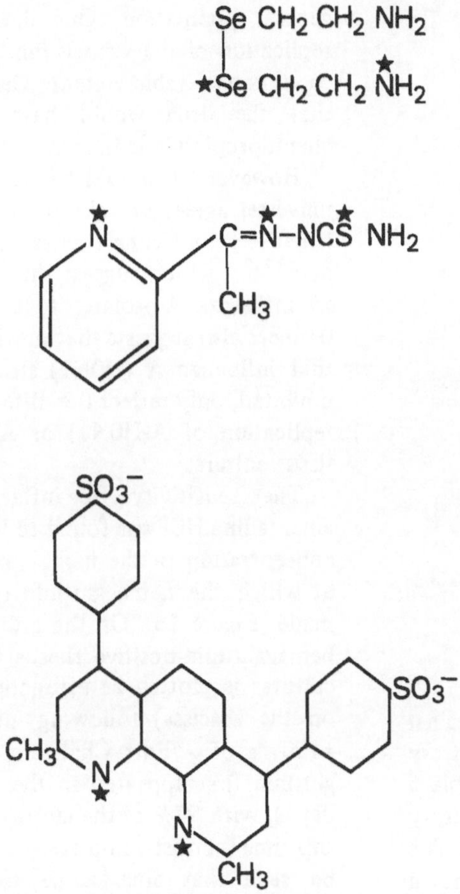

Selenocystamine

2–Acetyl Pyridine Thiosemicarbazone

Bathocuproine

FIGURE 14. Inhibitors of the virion-associated RNA-dependent RNA polymerase of influenza virus. (*) Illustrates the atoms active in metal chelation.

2. Certain compounds which act at a theoretically important stage on the RNA-dependent RNA polymerase of the virus, as described above, are currently under study. These compounds inhibit influenza A and B viruses[13,15] (Figure 14 and Table 4). Although the latter compounds have only been used as experimental models and have not been shown to inhibit influenza virus multiplication in laboratory animals, they do illustrate a theoretically good approach to the problem of selecting inhibitors of the virus. We suggest that the transcriptase enzyme assay be used in screening tests.

3. Ribavarin[11,12] (Figure 15), which has recently been shown to inhibit a wide range of influenza A and B viruses in tissue culture and in laboratory animals, acts at a most interesting stage of the nucleotide pathway. The compound is undergoing trials in volunteers at the present time but has toxicity problems, including teratogenic effects and immunosuppressive activity in animals.

TABLE 4

Inhibitors of Influenza A/PR8/34(H0N1) Virus-induced Polymerase

Compound	μμmol of GTP-C^{14} incorporated, 15 min/mg protein	% inhibition
None	314	–
Gliotoxin	87	72
Dehydrogliotoxin	100	63
Chetomin	130	59
Aranotin	103	64
Acetylaranotin	88	72

From Ho, P. P. K. and Walters, C. P., *Ann. N.Y. Acad. Sci.*, 173, 438, 1971. With permission

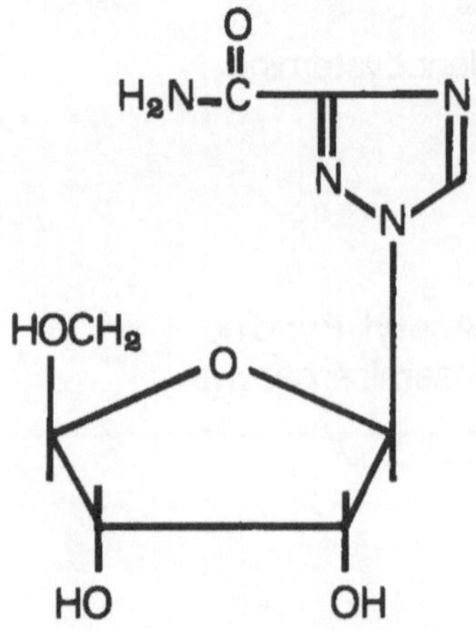

FIGURE 15. Molecular structure of ribavirin (RTCA), the nucleoside 1-β-D-ribofuranosyl-1,2,4-triazole-3-carboxamide.

IV. ANTIVIRAL ACTIVITY OF AMANTADINE

Amantadine is particularly active against influenza A viruses but has little or no inhibitory activity against influenza B or C viruses. Table 5 shows the degree of influenza virus inhibitory activity noted in tissue culture experiments. The first point of interest is the great variability shown when different influenza A strains are tested under similar conditions. Thus, influenza A/PR8 (H0N1), an old subtype of influenza A, is only marginally inhibited (0.5 $\log_{10}$ reduction in virus titer), whereas A/Scotland/49/57 (H2N2), an "Asian" influenza virus, is very significantly inhibited by the same concentration of amantadine. Fortunately, viruses having HA and NA of the H2N2 and H3N2 eras (that is, recent and current influenza viruses) are very susceptible to inhibition by amantadine in tissue culture. Because of the frequency of antigenic "drift" noted earlier, it is of great importance to investigate if the current influenza variants of epidemiological importance are inhibited. Our studies (Table 5) have indicated that all recently tested variants of influenza A are inhibited by amantadine. As expected, recent variants of influenza B virus are not inhibited and thus resemble the early B variants which were also resistant to inhibition. Of great potential interest are the data in the table which show the inhibition of A/New Jersey/76 (H_{swine}1N1) virus by amantadine. This virus, as discussed above, may have epidemiological impact if it spreads in the general population. Our data suggest that the replication of the virus is inhibited by amantadine to a considerable extent. One would anticipate that the drug would have some effect as a chemoprophylactic in man.

However, it should be noted that there is no universal agreement about subtype differences in inhibition. Grunert has recently presented data[97,98] which suggest that amantadine inhibits *all* influenza A isolates to a comparable degree. Grunert also suggests that previous data, indicating that influenza A (H0N1) strains were much less inhibited, only reflect the differences in the rate of replication of A(H0N1) or A (H3N2) viruses in tissue culture.

The sensitivity of influenza A viruses to amantadine HCl was found to be dependent on the concentration of the inoculum virus and the time at which the hemagglutinin determinations were made (Figure 16). On the ordinate the percent of hemagglutinin-positive rhesus monkey kidney cell cultures is plotted as a function of time (in days on the abscissa) following infection with $10^{2.7}$ EID_{50}'s of A/Port Chalmers/1/73 virus. Hemagglutinin first appeared in the control cultures on day 2, with 97% of the cultures being infected. At this time (almost complete virus production) it can be seen that amantadine HCl was inhibitory. Continued incubation, however, resulted in an overwhelming of the protective effect. The sensitivity of the virus can be expressed by the ED_{50} values at the bottom of the figure. The Port Chalmers strain was sensitive at 15 μg/ml, but only at the time of near maximum virus production.

With a tenfold higher concentration of inoculum virus, hemagglutinin production by the control cultures was already complete when first detected on day 2. There was no protective effect for amantadine HCl (Figure 17).

Following infection with $10^{4.7}$ EID_{50}'s, near maximum hemagglutinin production was evident 1 day sooner – on day 1. As shown in Figure 16, amantadine HCl proved to be inhibitory on day 1 with an ED_{50} of 7 μg/ml. Once again, the protective effect was overwhelmed by the next day. The importance of adjusting the conditions so that a strain's sensitivity to amantadine HCl can be determined at the point of near maximum virus production is suggested from these results. The data are summarized in Figure 17. The ED_{50}'s from a number of experiments are plotted on the ordinate as a function of the monkey kidney cell

TABLE 5

Inhibition of Human and Animal Influenza A Viruses by Amantadine

Virus		Log_{10} reduction in virus titer
Human		
A/PR8/34	(H0N1)	0.5
A/NWS	(H0N1)	1.3
A/FMI/47	(H1N1)	1.3
A/Singapore/57	(H2N2)	2.4
A/Tokyo/62	(H2N2)	4.0
A/HK/1/68	(H3N2)[a]	2.5 ± 0.5
A/Port Chalmers/73	(H3N2)[a]	2.3
A/Hannover/1/74	(H3N2)[a]	4.0
A/Puerto Rico/74	(H3N2)[a]	2.2
A/Victoria/3/75	(H3N2)[a]	3.5
A/England/75	(H3N2)[a]	4.0
Animal		
A/Mannitoba/67	(H_{swine}1N1)[a]	3.0
A/swine/30	(H_{swine}1N1)[a]	3.0
A/equine/Miami/63		3.0
A/New Jersey/8/76	(H_{swine}1N1)[a]	2.8

Note: Monkey kidney cells were infected with virus in the presence and absence of 25 μg/ml amantadine and virus growth estimated by hemadsorption after 72-hr incubation. Reduction in virus titer of 1.0 log_{10} or greater is significant.

[a]25 μg/ml amantadine in the egg piece system.[94]

$TCID_{50}$'s calculated at the particular time when each effective dose was determined. It can be seen that the Port Chalmers strain was consistently sensitive to amantadine HCl when the level of infection ranged from 2 to 10 $TCID_{50}$'s. At higher levels of infection there was little if any response to treatment. Furthermore, within the 2 to 10 $TCID_{50}$ range of infection there was a wide range of ED_{50v} values, from <1 to 18 μg/ml. Statistical analysis of these data indicates that this is due to a highly significant correlation between the two parameters.

The data developed with the seven other influenza A viruses were evaluated by this same procedure. In all cases the viruses were found to be consistently sensitive to amantadine HCl when the level of infection at the time of ED_{50} determination was in the 2 to 10 $TCID_{50}$ range, thus allowing for a direct comparison of relative sensitivities.

A. The Mode of Action of Amantadine

This has not been established definitely, partly because of the few details available on the early events in the replication of influenza virus in infected cells. Strains of influenza virus resistant to the inhibitory effects of amantadine, selected as described below, should prove useful for the elucidation of the precise point of action of the compound. Certainly amantadine acts early in the replication of influenza virus, since addition of the compound to cells later than 30 min after infection with a high multiplicity of a sensitive virus does not result in detectable virus inhibition.[56,57]

Experiments have been conducted showing that amantadine has no direct effect on free virus particles in suspension and has no effect on the absorption of influenza virus to susceptible cells.[56] Furthermore, we have recently demonstrated that amantadine does not inhibit the virion-associated RNA-dependent RNA polymerase activity (Table 6) and, by implication, the cell-associated RNA-dependent RNA polymerase which functions early in influenza virus-infected cells. In the control reaction mixture (Table 6) 97 × 10^3 dpm of H^3-UMP is incorporated into RNA

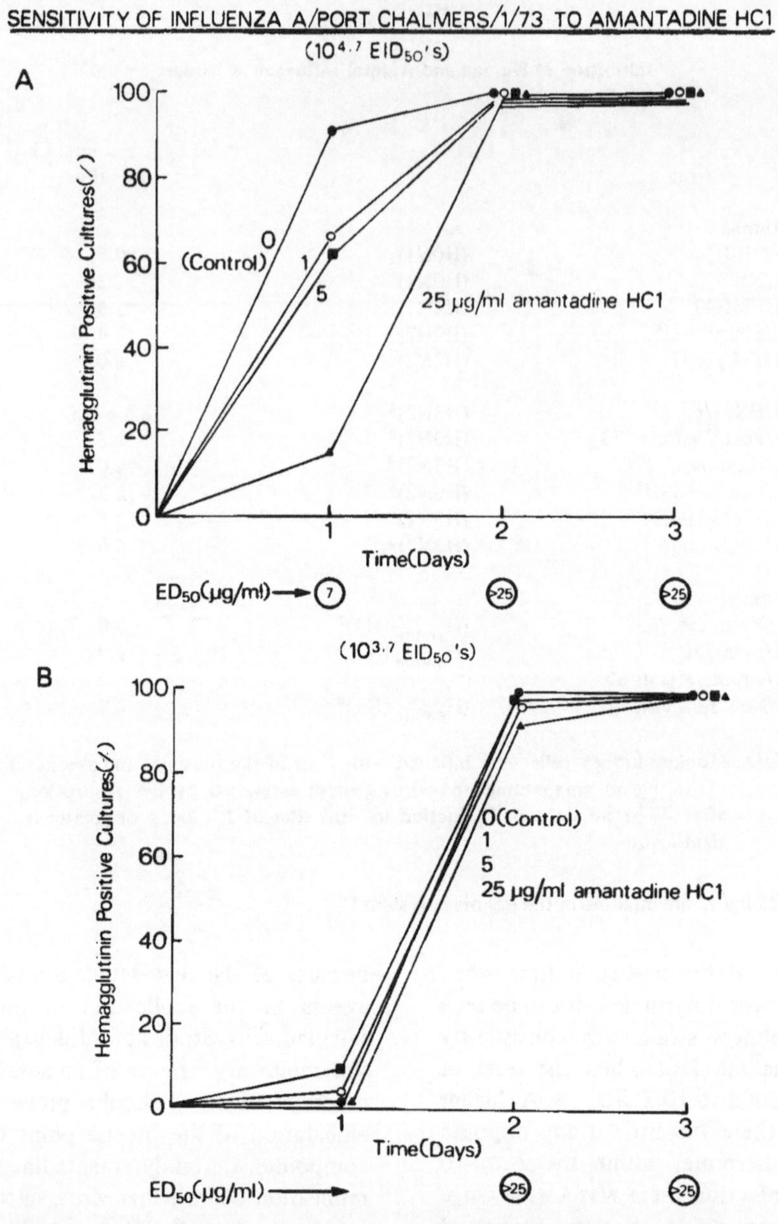

FIGURE 16. Assessment of inhibitory effect of amantadine on influenza A viruses. (Courtesy of Dr. R. R. Grunert.)

by the virion-associated RNA-dependent RNA polymerase enzyme, and this enzyme activity is completely inhibited by selenocystamine, a known inhibitor of the enzyme.[13]

In contrast, concentrations of amantadine 1000-fold greater than those which inhibit the growth of the same virus in tissue culture have no effect on the virus-associated RNA-dependent RNA polymerase activity. Amantadine may, however, inhibit the uncoating of penetrated virus particles.[58] A recent electron microscope study demonstrated that amantadine had no effect on viropexis and also suggested an effect on uncoating[33] (Figure 18).

B. Derivatives of 1-Aminoadamantane

A large number of derivatives of amantadine (Structure I, Figure 13) have been produced and tested (some are illustrated in Figure 13). At present, with the exception of Structure III,[59] there is little evidence that these are significantly more effective than 1-aminoadamantane itself.

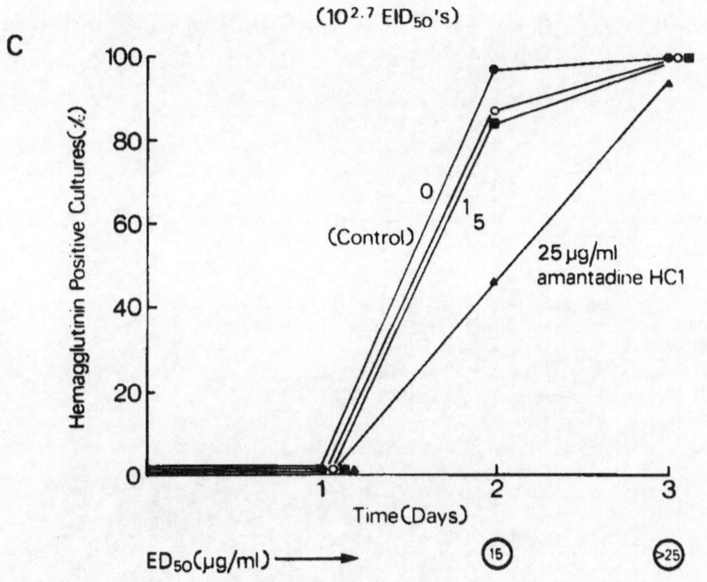

FIGURE 16C.

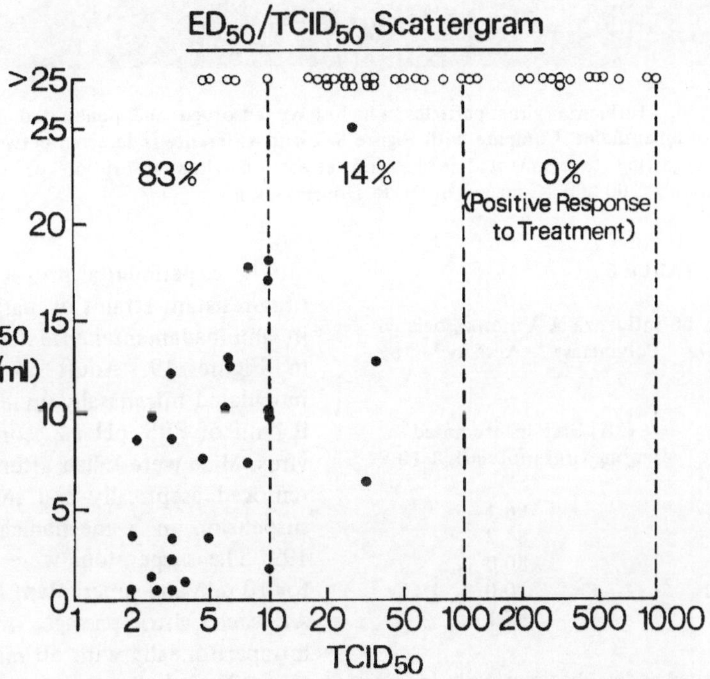

FIGURE 17. Summarized data of inhibitory effects of amantadine. Virus is inhibited by amantadine when 2 to 10 $TCID_{50}$ of virus is used as challenge dose. (Courtesy of Dr. R. R. Grunert.)

Any derivative which inhibited influenza B viruses as well as influenza A virus would be a definite advancement. Several such compounds have been reported,[60] but these must await confirmation from other laboratories.

C. Development of Amantadine-resistant Influenza Viruses

Features of the molecular biology of replication of influenza virus (discussed at the beginning of the paper), particularly the high rate of genetic

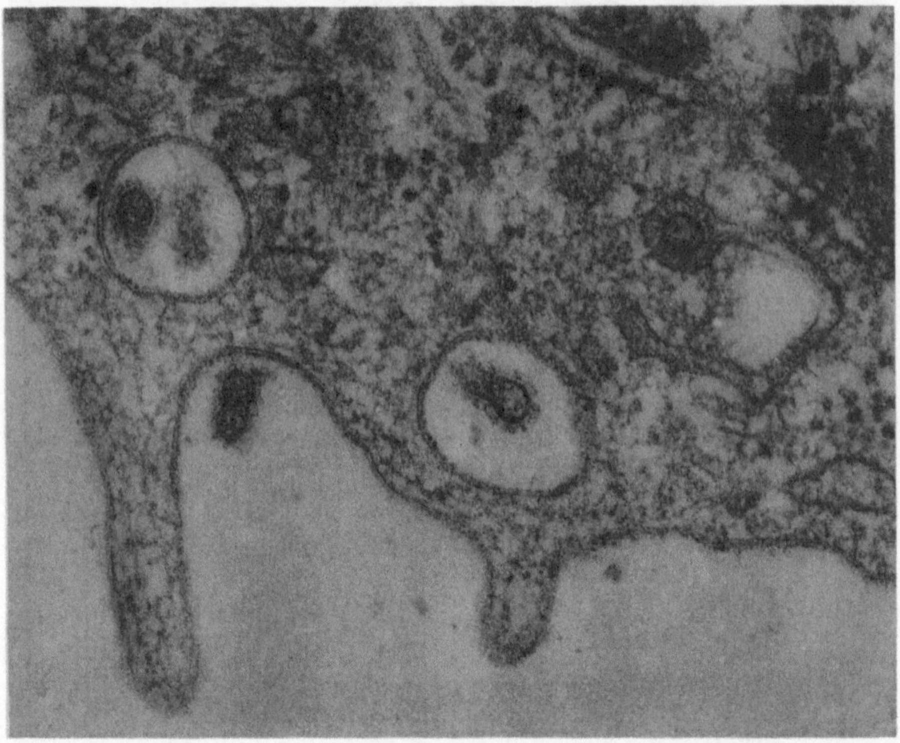

FIGURE 18. Influenza virus particles which have adsorbed and penetrated cells in the presence of amantadine. Compare with Figure 8. Little difference is detected between the two figures, suggesting that amantadine has no effect on virus adsorption or penetration. Magnification × 100,000. (Courtesy of Dr. R. Dourmashkin.)

TABLE 6

Absence of Inhibition of Influenza A Virion-associated RNA-dependent RNA Polymerase Activity[a] by Amantadine

Concentration of Compound (m*M*)		(3H) UMP-incorporated dpm/mg virus protein/h × 10^{-3}
Amantadine	100	96.1
	10	84.2
	1	80.0
Selenocystamine	0.1	0.0
Control (no drug)		97.1

[a]Enzyme activity estimated as described previously.[15]

recombination and reassortment, present theoretical problems for influenza-inhibitory compounds. Thus, we have demonstrated that the passage of an influenza A virus in mice in the presence of a high concentration of amantadine resulted in the selection of virus variants which required a 100-fold increase in amantadine concentration for inhibition. In fact, the variants were completely drug resistant.[50]

The experimental protocol for the selection of drug-resistant strains by passage of influenza virus in *aminoadamantane-treated mice* is summarized in Figure 19. Adult Swiss white mice were inoculated intranasally under ether anesthesia with 0.1 ml of PBS, pH 7.2, containing 10^4 EID_{50} of virus. Mice were killed after 72 hr. The lungs were removed aseptically and minced as a 10% (v/v) suspension in a mechanical blender in mixture 199. The suspensions were centrifuged at 2000 g for 10 min and supernatant fluids frozen at −80°C. At each virus passage, mice were inoculated intraperitoneally with 50 mg/kg of aminoadamantane 30 min before inoculation with 10^4 EID_{50} of virus and supplied with drinking water containing 1 mg/ml aminoadamantane. Separate tests established that mice received an approximate total of 150 mg/kg aminoadamantane per day with this treatment. The degree of inhibition of influenza viruses by aminoadamantane was estimated by quantitative hemadsorption (QH).

Influenza A2/Singapore/1/57 (H2N2) virus from a single egg pool was divided into 24 aliquots, and these were used to infect 24 mice in

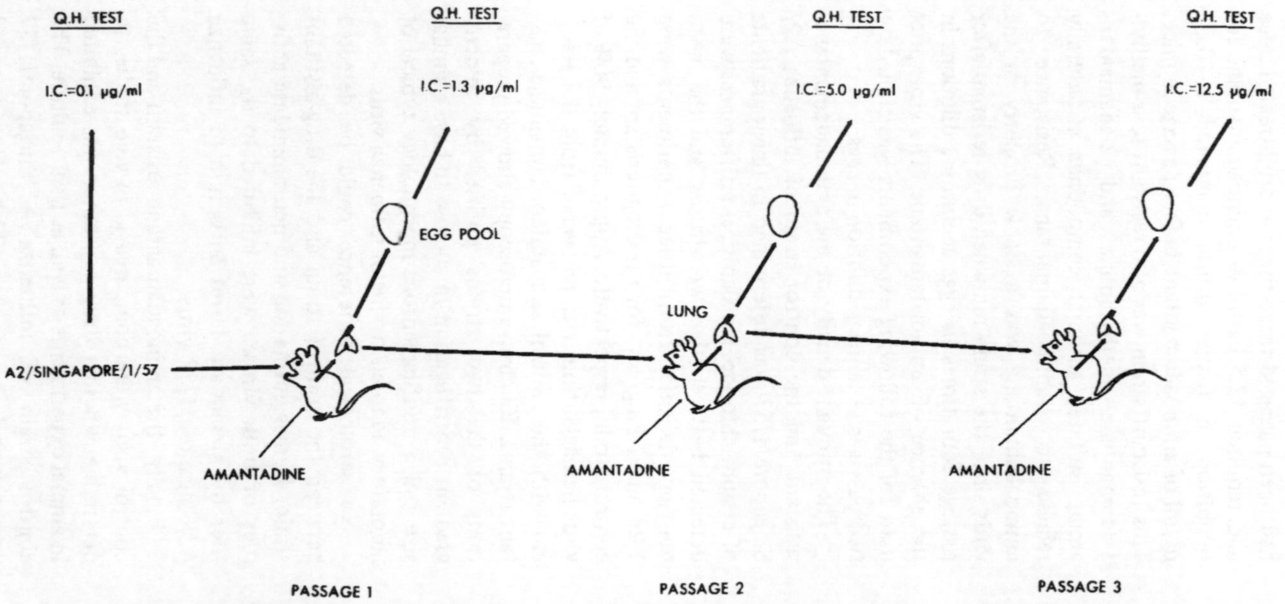

FIGURE 19. Protocol for experiment to select amantadine-resistant strains of influenza A virus. See also Figure 20. (From Oxford, J. S. and Potter, C. W., *J. Hyg.*, 71, 227, 1973. With permission.)

individual cages. Twelve of these mice were treated with aminoadamantane (150 mg/kg/day), and the remaining 12 were control animals which received no drug. All mice were inoculated intranasally with approximately 10^4 EID_{50} of virus. Lungs were harvested individually after 72 hr; after titration in eggs to determine the amount of infective virus, lung material was used to infect mice for the second passage (Figure 19). After each of the three passages in mice, egg allantoic fluid pools were prepared from the infective lung material, and the degree of inhibition by aminoadamantane of the isolated virus was estimated by the QH test.

The 36 virus pools from the control mice passage series were all inhibited by 0.02 to 0.2 μg/ml of aminoadamantane in the QH test. In contrast, influenza viruses resistant to inhibition by aminoadamantane were detected in a proportion of the 36 virus pools from drug-treated mice (Figure 20). Drug-resistant viruses were detected in the lungs of 1 of 12 mice on the first passage of virus in aminoadamantane-treated animals. After three passages in drug-treated mice, the proportion of aminoadamantane-resistant viruses increased, and resistant viruses were present in 8 of the 12 passage series. A single virus strain showed partial resistance to inhibition by aminoadamantane after one passage in mice: before passage in mice, growth of the virus in tissue culture was inhibited by 0.1 μg/ml of aminoadamantane, and after one pass in drug-treated mice, 1.3 μg/ml aminoadamantane was required for inhibition (Figure 20). After two and three passes in drug-treated mice, 5 and 12.5 μg/ml of aminoadamantane, respectively, were required for inhibition. Three virus strains required two passages in drug-treated mice before showing any resistance to inhibition by aminoadamantane, while in four strains drug resistance became apparent only after three passages in treated mice. Thus, strains of A2/Singapore/1/57 virus resistant to inhibition by aminoadamantane could be obtained with relatively high frequency after several passages of virus in mice treated with high concentrations of aminoadmantane.

In a second series of experiments, influenza A2/Singapore/1/57 (H2N2) virus was passed four times in groups of mice treated with varying concentrations of aminoadamantane. Drug-resistant viruses were recovered only from mice treated with large doses (150 mg/kg/day) of aminoadamantane; no drug-resistant viruses were recovered from mice treated with 15 or 1.5 mg/kg/day of aminoadamantane.

D. Biological and Physical Characteristics of an Aminoadamantane-resistant Virus Strain

Virus was examined from a single pool which had been passaged three times in drug-treated mice and required 12.5 μg/ml of aminoadamantane for inhibition in tissue culture compared to 0.02 μg/ml of aminoadamantane before passage in mice. This resistant strain was also resistant to α-methyl-1-adamantane methylamine and 2-adamantanamine sulfate, antiviral compounds structurally related to 1-aminoadamantane. Resistance to aminoadamantane was a stable property of the virus since the strain remained drug resistant after passage four times in eggs at limiting dilutions in the absence of aminoadamantane. The virus pool used for the following experiments was cloned by two passages at limiting dilution in eggs.

The buoyant density of the aminoadamantane-resistant and the control strain of influenza A2/Singapore/1/57 was determined in linear gradients of cesium chloride. No significant differences were detected between the two viruses, and the mean buoyant densities from three experiments were 1.24 and 1.23 g/cm^3 for the drug-resistant and the control strain, respectively. Approximately 90% of virus hemagglutinin was recovered from the gradient with the control and aminoadamantane-resistant strains. Electron microscopy showed preparations of the two strains purified by velocity gradient centrifugation in sucrose to have a similar size and a nonfilamentous morphology typical of laboratory-adapted strains of influenza virus.

No serological difference could be detected between the control strain and the drug-resistant strain. Hemagglutination and neuraminidase activities of both viruses were inhibited to the same titer by a standard ferret antiserum to influenza A2/Singapore/1/57 virus.

Finally, the aminoadamantane-resistant and the control virus were compared in in vivo studies to determine whether drug resistance was exhibited in animal experiments as well as tissue culture. The control strain of influenza A2/Singapore/1/57 caused lung lesions in mice 72 hr after intranasal inoculation under ether anesthesia. The amount of lung consolidation and the number of mice with lung lesions were both reduced significantly in mice treated with aminoadamantane. In contrast, aminoadamantane had no detectable effect on the

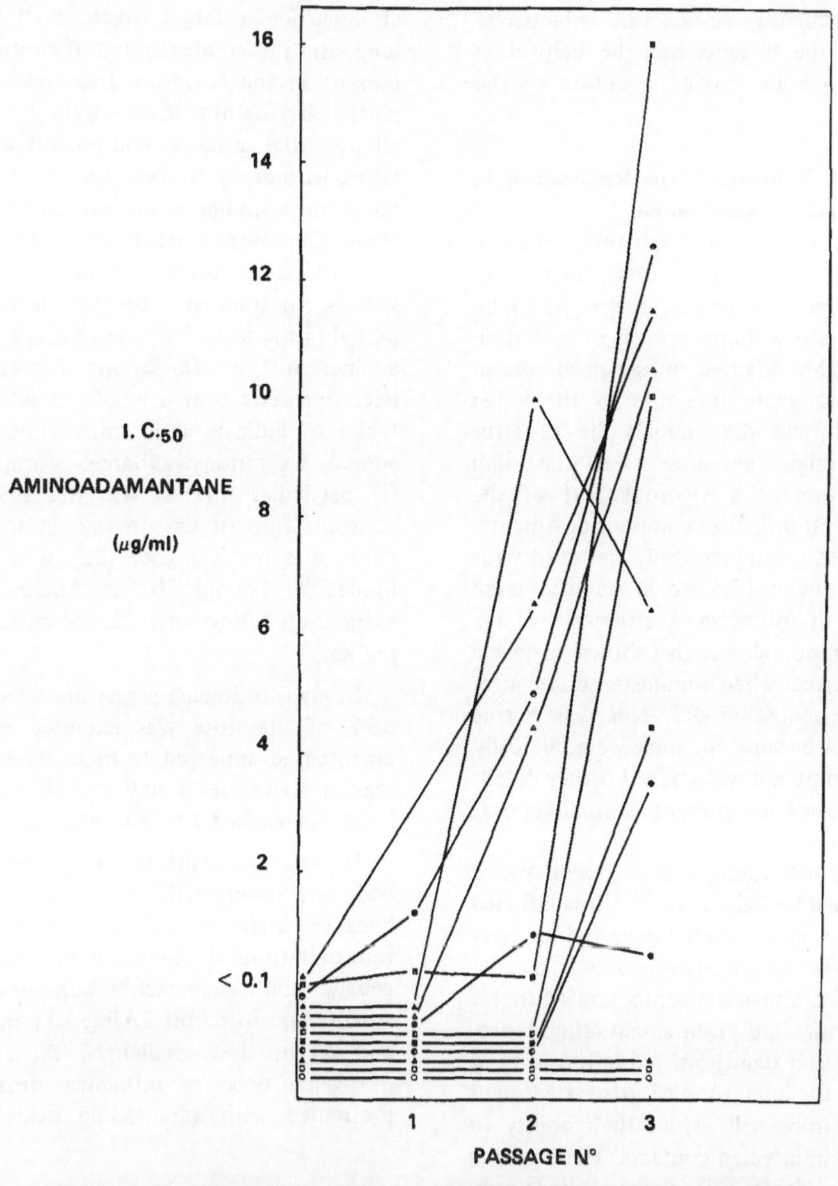

FIGURE 20. Analysis of influenza isolates from mouse lungs passaged successively in the presence of amantadine. Isolates from three successive passages were analyzed by QH and the results expressed as the concentration (IC_{50}) of amantadine required to inhibit virus hemadsorption by 50%. After three passages most of the isolates were resistant to the inhibitory effects of amantadine. (From Oxford, J. S. and Potter, C. W., *J. Hyg.*, 71, 227, 1973. With permission.)

incidence of lung lesions or on the degree of lung consolidation in mice infected with the amino-adamantane-resistant strain of virus. Thus, drug resistance was expressed in virus-susceptible animals as well as in tissue culture. Tuckova et al.[61] have demonstrated that in the genetic interaction between two viruses, resistance to amantadine can be transferred from one virus to another; this is a very important practical consideration.

We have attempted without success to recover amantadine-resistant variants from persons infected with influenza and undergoing treatment with amantadine.[62] However, the latter experiments involved very few persons, and similar experiments need to be carried out in future trials of this and any other anti-influenza compounds to monitor the possible emergence of any drug-resistant viruses. Continuing studies of amantadine-resistant influenza viruses are practically important; also, the variants may be helpful in establishing the precise mode of action of the compound.

E. Inhibition of Influenza Virus Replication in Laboratory Animals by Amantadine

In humans influenza is predominantly an acute infection of the upper respiratory tract, only rarely producing pneumonia and death (see Chapter 1). Most laboratory animal models of infection, however, use highly adapted influenza viruses in mice; these cause acute infection of the lower respiratory tract, and the animals die of virus pneumonia. It might be anticipated that such mouse models present a vigorous test of the efficacy of any anti-influenza compound. Amantadine, administered prophylactically, reduced virus pneumonia and the death rate in mice infected with a number of influenza A strains.[53] Unexpectedly, the compound was also shown to have a virus-inhibitory effect when administered late after infection in the mouse model. This was a true therapeutic effect because the mice were clinically ill before the compound was given. Further details of these experiments are presented in Chapter 1, Volume II.

A particularly interesting animal model system has been developed by Schulman.[63] Virus-infected mice are housed with uninfected mice and are later examined for infection by transmission. In a series of experiments Schulman has demonstrated that if the uninfected mice are given amantadine derivatives, then there is a significant reduction in their attack rate from the infector mice. Also, treatment of the infector mice will reduce their ability to transmit virus to uninfected contacts. These results are illustrated in Table 8 from Schulman's published data. Obviously this model is very close to the natural epidemiology of influenza in man and should be of great interest. The use of ferrets, in place of mice, with attached transmitters enabling individual temperature responses to be recorded from animals housed in groups should also improve the model.[64]

F. Pharmacology and Toxicology Studies with Amantadine

Analysis of tissues and body fluids for 1-aminoadamantane is done comparatively easily using a sensitive and specific analytical method based on vapor-phase chromatography.[65] This has enabled meaningful correlation between antiviral activity at a particular target organ, such as the mouse lung, and the concentration of compound actually present in the corresponding tissue. It would be particularly helpful if such data were provided for all potential antiviral compounds undergoing detailed laboratory studies prior to testing in man. Since most studies of the anti-influenza activity of amantadine were carried out in the mouse, details of tissue distribution will be given here for this species. In summary, the low level of the compound in the feces following a single oral dose over a range of 1 to 100 kg and the regularity of the excretion data over a whole range of doses were taken to indicate that amantadine was well absorbed. Excretion was largely complete in 12 hr. Of particular interest was the finding that the concentration of amantadine in the mouse lung tissue was much higher than that found in the blood, for example. No amantadine was found in expired air following intraperitoneal dosing of 186 mg/kg.

In man, following single oral doses of 2 mg/kg, 56% of the dose was excreted in 24 hr, and amantadine appeared to be excreted according to regular first-order kinetics. The maximum blood level was reached 1 to 4 hr after an oral dose.

It was important to determine if amantadine had any antipyretic effect in animal models, because such an effect in man would make interpretation of clinical data impossible. Grossgebauer and Langmaach[65] administered 60 mg of amantadine to rabbits. After 10 min, 20 min, 2 hr, and 24 hr they challenged the animals with a pyrogenic dose of influenza virus. All animals pretreated with amantadine exhibited a normal

TABLE 7

Effect of RTCA[a] on Replication of Influenza A and B Viruses in Egg Pieces

Virus	Inhibition $\log_{10}$ ID_{50}/ml
A/HK/1/68	4.0 ± 0.5
A/PC/1/73	3.5
B/HK/8/73	3.5
A/FPV	5.2
A/duck/England/62	4.5
A/equine/Miami/1/63	3.0
A/swine/Cambridge/1/39	4.5
A/New Jersey/76	4.0

[a]0.04 m*M* RTCA added at the time of virus infection.

fever response or an enhanced response, although amantadine, in the doses used, did not itself produce a febrile response.

G. Effect of Amantadine in Prophylactic and Therapeutic Trials in Man

Animal studies with amantadine, especially those of Schulman,[63] on the transmissibility of influenza in mice have suggested future means of administering influenza inhibitory compounds to man. Thus, of particular interest are the prophylactic trials of Galbraith et al.,[54] where amantadine was given to household *contacts* of persons with influenza. This epidemiological situation for the use of inhibitors of respiratory tract viruses, which could also be applied to schools and factories, is briefly summarized here. Details are given in Chapter 3, Volume II. Family doctors having diagnosed a case (the "index case") of influenza, administered amantadine or placebo to the remaining members of the family for 10 days under a double-bind protocol. Any spread of influenza from the untreated index case to the family was then monitored by laboratory serological and virus-isolation studies and by clinical observation. The incidence of subclinical influenza in household contacts was reduced from 14.1 to 3.6%, and the incidence of subclinical infections was reduced from 24.6 to 14.6% in the amantadine-treated group. In another controlled therapeutic trial in families, persons with clinically diagnosed influenza were treated with placebo or amantadine within 24 hr of onset; subsequently they were visited daily for detailed clinical observations. There was a significant reduction from 75 to 67 hr mean duration of fever in treated individuals compared to those receiving placebo.

Controlled field trials in a number of countries support the efficacy of amantadine, taken orally, as a chemical prophylactic or therapeutic agent against influenza A of subtypes H2N2 and H3N2.[54,55,67,68] A summary of 12 prophylactic trials in different parts of the world indicates that the protective effect of amantadine varied from 16 to 100%, with a mean of 63%.[69] The latter figure is approximately the same as estimates of protection following immunization with currently available inactivated influenza vaccines. At present, evaluation studies of amantadine and derivatives are continuing, particularly in field trials involving therapeutic administration.

V. INFLUENZA-INHIBITORY EFFECTS OF RIBAVIRIN

Ribavirin is a synthetic nucleoside analog, 1-β-D-ribofuranosyl-1,-2,4-triazole-3-carboxamide (RTCA, Figure 15) which inhibits (perhaps nonspecifically) the replication of a wide range of DNA- and RNA-containing viruses.[11] The compound also inhibits the replication of influenza virus in mice,[11] and ferrets, but has serious toxicity problems. These in vivo results are discussed in detail in Chapter 7. Ribavirin was discovered as part of a project searching for compounds which would inhibit RNA and DNA viruses by interfering with either nucleic acid or protein synthesis. Hence, it is included here as an example of a nucleoside analog rather than as a compound that has any great future as a chemoprophylactic against influenza virus in man. Obviously, such compounds would be likely to exert similar effects on the host cell, and at present it is not clear, for example, if ribavirin has a selective antiviral effect or whether it will be shown to have other unwanted effects on normal cell metabolism. In laboratory studies carried out with confluent, and hence nondividing, cell monolayers and in animal studies involving acute infections, antiviral effect is demonstrable against a wide range of viruses. In the past many related compounds have been investigated for antiviral activity. Thus, the ribofuranosyl nucleoside antibiotic pyrazomycin, 3-β-D-ribofuranosyl-4-hydroxypyrazole-5-carboxamide (Figure 21, compound 2) had been shown to have antiviral activity against rhinovirus, measles, herpes simplex, and vaccinia viruses in tissue culture. However, inhibition of viral replication in vivo and toxicity to the host could be only partially separated. Another ribonucleoside antibiotic, formycin, has also demonstrated in vitro antiviral activity. These considerations led the ICN group to concentrate on the synthesis of a substantial number of β-D-ribofuranosyl derivatives as potential wide-spectrum antiviral agents.

The ribonucleoside antibiotic showdomycin (Figure 21, compound 4) is an example of a biologically active nucleoside with the β-D-ribofuranose moiety attached to a five-membered heterocyclic ring. Further studies with other five-membered heterocycles have resulted in the synthesis of β-D-ribofuranosyl derivatives of various imidazoles and 1,2,4-triazoles.

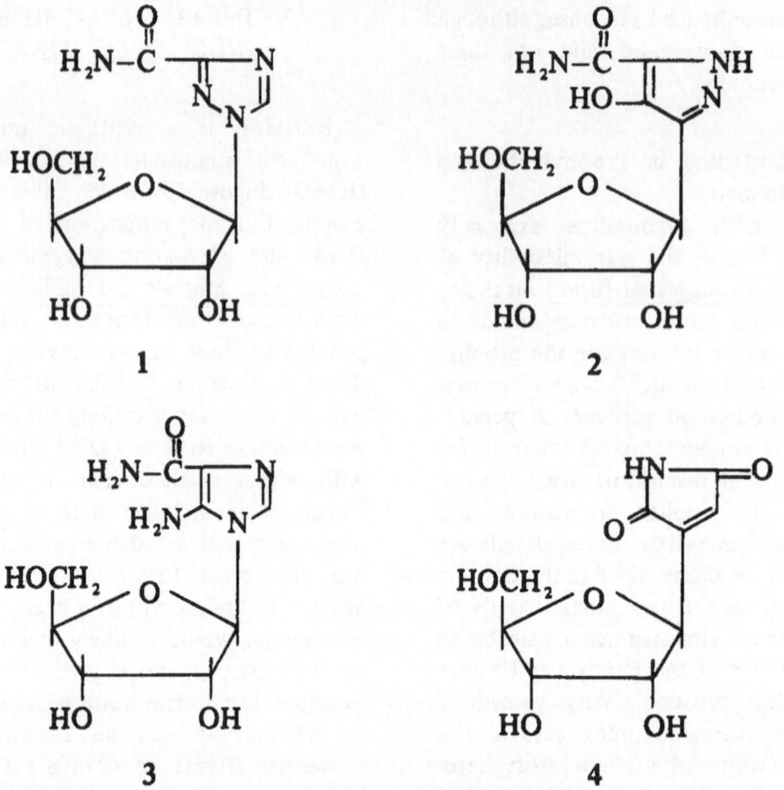

FIGURE 21. Molecular structure of compounds related to ribavirin: (1) ribavirin; (2) pyrazomycin; (3) 5-amino-1-β-D-ribofuranosylimidazole-4-carboxamide; (4) ribonucleoside antibiotic showdomycin.

The nucleoside 1-β-D-ribofuranosyl-1,2,4-triazole-3-carboxamide (1) is structurally related to 5-amino-1-β-D-ribofuranosylimidazole-4-carboxamide (3). The structural similarity of 1 and pyrazomycin (2) is also apparent (Figure 21).

Nucleosides related to ribavirin have been prepared and evaluated for antiviral activity. Several nucleosides which are isomeric with ribavirin, including 1,2,3-triazole nucleosides, were found to lack significant antiviral activity. Modification of the carbohydrate moiety and substitution on the heterocycle of ribavirin also reduced activity, in most cases. Changes in the glycosyl portion include the 2′-deoxy-β-D-ribofuranosyl, β-D-arabinofuranosyl, β-D-xylofuranosyl, and 5′-deoxy-5′-substituted-β-D-ribofuranosyl nucleosides. Substitution on the carboxamide group of ribavirin in a limited number of examples indicates that the primary carboxamide group is necessary for effective antiviral activity. Two nucleosides with structures closely resembling that of ribavirin 1-β-D-ribofuranosyl-1,2,4-triazole-3-thiocarboxamide and 1-β-D-ribofuranosyl-1,2,4-triazole-3-carboxamidine hydrochloride exhibit significant antiviral activity.

We have shown recently that ribavirin acts at an early stage in the influenza virus infected cell, possibly by inhibiting the production of essential nucleotides and hence RNA synthesis. The nucleoside analog caused a significant inhibition of the replication in egg pieces of a range of influenza A strains of human, equine, avian, and swine origin including the recent isolate A/New Jersey/76 (Table 7). In addition, influenza B/HK/1/73 virus was inhibited to a similar degree to the influenza A viruses. Separate toxicity studies indicated that 0.4 m*M* concentrations of RTCA had no effect on the uptake and incorporation of Se^{75}-methionine into acid-precipitable product after prior incubation with the compound for 72 hr. Thus, RTCA inhibited influenza virus replication in egg pieces at concentrations which had no detectable effect on cellular protein synthesis. It must be emphasized, however, that probably very few cells are active in cell division in this system, and any effect of the compound on DNA replication needs to be investigated.

TABLE 8

Effect of Amantadine and Rimantadine on the Susceptibility of Previously Uninfected Contact Mice to the Acquisition of Transmitted Influenza A_2 Virus Infection

		Contacts infected	
Infector mice	Contact mice	No.	%
NT[a]	Saline[b]	14/30	46.7
	Amantadine[c]	11/30	36.7
NT	Saline[b]	17/30	56.7
	Rimantadine[c]	8/30	26.7

[a]NT = no treatment.
[b]Paired contacts, one saline-inoculated and one drug-inoculated contact in each cage.
[c]A 20-mg/kg dose, i.p. 1 hr before and 12 hr after initiation of exposure to transmitted infection.

From Schulman, J. L., *Proc. Soc. Exp. Biol. Med.*, 128, 1173, 1968. With permission.

We investigated the effect of RTCA on influenza virus-induced polypeptide synthesis in infected cells. Most studies were carried out using FPV infected chick embryo fibroblast cells, where it is known that the virus undergoes a complete cycle of replication with production of all the virus structural antigens.[45] Virus-infected control or RTCA-treated (4 m*M*) CEF monolayers were pulsed with S^{35}-methionine for 20-min periods at carrying times after infection. The cells were lysed and solubilized, and total cell extracts were electrophoresed on slab polyacrylamide gels.[70] Examination of the autoradiographs (Figure 22) indicated that the nucleoside analog at a concentration of 4 m*M* had no effect on the polypeptide pattern in treated uninfected cells. However, the compound markedly inhibited the production of influenza virus-induced structural and nonstructural polypeptides. Thus, in the control CEF monolayer infected with A/FPV, new polypeptides corresponding in rate of migration to virus NP and virus P polypeptides are detected. In addition, a virus-coded nonstructural polypeptide (NS) is apparent. These three polypeptides are absent from extracts of FPV-infected monolayers treated with RTCA. Thus, it was apparent that RTCA had no selective effect on any individual virus-induced structural or nonstructural polypeptide, but inhibited the production of all to a comparable degree, as detected by isotopic labeling. Amantadine caused partial inhibition of

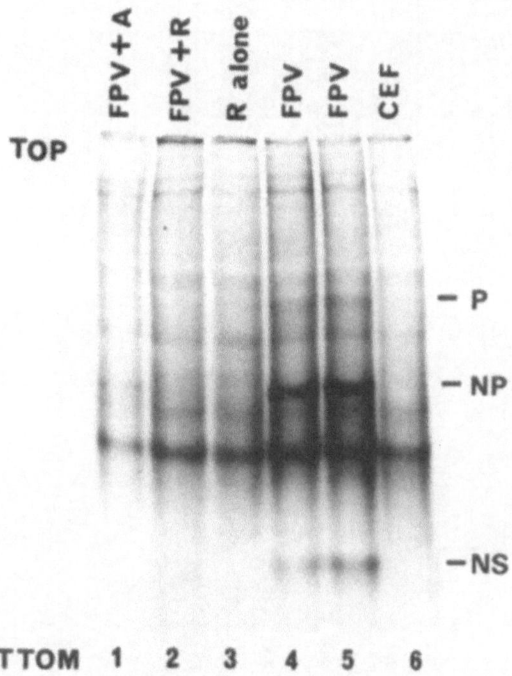

FIGURE 22. Autoradiograph of a slab polyacrylamide gel (SDS-PAGE) analyzing the inhibitory effect of ribavirin on virus and host cell polypeptide synthesis. Virus-induced polypeptides NS, NP, and P can be distinguished. (A) Amantadine. (R) Ribavirin (4 m*M*). (CEF) Chick embryo fibroblasts. CEF cells with or without drugs were infected with A/FPV (10 PFU/cell) and pulsed with S^{35}-methionine 6-hr postinfection. The cell monolayers were solubilized in SDS and β-mercaptoethanol and the polypeptides analyzed by SDS-PAGE.

the production of A/FPV-coded structural proteins (Figure 22), confirming earlier results obtained by immunofluorescence.[57]

A. Mode of Action Studies with Ribavirin

The effect of delayed addition of RTCA to virus-infected cells was investigated to establish the point of action of the compound in the virus replication. CEF cells were infected with 10 PFU/cell of FPV, and RTCA (4 m*M*) was added at varying times after infection. At 6-hr postinfection the monolayers were pulse labeled with S^{35}-methionine for 20 min, solubilized, and examined by electrophoresis in slabs of polyacrylamide for any inhibitory effect of the compound on influenza virus-induced polypeptide synthesis. The RTCA-sensitive step occurred within 1½ hr of infection of cells at this high multiplicity (Figure 23). In comparison, amantadine had to be added within 30 min of infection to obtain any inhibi-

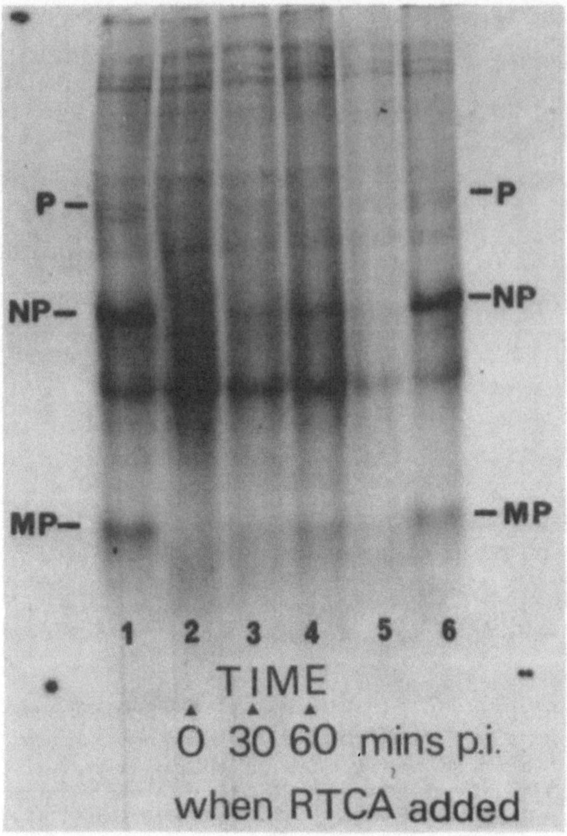

FIGURE 23. Effect of time of addition of ribavirin on influenza virus replication (SDS-PAGE). CEF cells were infected with A/FPV (10 PFU/cell) and RTCA (4 m*M*) added at infection or varying times thereafter. At 6-hr postinfection the monolayers were pulsed with S^{35}-methionine, and the cell monolayer was solubilized and analyzed for polypeptides by SDS-PAGE as in Figure 22.

tory effect in the same experiment. The nucleoside analog did not exert its inhibitory activity via an effect on virus adsorption, as shown by separate experiments in which RTCA was present only during the virus adsorption period. We attempted to reverse the inhibitory effect of RTCA by a variety of nucleosides to determine if the compound was exerting its inhibitory effects on the nucleotide pathways to RNA. The inhibitory effect of RTCA (0.2 m*M*) on plaque formation of FPV in CEF cells and on the growth of A/HK/1/68 virus in egg bit experiments was reversed by an equimolar concentration of the nucleosides guanosine and xanthine. In contrast, inosine and adenine were not able to reverse the virus-inhibitory effects of the compound. 1,β,D-ribofuranosyl-1,2,4-triazole-3-carboxamide-5′-phosphate is a potent competitive inhibitor of inosine-5′-phosphate dehydrogenase isolated from *E. coli* and Ehrlich ascites tumor cells; this suggests that the antiviral activity of RTCA may be mediated by inhibition of GMP biosynthesis at the stage of conversion of IMP to xanthosine-5′-phosphate.[70] The reversal results of the present study would be consistent with this hypothesis.

The apparent selective inhibitory effect on influenza virus-coded polypeptides compared to cell polypeptide synthesis could be a result of the relative instability of virus mRNA compared to cell mRNA. The synthesis of both may be inhibited, but host mRNA may be more stable; therefore, host polypeptide synthesis may continue while synthesis of virus-coded polypeptides ceases. Direct comparison of RNA species in infected and noninfected cells treated with RTCA may provide more precise data on the point of action and any selective inhibitory effect on virus RNA synthesis compared to cell RNA synthesis. We have recently established that RTCA and 5′-mono-PO_4 have no inhibitory effect on influenza transcriptase activity (Table 9).

B. Pharmacology and Toxicology of Ribavirin

As mentioned earlier with amantadine, it is important to carefully correlate the antiviral activity with the tissue distribution data if in vivo experiments are to be interpretable and meaningful. Similar absorption, distribution, and excretion

TABLE 9

Absence of Inhibition of Influenza Transcriptase by Ribavirin-5′-phosphate

Concentration of compound (m*M*)	cpm H^3-ATP incorporated/ mg virus/ hr × 10^2
Ribavirin 5′-PO_4	
10	30
5	58
2.5	43
1.25	37
0	40
Selenocystamine	
5	0

Note: Enzyme activity as described previously.[15] The monophosphate of ribavirin was supplied courtesy of Dr. D. Streeter, ICN, Irvine, California. Identical results were obtained with ribavirin itself.

data have been reported in mice, beagle dogs, and monkeys. In summary, ribavirin is approximately 90% absorbed, and 1 hr after dosing the majority of ribavirin is found in the tissues, probably intracellularly (Table 10). The compound is metabolized by phosphorylation, deribosylation, and amide hydrolysis. Ribavirin is recovered in the form of the 5′-monophosphate, diphosphate, and triphosphate and as the base, in the form of 1-β-D-ribofuranosyl-1,2,4-triazole-3-carboxylic acid and 1,2,4-triazole-3-carboxylic acid. The major components were the phosphates and the base. The terminal half-life of all components was 7 to 8 hr in the serum or urine, 3 hr in the lungs, 9 to 10 hr in the spleen, and 6 hr in the liver and kidneys. The acute toxicity level for the mouse is 864 mg/kg. Chronic toxicity studies in mice given 500 mg/kg twice weekly i.p. for 17 months failed to demonstrate carcinogenicity, but resulted in lethal anemia in 20% of animals. Relatively high oral doses of the compound (90 mg/kg) had no effect on fertility reproductive performance, perinatal and lactation performance, and thyroid function and showed no mutagenic effects. However, of considerable importance were the observations that ribavirin at relatively low doses (10 mg/kg) was teratogenic and embryocidal in the rat and embryocidal in the rabbit at 1 mg/kg. No data on teratogenicity are available to man, of course, but obviously clinical trials of the compound must be carried out with great care until the problem is more fully investigated. At higher concentrations, immunosuppressive activity is detected (see Chapter 1, Volume II).

C. In Vivo Inhibitory Effect of Ribavirin against Parainfluenza Virus

Ribavirin administered i.p. in multiple doses also reduced the mortality in mice infected i.n. with type 1 parainfluenza (Sendai) virus, as summarized in Table 11. To more realistically assess ribavirin's efficacy against a nonlethal, upper respiratory infection, hamsters were infected intranasally with type 3 parainfluenza (HA1 virus), and the drug was administered as a dry powder aerosol given in a freon gas-propelled spray device. Nasal washings were taken at varying times after infection prior to each day's treatment and assayed in tissue culture for virus. Treatment was effective in reducing the virus titer as much as 100%, and the lung consolidation produced by the virus was entirely eliminated by the ribavirin treatment. The latter was derived from five hamsters sacrificed on the days indicated. The drug was in a 5% concentration, with 4.5 mg of ribavirin administered in each spray. The sprays were given four times daily for 7 days, beginning 4-hr previrus exposure. However, oral gavage and i.p. treatment were not significantly effective against this parameter of the infection, although pooled HA1 antibody titers were reduced two-to fourfold in the treated animals.

TABLE 10

Distribution of Radioactivity in Rats at Various Times after Oral Dosing[a] with [^{3}H]-Ribavirin

	% of administered radioactivity			
	1 hr	4 hr	6 hr	24 hr
Serum	0.72	0.28	0.11	0.006
Urine	13	31	63	83
Liver	44	26	6.0	0.85
Kidney	10	3.5	0.7	0.10
Spleen	6.4	2.5	0.55	0.13
Lung	8.7	3.2	0.39	0.20

[a] 38 mg/kg, 1.2×10^4 cpm/μg, in 0.5 ml.

From Sidwell, R., personal communication.

TABLE 11

Activity of Ribavirin Against Lethal Type 1 Parainfluenza Virus Infections in Mice

	Percent survivors		Mean day of death	
Drug dose (mg/kg)	Ribavirin treated	Virus control	Ribavirin treated	Virus control
37.7	90[a]	0	19.9[a]	9.8
18.8	70[a]	0	18.3[a]	9.8

[a] $P < 0.001$.

From Sidwell, R., personal communication.

VI. INHIBITORS OF INFLUENZA RNA-DEPENDENT RNA POLYMERASE

As described above, inhibitors of virus RNA-dependent RNA polymerase activity would have potential application as chemoprophylactic agents against RNA-containing viruses. Ho and Walters[13] initiated studies using this approach and have described the inhibition of cell-associated RNA-dependent RNA polymerase of influenza A/PR8 (H0N1) virus by selenocystine. The compound also has virus-inhibitory effects in vivo. More recently, we have described the in vitro inhibition of influenza virion-associated RNA-dependent RNA polymerase by selenocystamine dihydrochloride, bathophenanthroline disodium disulfonate, and certain heterocyclic thiosemicarbazones (Figure 14). A property common to these compounds is the ability to chelate soft, heavy, metal ions such as zinc and copper. Conversely, similar types of compounds, in which the possibility of chelation was diminished, showed significantly less inhibitory activity against influenza virus RNA-dependent RNA polymerase. This and other data[15] suggest that the RNA polymerase of influenza is a zinc metalloenzyme. However, the hypothesis that the RNA-dependent RNA polymerase enzyme of influenza virus is a zinc-activated enzyme or a zinc metalloenzyme[15] requires the demonstration of zinc association with the enzyme polypeptide; this is not possible at present because the exact polypeptide composition of the polymerase enzyme is not known. As an initial approach to this problem, we have determined the zinc content of influenza virus, which has been carefully purified using metal-free buffers and equipment, and then extensively dialysed vs. tris buffer at pH 8.0 to remove any loosely bound zinc atoms.[72] Examination of both influenza A and B viruses by atomic absorption and atomic mass spectroscopy has indicated the presence of approximately 6×10^3 atoms of zinc per influenza virus particle. This zinc is relatively tightly bound and can only be removed by disruption of virus with SDS and dialysis against tris buffer, pH 8.0, containing 1 m*M* EDTA and 0.1% SDS (Figure 24). Since the polymerase enzyme most likely consists of polypeptides NP, P1, and P2, and the maximum number of molecules of NP is around 1000, it is obvious that there could be as many as six atoms of zinc per NP polypeptide. With many other zinc metalloenzymes, one or two atoms of zinc have been reported per enzyme polypeptide. In this study we were interested in investigating particular characters of the enzyme, such as metal ion dependency, to use as a probe for future more logically based syntheses of influenza-inhibitory compounds. We considered that if selective inhibitors of the enzyme could be found, then the other problems of cell toxicity, pharmacology, etc. could be approached later.

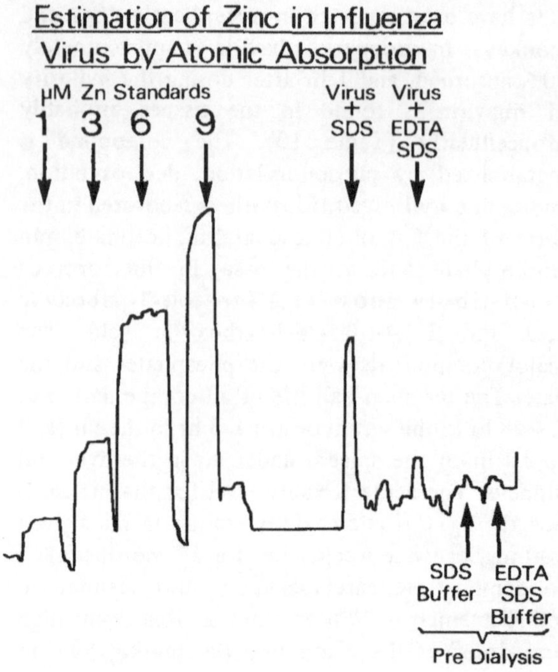

FIGURE 24. Quantitation of zinc associated with purified influenza A virus by atomic adsorption. The tracing of the peaks of zinc standards is given on the left of the figure. Approximately 0.004 m*M* zinc was associated with the purified virus (4 mg protein), and this could only be removed by disruption of virus with SDS and extensive dialysis against EDTA, suggesting strong binding of zinc to an internally situated virus protein.

Table 12 shows the in vitro inhibition of RNA-dependent RNA polymerase of influenza A and B viruses by selenocystamine dihydrochloride and other chelating agents. Selenocystamine also inhibited the DNA-dependent RNA polymerase of *Escherichia coli* at a concentration of 0.04 m*M*, but had little effect on the DNA-dependent DNA polymerase of *Micrococcus lysodeikticus.* Thus, a certain degree of selectivity was detected, particularly between influenza B/Lee RNA polymerase and *E. coli* RNA polymerase, although at present we have no data on the effect of this compound on mammalian DNA-dependent RNA polymerase.

The RNA-dependent RNA polymerase of influenza A and B viruses was also inhibited by bathophenanthroline (disodium disulfonate) and bathocuproine (disodium disulfonate), as shown in Table 12. Bathocuproine also inhibited the DNA-depend-

TABLE 12

Effect of Chelating Agents on RNA-dependent RNA Polymerase Activity of Influenza B/Lee/40 Virus

Chelating agent	Concentration (m*M*) inhibiting B/Lee/40 virus RNA polymerase activity by 50%[a]
With donor N atoms	
Bathophenanthroline disulfonic acid disodium salt	0.05 ± 0.02
Bathocuproine disulfonic acid disodium salt	0.08 ± 0.02[b]
Triethylenetetramine	0.93 ± 0.1
Diethylenetriamine	4.66 ± 2.18
2:2′-Dipyridyl	6.4 ± 0.10
With donor S atoms	
Isatin-3-thiosemicarbazone	0.02 ± 0.004
Isatin-3-selenosemicarbazone	0.03 ± 0.008
Methylisatin-3-thiosemicarbazone	0.10
Isatin-4′-Methylthiosemicarbazone	0.10
Isatin-4′-Ethylthiosemicarbazone	0.04
Isatin-3-semicarbazone	0.5
Isatin-3-amidinohydrazone	0.5
2-Acetylpyridine thiosemicarbazone	0.002 ± 0.001
3-Acetylpyridine thiosemicarbazone	1.0
4-Acetylpyridine thiosemicarbazone	1.0
2-Acetylpyridine semicarbazone	1.0
With mixed donor atoms	
Solochrome black	0.13 ± 0.03
Eriochrome blue	0.15 ± 0.05
Zincone	0.19 ± 0.07
PAR	0.26 ± 0.08
8-Hydroxyquinoline	1.30

[a]The RNA-dependent RNA polymerase enzyme associated with the virus preparations incorporated approximately 640,000 dpm of (^{3}H)-UMP into acid-insoluble product/mg virus protein/30 min/37°C.[15]
[b]Standard errors in this table are given for five or more determinations. Other figures represent mean values of two or three determinations.

ent RNA polymerase of *E. coli,* although at concentrations higher than those required for the inhibition of influenza A/R1-5$^+$ virus-associated RNA polymerase, again indicating some selective activity. Bathophenanthroline and bathocuproine chelate zinc strongly,[73] forming a series of complexes such as the 1:1 species.

Certain heterocyclic thiosemicarbazones, which are effective chelators of heavy metal ions,[74] were also demonstrated to be inhibitors of the RNA polymerase activity associated with influenza A and B viruses (Table 12). 2-Acetylpyridine thiosemicarbazone exerts its maximum binding effect when it forms a terdentate chelate with the metal ion by loss of a proton from the thioenolic form, with coordination of the metal through the sulfur and two of the nitrogen atoms. In 3- and 4-acetylpyridine thiosemicarbazone, only bidentate chelation (by the thiosemicarbazone moiety) is possible, leading to appreciably less stable complex formation. Similarly, isatin-3-semicarbazone and isatin-3-amidinohydrazone, in which the sulfur atom of the thiosemicarbazone is replaced by oxygen and nitrogen, respectively, should also have markedly reduced chelating ability for Cu(I) and Zn(II).[75] These analogs were synthesized and found to be significantly less inhibitory towards influenza B/Lee/40 virus-associated RNA polymerase (Table 12). A correlation was thus established between chelating ability and inhibitory effect on influenza virus RNA polymerase.

We have proposed, therefore, that the compounds described above inhibit the virus-associated RNA polymerase enzyme by formation of a ternary enzyme-metal ligand complex. We suggest that in the RNA polymerase the zinc atom show its usual stereochemistry with three to four sites occupied by binding groups on the enzyme. Configuration of the enzyme is maintained by zinc, and enzyme functioning is inhibited either when zinc is completely removed or when a ternary complex is formed. Our dialysis results with 2-acetylpyridine would suggest that zinc is not removed from the virus; therefore, the formation of an inhibitory ligand zinc enzyme complex is more likely.

However, other modes of action of the compounds cannot yet be excluded. Thus, the inhibition of RNA-dependent DNA polymerase of Rous sarcoma virus by heterocyclic thiosemicarbazones, described by Levinson et al.,[76] was attributed to the formation of ligand-metal complex involving metal ions already in the solution. Analogous metal-ligand complexes are known which inhibit microbial growth and have virucidal activity.[77,78] Further, Ohki and Hori[79] have described the inhibition of QB replicase by sulfhydryl-blocking reagents, and this mode of action is a possible explanation for the inhibitory effects of selenocystamine in particular. Billard and Peets[80] reported the inhibition of the cell-associated influenza RNA-dependent RNA polymerase by selenocystine, 4-(2-propinyloxy)-β-nitrostyrene and acetylaranotin (Figure 25). Since the inhibitory effects were reversed by an excess of dithiothreitol, the authors concluded that the influenza polymerase was very sensitive to compounds reacting with enzyme SH groups. It was postulated that since these compounds had either no effect or a very reduced inhibitory effect on cell polymerases, then the viral enzyme might possess a much more vulnerable configuration with respect to the chemical assessibility of various critical SH groups. If correct, of course, this property could be exploited further for the selection of virus-inhibitory compounds. Studies with purified RNA-dependent RNA polymerase and exogenous influenza RNA template will be required if the mode of action of these inhibitors is to be established unequivocally. The isolation and purification of influenza RNA polymerase will also allow direct determination of any heavy metal content, as described for the demonstration of *E. coli* RNA polymerase,[81] *E. coli* DNA polymerase,[82] terminal deoxy nucleotidyl transferase from calf thymus gland, and reverse transcriptase of AMV[83] as zinc metalloenzymes. Our results, although using carefully purified virus, do not exclude the possibility that the zinc detected is a strongly bound contaminant.

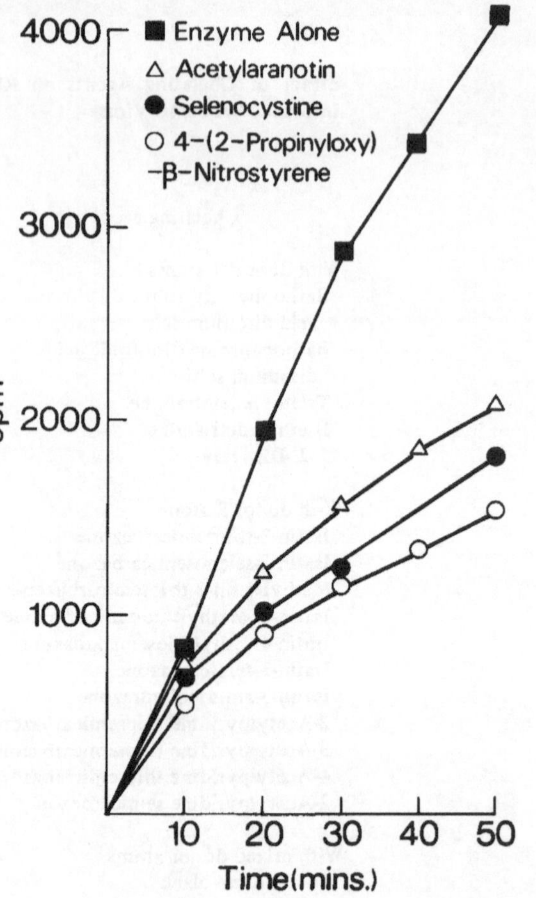

FIGURE 25. Inhibitors of influenza RNA transcriptase enzyme. Inhibitors were added at the beginning of the experiment, and their effect on enzyme activity was estimated by removing samples from the RNA polymerase enzyme reaction mixture, precipitating newly labeled RNA by TCA, and counting in a liquid scintillation counter. (From Billard, W. and Peets, E., *Antimicrob. Agents Chemother.*, 5, 19, 1976. With permission.)

More information is required on the possible virustatic effects of the inhibitors of influenza RNA polymerase. At present, inhibitory effects on influenza A and B virus growth in vitro have been detected only with concentrations of chelating agents close to levels at which cellular protein synthesis is adversely affected. However, if the

RNA polymerase of influenza virus is a metalloenzyme, more active and more selective chelating agents could be designed and tested with advantage. The compounds tested at present also inhibit *E. coli* RNA polymerase, which is a zinc metalloenzyme,[81] but any inhibitory effect on mammalian cell polymerases would depend on the configuration of RNA polymerase enzyme polypeptides near the zinc binding site and on the relative stability constants of zinc for the polypeptide ligand and any competing ligand.

A more objective method for assessing the metal-binding ability of a chelating agent in a biological situation is provided by the use of the program COMICS,[72] (see Chapter 8, Volume II) which computes equilibrium concentrations of metal ions and complex species in mixtures containing several kinds of metal ions and many kinds of chelating agents. Using this approach and stability constants for copper and zinc complexes of 2-acetylpyridine thiosemicarbazone,[99] we have computed that the concentrations of this reagent would need to lie in the region 0.05 to 0.1 m*M* if the level of intracellular free zinc ion was to be significantly depressed. Alternatively, much lower concentrations of ligand would be sufficient if the added ligand showed a relative preference for ternary complex formation through any metal associated with the influenza RNA polymerase enzyme rather than its removal from the enzyme polypeptide. More details are given in Chapter 8, Volume II. It is noteworthy that some isatin isothiosemicarbazones have recently been shown to inhibit RNA-dependent RNA polymerase activity of mengovirus,[84] although the compounds tested had no effect on A/FPV plaquing. Approximately 15 μmol of 1-ethylisatin-S-*n*-butylisatin isothiosemicarbazone caused a 99% plaque reduction in mengovirus-infected FL cells and inhibited RNA polymerase activity in a cell-free system. Preliminary spectroscopical studies indicated that the compound formed a complex with standard proteins rather than nucleic acids.

VII. ALTERNATIVE APPROACHES BY MODIFYING THE HOST RESPONSE TO INFECTION

An alternative approach to the problem of specific inhibition of influenza virus replication is that of altering the immunological response of the host. However, at present the details of the immunopathology of influenza in humans and experimental animals are not clear. It is known, for example, that serum IgG immunoglobulin to the virus HA can give complete protection to infection with a virulent influenza virus in animal models (reviewed by Schild et al.[85]) Also, recent studies have demonstrated that antibody response to the virus HA is thymus dependent, but immunological memory is thymus independent.[85]

The role of cell-mediated immunity, for example, in recovery from influenza infection is not at all clear, and conflicting results are found in the literature. In an avian model an elegant study involving bursectomy and thymectomy suggested that cell-mediated immunity is less important than humoral immunity in recovery from infection. Our own results[85] suggest that humoral immunity is more important than cellular immunity in protection against influenza A virus infection in mice. Recipients of immune spleen cells were very poorly protected against a lethal challenge with A/PR8 virus. This was not due to inefficient transfer of immunity, since the presence of both thymus-derived (T) and bone marrow-derived (B) memory cells in the recipients was confirmed by the typical secondary antibody response obtained after "boosting" the host with a dose of purified HA which was not immunogenic in virgin animals. Such boosted recipients were shown to be fully protected, indicating that immune spleen cells are fully protective only when they secrete antibody. Conversely, we have found that cyclophosphamide (CY) treatment prevents the anti-influenza antibody response and potentiates influenza infection. This finding apparently conflicts with reports that CY[86] or antilymphocytic serum treatment[87] may lead to a less severe disease in influenza-infected mice. However, experimental conditions may be critical, and immunosuppression may decrease antibody production in some systems and decrease the cellular recruitment and minimize the resultant harmful lung consolidation in others. Indeed, a recent publication[88] reconciles these different data by showing that CY increases the mortality of mice infected with low concentrations of influenza virus, but delays the time of death of mice infected with high concentrations of the same variant.

Our finding that transfer of antibody, resulting in serum titers in the recipients comparable to those found in mice which later recover from infection, can protect both normal and immunosuppressed mice strongly suggests that humoral immunity plays a major role in protection under natural conditions. Since the analysis of our hyperimmune rabbit antisera by rate centrifugation in sucrose gradients showed that they contained exclusively 7S anti-influenza immunoglobulins, it can be assumed that IgG antibody was in fact protective in our system. While the role of respiratory IgA remains speculative, our results suggest that IgG molecules are efficient against a pulmonary infection. This is in agreement with reports that the nasal antibody to influenza virus appears to be a 7S IgG protein in ferrets[89] as well as horses.[90] This finding is in accordance with the well-documented relative inefficiency of vaccination in man when a previously prevalent cross-reacting strain of influenza, rather than the current one, is used.

Fujita et al.[91] have reported most interesting data on the inhibition of influenza A and B virus-induced lung inflammation in mice by benzamidine derivatives (Figure 26). The benzamidines are less inhibitory than amantadine against influenza A in vitro virus, but are more effective in vivo especially when given therapeutically 3 days after virus infection. This suggested some effect on virus-induced inflammation in the lungs, perhaps as well as some inhibitory effect on virus replication. In a most interesting experiment mice were inoculated intranasally with 3,200 HA units of noninfective merzonine-inactivated virus. Five days after inoculation lung consolidation was apparent in these mice. However, in comparably inoculated mice given benzamidine derivatives at the time of inoculation and for 4 subsequent days, there was a 50% reduction in lung consolidation. Further, mates were inoculated in the foot pad with 450 μg of crude protein extracted from virus-induced lung lesions. Thirty minutes before some animals had been inoculated i.p. with benzamidine derivations, they were examined to determine the amount of edema by measuring foot pad volumes 0.5, 1, 2, and 3 hr after injection. Benzamidine derivatives exerted a significant effect on viral inflammation (50% inhibition of foot pad swelling). One reservation is that very large quantities of benzamidine were administered (as much as 0.2 mean lethal dose in the latter experiment) to produce an effect. However, the mouse model could be used with ease to screen for such activity in other derivatives and molecules. Presumably such an action would be independent

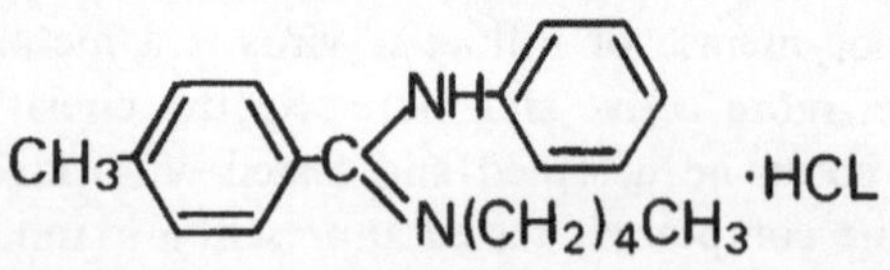

FIGURE 26. An inhibitor of host cell response to influenza infection.

of influenza subtypes and would circumvent the problem of influenza HA and NA antigenic changes during antigenic "drift" or "shift."

VIII. SUMMARY AND CONCLUSIONS

Amantadine is presently the most studied influenza-inhibitory compound and is a basic standard against which other compounds can be compared in the laboratory. Clinical trials of amantadine have established for the first time that prophylactic administration of a specific inhibitor of virus replication can prevent the spread of influenza in the community. Even more important, studies in both laboratory animals and in man have established the therapeutic activity of amantadine and have thus opened up the completely new and unexpected possibility of the treatment of a partly established upper respiratory tract infection. Clinical assessment of amantadine as an anti-influenza compound is continuing, and critical analysis of the accumulated data compared to, for example, influenza vaccine trials will give an indication of the relative value of the two approaches. The compound may be useful in any program against the recent virus A/New Jersey/76 (H_{swine}lNl).

Ribavirin is at present under extensive experimental investigation in the laboratory and is also undergoing limited clinical trials. However, preliminary data suggest only very weak or no prophylactic activity against influenza A infections in man,[100] and the possible teratogenic and immunosuppressive activity of the compound is worrisome. It is possible that administration of the compound in small doses by aerosol may partially overcome the toxicity problems, and such studies are being carried out at present. The compound is particularly interesting because of its wide spectrum of antiviral activity and because it may depend for its mode of action on quantitative differences in the levels of various cellular enzymes following virus infection. Certainly the inhibitory effects of ribavirin in the laboratory against influenza B as well as influenza A viruses would make it a useful inhibitor for in vitro studies of virus replication.

Finally, inhibitors of unique virus-specified enzymes such as RNA-dependent RNA polymerase, as suggested earlier by Ho and Walters,[13] and neuraminidase are worthy of continued and much more extensive investigation, particularly since the virion enzymes can be obtained in a high degree of purity. Assays using the virion-associated transcriptase can be easily established and are subject to a great degree of automation and computerization (see Appendix). Thus, it is possible to screen a large number of compounds in a few hours and to analyze the results in a computer. Without doubt, advances in our knowledge of the molecular biology of influenza replication will suggest further points of action for potential inhibitors, although the applied techniques have not yet resulted in the selection of a compound with both in vitro activity in the laboratory and in vivo activity in man, perhaps because of their present limited use in commercial laboratories. At present it would seem sensible to continue with a combination of empirical and random screening and applied molecular biological methods in the search for inhibitors in influenza virus. It should be clearly realized that biological screening methods using tissue cultures and animals are attempting to simultaneously resolve two problems, virological and pharmacological. Thus, compounds may have been discarded which possessed virus-inhibitory effects; because of abnormal or curious pharmacokinetics; however, they never reached the target organ in the animal or penetrated the plasma membrane in the tissue culture. The point of this chapter has been that by screening using the RNA transcriptase or neuraminidase test, for example, inhibitors with a precise mode of action will be selected. The pharmacology can then be considered as a separate problem to be solved. Furthermore, the RNA transcriptase screen in particular could be made more sophisticated by analyzing the effect of compounds on the enzyme-produced RNA. It is quite possible that compounds may be discovered which cause mistranscription of one vital fragment of the influenza genome, and this might be missed if only compounds causing a 50% inhibition of enzyme activity are thought to be of interest.

APPENDIX

Current Classification of Influenza Virus

The system of influenza nomenclature proposed by a World Health Organization expert committee[91] designates viruses as type A, B, or C and contains the following information.

1. A description of the antigenic type of ribonucleoprotein (A, B, or C).
2. The host of origin. This is not indicated for strains isolated from man, but is indicated for all strains isolated from nonhuman hosts, e.g., swine, horse (equine), duck, chicken, turkey, quail, tern, etc.
3. Geographical origin.
4. Strain number.
5. Year of isolation.

For influenza A viruses the antigenic description follows the strain designation and includes, in parentheses, the following information.

1. An index describing the antigenic character of the hemagglutinin subtype. Some examples are

Human	H0, H1, H2, H3
Equine	Heq1, Heq2
Swine	Hsw1
Avian	Hav1, etc.

2. An index describing the antigenic character of the neuraminidase subtype, for example

Human	N1, N2
Equine	Neq1, Neq2
Avian	Nav1

The host designation for the H and N (e.g., Heq1, Nav1) antigens refers to the source of the virus where the antigen was first characterized. Such designations are intended to indicate the possession of a common subtype of hemagglutinin or neuraminidase antigen, but it is implicit that a given H or N subtype encompass strains showing a considerable degree of antigenic variation within the subtype.

The strain designation describes the origin of the virus and is essentially the same as in the previous system but omits the subtype designation. The second part of the designation (given in parentheses) describes the antigenic composition (H and N).

There is no provision for describing H and N

subtypes of influenza B and C viruses. The existence of antigenic variation among their strains is known, but the information is inadequate to enable a division into subtypes. The description of these viruses is, therefore, limited to strain designation, e.g., B/England/5/66, C/Paris/1/67.

Genetically stable recombinants of influenza A viruses may be produced containing the hemagglutinin derived from one parent and the neuraminidase from the other (antigenic hybrids). Consequently, the antigenic composition of a recombinant virus containing the hemagglutinin antigen from A/BEL/42(H0N1) and the neuraminidase from A/Singapore/1/57(H2N2) would be: A/BEL/42(H0)-A/Singapore/1/57(N2). For tabular material or for reference in the text of publications, this description may be abbreviated to A/BEL(H0)-Singapore(N2).

Virion-associated RNA-dependent RNA Polymerase Enzyme Assay for Drug Screening

Briefly, the influenza virion-associated RNA-dependent RNA transcriptase enzyme has to be activated by incubation of the virus with mild detergent. With a supply of nucleoside triphosphates, one of them labeled with H^3 or C^{14} and the correct cations (Mg^{++} and Mn^{++}), the enzyme can transcribe the virion RNA into an RNA copy, precipitable by TCA. For the test, therefore, enzyme activity is proportional to the radioactive counts incorporated into TCA precipitable material. It is easily possible to screen 20 compounds a day at varying compound concentrations. The results from the scintillation counter can be directly printed on tape and fed into a computer to enable detailed analysis of the results. The test is technically simple, the most care being required to remove the possibility of reagents being contaminated with RNAase, an extremely heat-stable enzyme. The presence of very low quantities of RNAase would result, of course, in the digestion of single-stranded RNA; therefore, the virus sample would appear to have little or no RNA polymerase activity. It is particularly important to ensure that any inhibition of RNA polymerase enzyme activity by a particular chemical "inhibitor" is not, in fact, simply due to contamination of the compound with RNAase.

Influenza virus purified from infected egg allantoic fluids has proven satisfactory for this test (Figures 27 and 28). Approximately 1000 10-day-old embryonated hen's eggs are inoculated allantoically with 0.05 ml of 10^3 EID_{50} of the appropriate virus, commonly a high-yield recombinant such as X-31 (A/Hong Kong/1/68(H3N2)) or MRC-11 (A/Port Chalmers/1/73(H3N2)). After 2 days incubation at 37°C the eggs are cooled to 4°C; the allantoic fluid (approximately 8 l) is harvested using an aspirator and centrifuged at 2000 rpm for 20 min to deposit unwanted egg membranes, etc. The supernatant is centrifuged at 19,000 rpm for 2 hr in an MSE 6 × 250 ml rotor to deposit virus. The deposit is taken up in a small volume of tris-HCl buffer pH 8.0 and carefully resuspended. From this point in the virus purification, precautions should be taken to avoid further RNAase contamination of the virus sample. All buffers, pipettes, sucrose centrifuge tubes, etc. are autoclaved at 10 lb/10 min, and an aseptic technique is used. The virus suspension (10 ml) is divided, placed on six gradients of 10 to 40% w/v sucrose in tris-HCl buffer pH 8.0 and centrifuged using an SW27 rotor in a Beckman ultracentrifuge for 1 hr at 20,000 rpm. The visible band of virus (Figure 29) (if two bands are present, the top band should be collected) is removed by side puncture of the tube and suspended in 100 ml of tris-HCl buffer pH 8.0 to dilute out the sucrose. Finally the dilute suspension is centrifuged to deposit the virus in an MSE 8 × 50 ml fixed-angle rotor at 30,000 rpm for 1 hr. The virus pellet is resuspended in tris-HCl buffer pH 8.0 to a protein concentration of 15 mg/ml and frozen in small aliquots at −70°C until used. Approximately 50 mg of virus protein would be purified from 1000 eggs and can be used for 2000 reaction tubes. Several different virus preparations are screened in the RNA polymerase test to select a preparation with high enzyme activity and low RNAase contamination.

RNA Polymerase Assay

All reagents are stable and can be autoclaved (10 lb for 10 min) to destroy any RNAase enzyme

1. 50 m*M* tris HCl buffer pH 8.0.
2. 50 m*M* $MgCl_2$.
3. 20 m*M* $MnCl_2$.
4. 2% v/v Nonidet P40.

The nucleoside triphosphates ATP, CTP, GTP, and UTP are prepared at 10 times concentration (approximately 2 mg/ml) using sterile tris buffer and sterile glassware; they are not heat sterilized. The

FIGURE 27. Medium-scale infection of embryonated hens' eggs with influenza virus. The apparatus (designed at the Statens Institutt for Folkehelse, Oslo) can inoculate batches of eggs and is suitable for the processing of about 8000 eggs per week. All parts in contact with virus can be sterilized by autoclaving. (Courtesy of Dr. A. Harboe.)

reaction mixture is prepared to the following composition and can be frozen at -80°C in aliquots without addition of virus and used for different experiments.

1. 50 m*M* tris-HCl pH 8.0.
2. 5 m*M* $MgCl_2$.
3. 2 m*M* $MnCl_2$.
4. 0.5 m*M* GTP.
5. 0.5 m*M* ATP.
6. 0.5 m*M* CTP.
7. 0.005 m*M* H^3-UTP.
8. 0.2% v/v Nonidet P40.
9. 500 μg/ml influenza virus protein.

The mixture is kept in ice and water, and influenza virus is added last. Small aliquots are then distributed (usually 60 μl) to Dreyer tubes for the actual experiment. A small volume (10 μl) of varying dilutions of the inhibitor under test can then be added to the appropriate Dreyer tubes. (Separate experiments can be carried out to check that the "inhibitor" compound under test contains no RNAase. A simple test is to incubate different concentrations of the compound with H^3-uridine-labeled ribosomal RNA for 30 min at 37°C and examine the mixture for any decrease in TCA-precipitable counts.)[15] Samples (20 μl) are removed from the polymerase reaction mixture at zero time using a sterile "microcap" pipette and pipetted onto small numbered filter paper disks. The Dreyer tubes are incubated at 37°C for 30 min, cooled to 4°C, and further 20-μl samples are pipetted onto filter disks. Duplicate disks are used. The disks are air dried and then washed in 6% trichloracetic acid (TCA) containing 0.1 *M* pyrophosphate (approximately 200 disks can be washed in 500 ml of 6% TCA) and then in 6% TCA, ethanol, and finally ether. The disks are air dried for 20 min at 37°C, placed in vials containing 10 ml of toluene solution containing 0.5% PPO and 0.3% POPOP, and counted for radioactivity in a liquid scintillation spectrometer.

In a typical experiment sample disks at time

FIGURE 28. Apparatus for inoculation of eggs with influenza virus. (Courtesy of Dr. A. Harboe.)

zero would have around 50 cpm; after incubation at 37°C a five- to tenfold increase in TCA precipitable counts would be expected with an influenza virus preparation with an active polymerase enzyme.

Compounds causing 50% inhibition of enzyme activity at low molarity are of interest. Further tests can then be set up to examine any differential effect on host DNA-dependent DNA polymerase or DNA-dependent RNA polymerase enzymes and also any effect on influenza B polymerase.

A further sophistication would be to analyze the RNA product of the enzyme reaction. Since the influenza genome is fragmented, it is possible that inhibitors may stop transcription on a vital RNA fragment while allowing transcription of the remaining 80% of the genome. Such a compound might be missed if only inhibitors causing 50% or greater inhibition of TCA-precipitable material are selected for further study.

Neuraminidase-inhibition Test for Drug Screening

Since at present most evidence suggests that the influenza virus neuraminidase enzyme performs an important function in virus release from cells, inhibitors of the enzyme would have potential as antiviral agents. Furthermore, the assay is relatively simple, inexpensive, and quick to perform. Other neuraminidase enzymes can be assayed to give an indication of any selectivity of the inhibitors for specific viral neuraminidases. Controls are very important in the assay, including incubation of the compound under test with NANA to ensure that the compound does not only *appear* to inhibit the enzyme by virtue of an inhibitory effect on the color development of NANA in the thiobarbituric acid assay procedure[48] (see Chapter 6 for further details).

Preparation of Working Reagents for Neuraminidase Assay

Use analytical grade reagents or their equivalent.

1. Periodate reagent	
Sodium metaperiodate	4.28 g
Dissolved in distilled water	38 ml
Add syrupy *o*-phosphoric acid; mix well, store in glass-stoppered brown bottle	62 ml
2. Arsenite reagent	
Sodium metaarsenite	10 g
Anhydrous sodium sulfate	7.1 g
Dissolve in distilled water	100 ml
Add concentrated sulfuric acid	0.3 ml

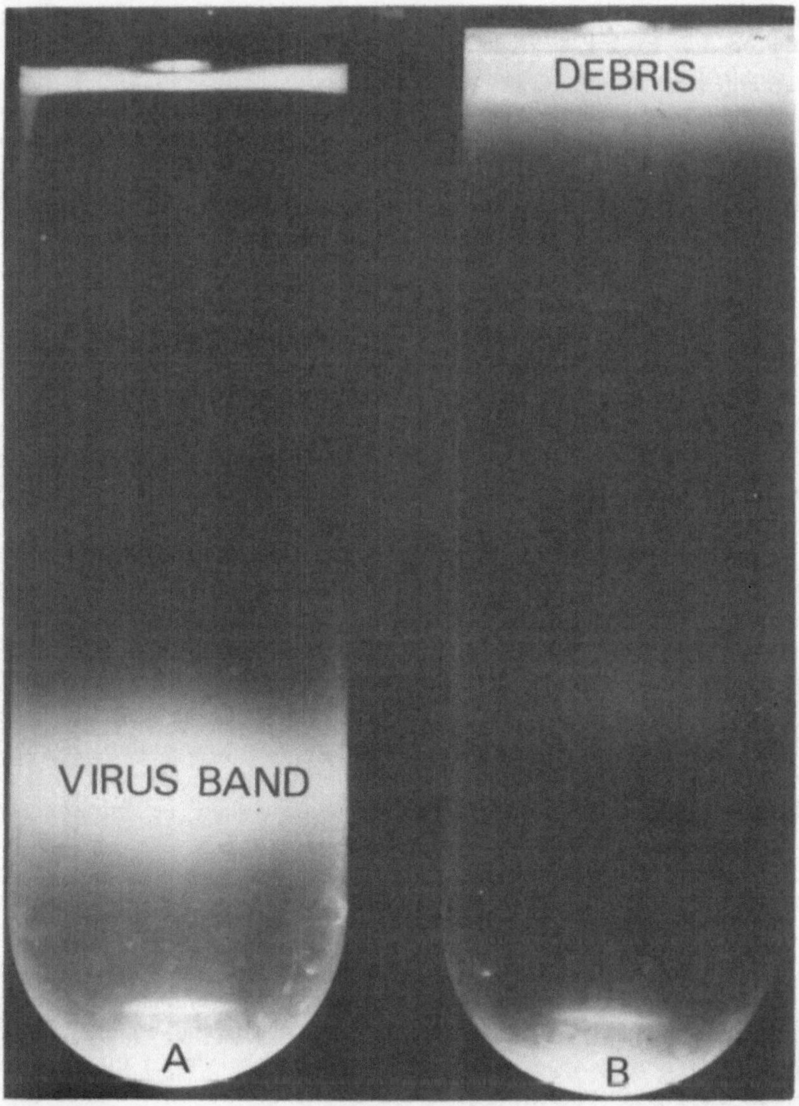

FIGURE 29. Purification of influenza virus for RNA polymerase inhibition screening test by sucrose gradients. (A) Dense virus band near in bottom third of the sucrose gradient. Satisfactory purification. (B) Weak virus band and large quantity of cell debris at the top of the tube. Unsatisfactory virus preparation.

3. Thiobarbituric acid 1.2 g
   Anhydrous sodium sulfate 14.2 g
   Dissolve in distilled water by heating in boiling water bath 200 ml
4. Warrenoff reagent (make up fresh)
   β-butanol 100 ml
   Concentrated HCl 5 ml
5. Phosphate buffer
   A. 0.4 *M* disodium hydrogen orthophosphate (Na_2HPO_4)
   B. 0.4 *M* sodium dihydrogen orthophosphate (NaH_2PO_4)

   Mix 19 ml solution A and 81 ml solution B to give 0.4 *M* buffer with pH 5.9.

   Adjust pH if necessary. Addition of "B" (acidic) buffer) lowers the pH.
6. Fetuin (prepared by ammonium sulfate precipitation from fetal calf serum-reconstitute 1 vial with 2.5 ml sterile H_2O.

Methods

Assay of neuraminidase activity:

1. Make serial twofold dilution of virus in

0.15 *M* saline. For allantoic fluids a suitable dilution series would be 1:2 to 1:128. Higher dilutions will be required for virus concentrates. The reference source of neuraminidase (A/eq/Heq1)-HK(N2)) and unknown test strains should be assayed in the same test.

2. Transfer 0.05 ml of each virus dilution into each of a series of test tubes (ca. 16 × 100 mm). Include duplicate *blank* tubes which contain saline only.

3. To each tube (including the blanks) add 0.05 ml of saline, 0.05 ml of pH 5.9 buffer, and 0.05 ml fetuin. (These can be mixed in equal volumes and added as 0.15 ml of mixture.) Mix well and place at 37°C for 1 hr.

4. Cool tubes to 20°C and add 0.1 ml periodate reagent to each tube, mix thoroughly, and place at 20°C for exactly 20 min.

5. Add 1.0 ml arsenite reagent. A brown color will form resulting from the release of iodine. Shake until the brown color disappears and then add 2.5 ml of thiobarbituric acid reagent. Mix thoroughly.

6. Immediately place tubes in a boiling water bath and hold for 15 min. The red color which develops indicates neuraminidase activity.

7. Cool tubes to room temperature and add 4.0 ml of Warrenoff reagent. Mix well using a mechanical mix (Vortex or Whirlmixer, if available) or vigorously shake by hand. This procedure extracts the color into the organic (butanol) phase.

8. Centrifuge tubes for 2 to 5 min at 500 to 1000 rpm. At this stage the butanol layer should be free of turbidity. If it has a misty appearance due to H_2O droplets in the organic phase, place the tubes in warm water (30°C) until the organic phase becomes clear. Carefully pipette the upper (butanol) phase into a colorimeter tube.

9. Read the optical density at 549 nm on a spectrophotometer using cells with 1-cm light path. The fetuin blanks are used to equilibrate the machine, and tests are read against one fetuin blank. Determine virus concentrations giving an OD_{549} of 0.45 to 0.85, and these are then used for the neuraminidase inhibition test.

Neuraminidase-inhibition (NI) Test

1. For the assay of NI_{50} concentration of an inhibitor, use a reference rabbit anti-neuraminidase serum as a control. Use test viruses at appropriate dilutions (i.e., those giving OD_{549} readings of 0.45 to 0.85 after 18 hr incubation with fetuin).

2. Mix in test tubes 0.05 ml of virus dilution with 0.05 ml of control serum or inhibitor in half log (i.e., 1:10, 30, 100, . . .) dilutions. Control and blank tubes include the fetuin alone, virus plus fetuin, drug plus fetuin alone.

3. The enzyme inhibitor reaction is carried out at 20°C for 3 hr.

4. Now add 0.05 ml of phosphate buffer pH 5.9 and 0.05 ml fetuin; shake the tubes and incubate at 37°C for 1 hr.

5. Remove from 37°C and assay for residual neuraminidase activity by adding the reagents as previously described for the assay beginning with step 4.

6. The NI_{50} titer of an inhibitor is defined as the concentration of compound giving 50% inhibition of neuraminidase activity. In practice this is determined using standard graph paper and plotting OD against compound concentration. The NI_{50} end point is read from the graph as the compound concentration giving a twofold reduction in OD compared to that of the appropriate control tube, i.e., virus plus normal buffer diluent.

In Vitro Methods with Influenza Viruses Suitable for Biological Screening

A feature of the virology of influenza is the great variability among the influenza A strains of human, avian, and equine origin regarding the growth of the viruses in tissue cultures. On the other hand, the embryonated hen's eggs can be used successfully to propagate and quantitate all known influenza A and B viruses. Nevertheless, tissue culture experiments are required for some studies; an egg piece method and methods of quantitative hemadsorption (QH test) are described below that can be used with most influenza strains regardless of whether they go through a complete or incomplete cycle of replication in the tissue culture cells used.

Quantitative Hemadsorption (QH) with Amantadine and Influenza A/Singapore/1/57 (H2N2)

The method described by Finter[93] was followed with some modifications. Aminoadamantane was diluted in twofold steps from 0.02 to 25 μg/ml in mixture 199, and 1.5 ml was added per cell culture tube of the continuous line of monkey kidney (MK) cells BSC-1. Alternative MK lines have been used including VERO and secondary MK cultures. After a 15-min incubation, 0.2 ml of influenza virus, diluted to give a suitable level of hemadsorption, was added to each tube. (For influenza A2 viruses this was approximately 10 EID_{50} per cell.) After 19 hr of incubation at 35 to 36°C, the tubes were hemadsorbed with a 0.5% suspension of guinea pig cells for 15 min at 4°C. After two washings with phosphate buffered saline at 4°C, 1.5 ml of deionized water at 37°C was added per tube to lyse the adhering red blood cells. Eight BSC-1 tissue culture tubes were used for each dilution of aminoadamantane, and the relative hemoglobin concentrations of the lysed guinea pig cells were determined from the optical density readings for pairs of tubes using a Unican SP600 spectrophotometer at 410 nm (Figure 30). The results were expressed as micrograms per milliliter at QH_{50}; this was the concentration of aminoadamantane required to inhibit hemadsorption of a virus by 50%. To determine variation between individual tubes, 26 tubes were infected with $10^{6.5}$ EID_{50} of fowl plague virus. The tubes were hemadsorbed after 19 hr, and the mean optical density reading was 0.42 with a standard deviation of 0.08. In experiments to determine reproducibility of the technique, multiple experiments using a single pool of virus and aminoadamantane showed a fivefold variation in the concentration of aminoadamantane required to inhibit hemadsorption by 50%.

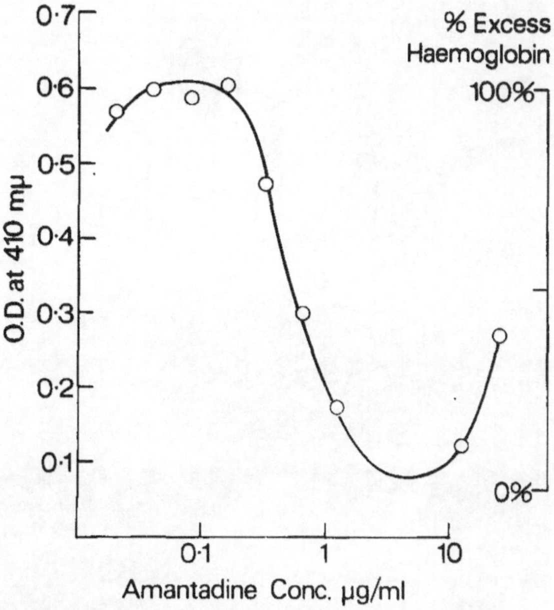

FIGURE 30. Quantitative hemadsorption test to estimate inhibitory effects of drugs on influenza virus replication. Tissue culture monolayers were treated with varying concentrations of amantadine and then infected with influenza virus. Red blood cells were added to the monolayers 24-hr later, and after a period for hemadsorption were washed off. Hemadsorbed cells were lysed in water and OD readings taken on a spectrophotometer. An S-shaped curve is commonly obtained. (From Oxford, J. S., Potter, C. W., and Logan, I., *Ann. N.Y. Acad. Sci.*, 173, 300, 1970. With permission.)

Egg Bit Technique of Fazekas de St. Groth and White[94]

This method is used routinely in our laboratory with very reproducible results. The WHO plastic trays are cleaned in a bath of 4% NaOH for an hour or two, then rinsed thoroughly under the tap in running water. Finally, they are mounted in the spacing frames and dried, face down, in an incubator at 36°C, where they can be stored ready for use. The antibiotic in the standard medium will easily cope with the small number of bacteria which might be present in the tap water. Embryonated eggs were opened with scissors at the albumen end, below the reflection of the chorioallantois (Figure 31). The embryo was tipped out, and the membrane adhering to the shell was rinsed twice with standard medium. In those few cases where the membrane lifted off, the egg was discarded.

Parallel strips (about 0.6-cm wide) running from the albumen end to the air space were cut first, and these were then cut crosswise into squares and dropped into a Petri dish half filled with standard medium. (Hesitant, mincing use of the scissors will crack the shell; however, this is not detrimental to the outcome of the test.) One of these small squares of membrane-on-shell was then transferred with a pair of forceps into each cup of the plastic trays. The size of host tissue was varied by the above authors, and it was found that whereas the final yield of virus varied in direct proportion to the surface area, the susceptibility to infection was the same for all sizes.

When testing whether orientation of the bits of tissue had any effect on the final infectivity titers, only a trivial difference was found in favor of membranes facing upward. This, fortunately, is the position they naturally tend to take in the cups.

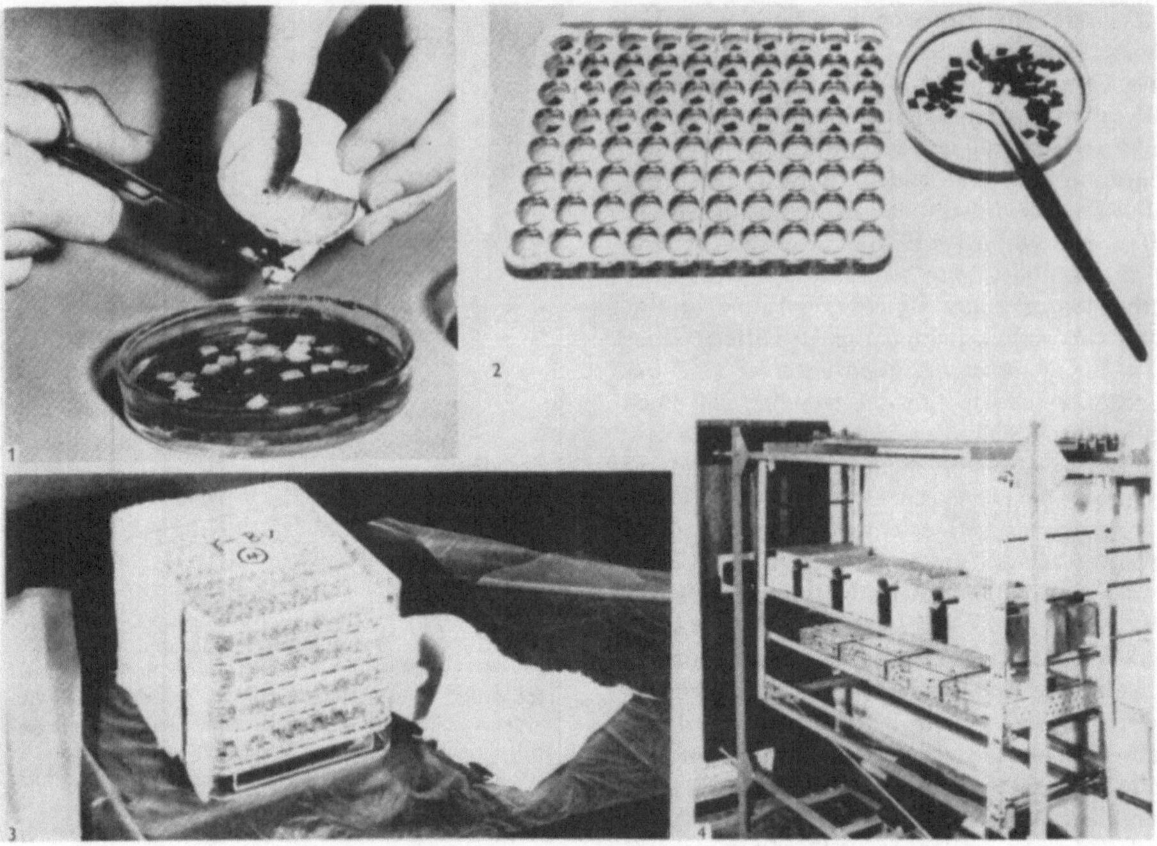

FIGURE 31. Technical details of the egg piece technique.[96]

All of the experiments in this study were made on eggs previously incubated for 11 days at 38.4°C. There is an important correlation between age and susceptibility, which also varies from strain to strain of virus.

Dilutions of the material to be tested were made up in tubes of standard medium kept in an ice bath, usually while an assistant prepared and distributed the host tissue in the trays. Calibrated sterile Pasteur pipettes were used to deliver the drug or virus inoculum as a drop of 0.025 ml. If for any reason the prepared trays cannot be inoculated immediately, the sensitivity of the method does not necessarily suffer. After cutting up, the bits of tissue may be left in either the Petri dishes or the trays for at least 2 hr, either on the bench or in the incubator. If the trays are agitated at 35°C, the sensitivity to infection does not drop more than 0.1 $\log_{10}$ unit in the first 24 hr. Such stability of the host tissue allows interruption of a set of titrations whenever convenient. Indeed, samples from a time series, such as growth experiments, can be titrated on tissue derived from the same host.

After inoculation, the trays were mounted, as described above, and placed on the shaking machine. The temperature of incubation makes no difference to the titers obtained as long as it lies within the range of 34 to 37°C. The degree of agitation was tested on the reciprocating shaker run at 120 oscillations per minute, with the thrust varied between 2 and 10 cm. This was not reflected in any change of susceptibility to infection. Since other combinations of frequency and amplitude can give the same degree of agitation, these results can be extrapolated to other rates of shaking. Shaking was essential only during the first day of incubation.

When testing whether 3 days' incubation gives higher titers than two, the average rise was found to be only 0.04 $\log_{10}$ units, not significant. This behavior characterizes BEL and other fast-growing influenza strains; some, however, multiply more slowly and hence have to be incubated for 3 days.

When the trays are taken down, the medium is clear and uniformly pink, due to the pH change; turbidity caused by bacterial and fungal contamination is obvious when present. The bits of tissue were picked out with a pair of fine forceps, moving from higher dilution to lower. At first the forceps were sterilized by dipping into boiling water between each cup. Later, after extensive control tests had demonstrated that the volume of fluid carried on the forceps is never sufficient to cause false positive reaction in an otherwise negative cup, this practice was abandoned.

When the host tissue has been removed, a standard volume (0.25 ml) of a 0.5% fowl red-cell suspension is added to each cup. The trays are well shaken and left standing for ½ hr at room temperature. The pattern of settled cells is read as positive (complete agglutination) or negative (no agglutination). Intermediate patterns rarely occur, and then only with slowly growing strains; they are read as positive.

The new test, developed above, was compared with the orthodox allantoic infectivity test using nine strains of influenza virus besides BEL by Fazekas de St. Groth and White. The most striking difference is the steepness of the dose-response curves with the new method as opposed to the varying degree of deviation from the Poissonian observed in intact eggs.

Influenza Virus Infectivity Assay: Conclusions on The Assay of Infectivity

1. Prepare standard medium (SM).

NaCl	8.0 g
KCl	0.6 g
$CaCl_2$	0.8 g
$MgCl_2$	0.05 g
Glucose	0.3 g
Gelatine (acid free)	2.0 g
Chloramphenicol	0.1 g
Phenol red (= 25 ml of 0.01% stock solution)	0.0025 g
H_2O (glass distilled or deionized)	to 1000 ml

Adjust pH with normal NaOH to about 7 (yellowish-orange color of indicator). Sterilize by autoclaving at 115°C for 30 min.

2. Deliver 0.3 ml of SM into each cup of clean plastic tray.

3. Deembryonate 11-day-old eggs at the sharp end, rinse membrane twice with SM, and cut shell into 6 × 6 mm squares into Petri dish half filled with SM. Surviving allantoic epithelium serves as host tissue in the form of small squares of membrane left attached to the eggshell. One 11-day egg yields up to 100 bits of tissue, each the equal in sensitivity of a whole egg.

4. Transfer one square of tissue-on-shell into each cup of prepared trays.

5. Inoculate cups by adding a 0.025-ml drop of dilutions prepared from test material.

6. Build battery of four trays separated by spacers and covered by blank on top; surround it by strip of moistened lint and wrap in polythene sheet (Figure 31).

7. Mount pairs of batteries in metal supports, place on horizontal shaker working in warm room, and incubate for 48 hr at 35°C.

8. Remove trays, pick out bits of tissue with fine forceps, add one drop of 10% fowl red cells to each cup, and shake.

9. Read test after 30 min; positive hemagglutination equals infection.

10. Clean trays and spacers by immersing in 4% NaOH for 1 hr, thoroughly rinse in running tap water, fit trays in spacers, and dry in incubator at 35°C.

ACKNOWLEDGMENT

I would like to thank Miss Sue O'Connor and Miss Joan Watkins for good technical assistance and Alan Hodge for drawing the figures.

REFERENCES

1. **Stuart-Harris, C. H.,** *Influenza and Other Virus Infection of the Respiratory Tract,* 2nd ed., Edward Arnold, London, 1965.
2. **Buthala, D. A.,** Experience gained in screening for synthetic antiviral compounds, *Ann. N.Y. Acad. Sci.,* 130, 17, 1965.
3. **Smith, W., Andrewes, C. H., and Laidlaw, P. P.,** A virus obtained from influenza patients, *Lancet,* p. 66, 1933.
4. **Hoyle, L.,** The influenza virus, *Virol. Monogr.,* p. 4, 1968.
5. Ackermann, W. W., The role of 1-methionine in virus propagation, *J. Exp. Med.,* 93, 337, 1951.
6. **Pilcher, K. S., Soite, K. F., Smith, V. H., Trosper, F., and Fulston, B.,** Inhibition of multiplication of Lee influenza virus by canavanine, *Proc. Soc. Exp. Biol. Med.,* 88, 79, 1955.
7. **Dickinson, L.,** The formaldehyde derivatives of amino acids and the mode of action of a formaldehyde-tyrosine derivative against influenza A virus on ovo, *Br. J. Pharmacol.,* 10, 56, 1955.
8. **Broncova, O. and Bydzovsky, V.,** Effect of some 5-arylpyrimidines on influenza, vaccinia and Newcastle disease viruses, *Czech. Epidemiol.,* 11, 179, 1962.
9. **Rolly, H.,** Substituted isonicotinic acid diphenylalklamides with antiviral activity, *Zentralbl. Bakteriol. Parasitenkd. Infektionskr. Hyg. Abt. 1 Orig.,* 188, 335, 1963.
10. **Tamm, I., Folkers, K., Shunk, C. H., and Horsfall, F. L.,** Inhibition of influenza virus multiplication by N-glycosides of benzimidazoles, *J. Exp. Med.,* 99, 227, 1954.
11. **Sidwell, R. W., Huffman, J. H., Khare, G. P., Allen, L. B., Witkowski, J. T., and Robbins, R. K.,** Broad spectrum antiviral activity of virazole: 1-β-D-ribofuranosyl-1,2,4-triazole-3-carbox-amide, *Science,* 177, 705, 1972.
12. **Oxford, J. S.,** Specific inhibitors of influenza virus replication as potential chemophrophylactic agents, *J. Antimicrob. Chemother.,* 1, 7, 1975.
13. **Ho, P. P. K. and Walters, C. P.,** Inhibitors of influenza virus induced RNA polymerase, *Ann. N.Y. Acad. Sci.,* 173, 438, 1971.
14. **Bauer, D. J., Apostolov, K., and Selway, J. W. T.,** Activity of methisazone against RNA viruses, *Ann. N.Y. Acad. Sci.,* 173, 314, 1970.
15. **Oxford, J. S. and Perrin, D. D.,** Inhibition of the particle associated RNA-dependent RNA polymerase activity of influenza viruses by chelating agents, *J. Gen. Virol.,* 23, 59, 1974.
16. **White, D. O. and Harris, A. W.,** Actions of metal chelates of substituted 1,10 phenanthrolines on viruses and cells. I. Inactivation of viruses, *Aust. J. Exp. Biol.,* 41, 517, 1963.
17. **Perrin, D. D.,** Personal communication.
18. **Barry, R. D., Ives, D. R., and Cruickshank, J. G.,** Participation of DNA in the multiplication of influenza virus, *Nature,* 194, 1139, 1962.
19. **Mahy, B. W. J., Cox, N. J., Armstrong, S. J., and Barry, R. D.,** Multiplication of influenza virus in the presence of cordycepin, an inhibitor of cellular RNA synthesis, *Nat. New Biol.,* 243, 172, 1973.
20. **Ackermann, W. W. and Maassaba, H. F.,** Growth characteristics of influenza virus: the influence of a sulphonic acid, *J. Exp. Med.,* 99, 105, 1956.
21. **Eaton, M. D., Low, I. E., Scala, A. R., and Vretsley, S.,** Inhibition by ammonium ion of the growth of influenza virus in chorioallantoic tissue, *Virology,* 18, 102, 1962.
22. **Oxford, J. S. and Schild, G. C.,** Inhibition of the growth of influenza and rubella viruses by amines and ammonium salts, *Br. J. Exp. Pathol.,* 48, 235, 1967.
23. **Davies, W. L., Grunert, R. R., Haff, R. F., McGahen, J. W., Neumayer, E. M., Paulshock, M., Watts, J. C., Wood, T. R., Hermann, E. C., and Hoffman, C. E.,** Antiviral activity of 1-adamantanamine (amantadine), *Science,* 144, 862, 1964.
24. **Swallow, D. L., Edwards, P. N., and Finter, N. B.,** Decalins – a new class of compounds active against influenza A viruses, *Ann. N.Y. Acad. Sci.,* 173, 292, 1970.
25. **Johnstone, E. J. D., Kidd, R. G., Rylance, H. J., and Sommerville, R. G.,** The inhibition of neuraminidase and antiviral action, *Br. J. Pharmacol.,* 27, 415, 1966.
26. **Palese, P., Schulman, J. L., Bodo, G., and Meindl, P.,** Inhibition of influenza and parainfluenza virus replication in tissue culture by 2-deoxy-2,3-deoxy-2,3-dehydro-N-trifluoracetyl neuraminic acid (FANA), *Virology,* 59, 490, 1974.
27. **Brammer, K. W., McDonald, C. R., and Tute, M. S.,** Antiviral properties of 1-phenoxymethyl-3,4-dihydro and 1,2,3,4-tetrahydroisoquinolines, *Nature,* 219, 515, 1968.
28. **Link, F., Blaskovic, D., and Raus, J.,** Studies on the effect of antiviral substances on experimental virus infections. I. The effect of urethane on influenza virus infection, *Acta Virol.,* 5, 89, 1961.
29. **Toyoshima, S., Kano, S., and Veda, T.,** Search for effective synthetic agents against influenza virus, *Keio J. Med.,* 12, 149, 1963.
30. Sidwell, R. W., Dixon, G. J., Sellers, S. M., and Schabel, F. M., In vivo antiviral properties of biologically active compounds. II. Studies with influenza and vaccinia viruses, *Appl. Microbiol.,* 16, 370, 1968.
31. **Almeida, J. D. and Brand, C. M.,** A morphological study of the internal component of influenza virus, *J. Gen. Virol.,* 27, 313, 1975.

32. **Marchesi, V. T., Tillack, T. W., Jackson, R. L., Segrest, J. P., and Scott, R. E.,** *Proc. Natl. Acad. Sci. U.S.A.*, 69, 1445, 1972.
33. **Dourmashkin, R. R., and Tyrrell, D. J.,** Electron microscope observation on the entry of influenza virus into susceptible cells, *J. Gen. Virol.*, 24, 129, 1974.
34. **Aoyagi, T., Suzuki, J., Nerome, K., Nishizawa, R., Takeuchi, T., and Umezawa, H.,** Sialic acid residues exposed on mammalian cell surface: the effect of adsorption of denatured virus particles, *Biochem. Biophys. Res. Commun.*, 57, 271, 1974.
35. **Pons, N. W.,** Influenza virus messenger ribonucleoprotein, *Virology*, 67, 209, 1975.
36. **Bishop, D. M. L. and Flamand, A.,** Transcriptional processes of animal RNA viruses, in *Control Processes in Virus Multiplication*, Burke, D. C. and Russell, W. C., Eds., Cambridge University Press, New York, 1975.
37. **Lee-Huang, S. and Ochoa, S.,** Specific inhibition of M52 and late T4 RNA translation in *E. coli*, *Biochem. Biophys. Res. Commun.*, 49, 371, 1972.
38. **Nudel, U., Lebleu, B., and Revel, M.,** Discrimination between messenger ribonucleic acids by a mammalian translation initiation factor, *Proc. Natl. Acad. Sci. U.S.A.*, 70, 2139, 1973.
39. **Wigle, D. I. and Smith, A. E.,** Specificity in initiation of protein synthesis in a fractionated mammalian cell free system, *Nat. New Biol.*; 242, 136, 1973.
40. **Downey, K. M., Byrnes, J. J., Jurmark, B. S., and So, A. G.,** The effect of 6-mercaptopurine and its derivatives on mammalian RNA-dependent RNA synthesis, *Biochem. Biophys. Res. Commun.*, 56, 227, 1974.
41. **Kuo, C. H. and August, J. T.,** Histone and bacterial basic protein required for replication of bacteriophage RNA, *Nat. New Biol.*, 237, 105, 1970.
42. **Kelly, D. C., Avery, R. J., and Dimmock, N. J.,** Failure of an influenza virus to initiate infection in enucleate BHK cells, *J. Virol.*, 13, 1155, 1974.
43. **Krug, R. M. and Etkind, P. R.,** Cytoplasmic and nuclear virus specific proteins in influenza virus infected MDCK cells, *Antimicrob. Agents Chemother.*, p. 334, 1973.
44. **Cooper, P. D., Steiner-Pryor, A., and Wright, P. J.,** A proposed regulator for poliovirus: the equestron, *Intervirology*, 1, 1, 1973.
45. **Hay, A. J. and Skehel, J. J.,** Studies on the synthesis of influenza virus proteins, in *Negative Strand Viruses*, Vol. 2, Mahy, B. W. J. and Barry, R. D., Eds., Academic Press, London, 1975, 635.
46. **Bachi, T., Gerhard, W., Lindenmann, J., and Muhlethales, K.,** Morphogenesis of influenza A virus in Ehrlich ascites tumour cells as revealed by thin sectioning and freeze etching, *J. Virol.*, 4, 769, 1969.
47. **Dowdle, W. R., Downie, J. C., and Laver, W. G.,** Inhibition of virus release by antibodies to surface antigens of influenza virus, *J. Virol.*, 13, 269, 1974.
48. **Shinkai, K. and Nishimura, T.,** Inability of isoquinoline derivatives to inhibit virus neuraminidase activity, *J. Gen. Virol.*, 16, 227, 1972.
49. **Schild, G. C., Oxford, J. S., and Virelizier, J. L.,** The immune response to influenza-methods of analysis for antibodies and antigens, *Proc. Miles Symp. London*, in press.
50. **Oxford, J. S., Potter, C. W., and Logan, I.,** Passage of influenza strains in the presence of amino-adamantane, *Ann. N.Y. Acad. Sci.*, 173, 300, 1970.
51. **Subak-Sharpe, J. H., Timbury, M. C., and Williams, J. F.,** Rifampicin inhibits the growth of some mammalian viruses, *Nature*, 222, 341, 1969.
52. **Bauer, D. J.,** Antiviral chemotherapy: the first decade, *Br. Med. J.*, 1, 275, 1973.
53. **Grunert, R. R., McGahen, J. W., and Davies, W. L.,** The in vivo antiviral activity of 1-adamantanamine (amantadine). I. Prophylactic and therapeutic activity against influenza viruses, *Virology*, 26, 262, 1965.
54. **Galbraith, A. W., Oxford, J. S., Schild, G. C., and Watson, G. I.,** Protective effect of amino-adamantane on influenza A2 infections in the family environment, *Ann. N.Y. Acad. Sci.*, 173, 29, 1970.
55. **Togo, Y., Hornick, R. B., and Dawkins, A. T.,** Studies of induced influenza in man. I. Double-blind studies designed to assess prophylactic efficacy of amantadine hydrochloride against A2/Rockville/1/65 strain, *JAMA*, 203, 1089, 1968.
56. **Hoffman, C. E., Neumayer, E. M., Haff, R. F., and Goldsby, R. A.,** Mode of action of the antiviral activity of amantadine in tissue culture, *J. Bacteriol.*, 90, 623, 1965.
57. **Oxford, J. S. and Schild, G. C.,** Immunofluorescent studies on the inhibition of influenza A and B viruses in mammalian cell cultures by amines and ammonium compounds, *J. Gen. Virol.*, 2, 377, 1968.
58. **Kato, N. and Eggers, H. J.,** Inhibition of uncoating of fowl plague virus by 1-adamantanamine hydrochloride, *Virology*, 37, 632, 1969.
59. **Beare, A. S., Hall, T. S., and Tyrrell, D. A. J.,** Protection of volunteers against challenge with A/Hong Kong/68 influenza virus by a new adamantane compound, *Lancet*, 1, 1039, 1972.
60. **Indulen, M. K., Kanele, I. A., Dzeguze, D. P., Ryasantseva, G. M., and Kalninyl, V. A.,** Selection of adamantane derivatives, in Proc. Symp. Antiviral Substances, Slovak Academy of Sciences, 1974.
61. **Tuckova, E., Vonka, V., Zavadova, H., and Kutinova, L.,** Sensitivity of 1-adamantanamine as a marker in genetic studies with influenza viruses, *J. Biol. Stand.*, 1, 341, 1973.
62. **Oxford, J. S. and Potter, C. W.,** Aminoadamantane resistant strains of influenza A2 virus, *J. Hyg.*, 71, 227, 1973.
63. **Schulman, J. L.,** Effect of amantadine-HCl and rimantadine-HCl on transmission of influenza virus infection in mice, *Proc. Soc. Exp. Biol. Med.*, 128, 1173, 1968.

64. **Squires, S.**, The evaluation of compounds against influenza viruses, *Ann. N.Y. Acad. Sci.*, 173, 239, 1970.
65. **Bleidner, W. E., Harmon, J. B., Hewes, W. E., Lynes, T. E., and Hermann, E. C.**, Adsorption, distribution and excretion of amantadine hydrochloride, *J. Pharm. Exp. Ther.*, 150, 684, 1965.
66. **Grossgebauer, K. and Langmaack, H.**, Failure of 1-adamantanamine (Symmetrel) to modify influenza virus induced pyrogenicity, *Arch. Gesamte Virusforsch.*, 31, 385, 1970.
67. **Knight, V., Fedson, D., and Baldini, J.**, Amantadine therapy of epidemic influenza A2/Hong Kong, *Antimicrob. Agents Chemother.*, p. 370, 1969.
68. **Likar, M.**, Effectiveness of amantadine in protecting vaccinated volunteers from an attenuated strain of influenza A/Hong Kong virus, *Ann. N.Y. Acad. Sci.*, 173, 108, 1970.
69. Influenza prophylaxis with amantadine, *WHO Tech. Rep. Ser.*, 8, 84, 1972.
70. **Oxford, J. S.**, Inhibition of the replication of influenza A and B viruses by a nucleoside analogue (Ribavirin), *J. Gen. Virol.*, 28, 409, 1975.
71. **Streeter, D. G., Witkowski, J. T., Khare, G. P., Sidwell, R. W., Bauer, R. J., Robbins, R. K., and Simon, L. N.**, Mechanism of action of virazole, a broad-spectrum antiviral agent, *Proc. Natl. Acad. Sci. U.S.A.*, 70, 1174, 1973.
72. **Oxford, J. S. and Perrin, D. D.**, Inhibitors of influenza virus associated RNA dependent RNA polymerase, in *Negative Strand Viruses*, Barry, R. D. and Mahy, B. W., Eds., Academic Press, London, 1974.
73. **Schilt, A. A.**, *Analytical Application of 1,10-Phenanthroline and Related Compounds*, Pergamon Press, Oxford, 1969, 59.
74. **Perrin, D. D.**, *Masking and Demasking of Chemical Reactions*, Interscience, New York, 1970, 183.
75. **Perrin, D. D. and Joyce, I. G.**, Computer calculation of equilibrium concentrations in mixtures of metal ions and complexing species, *Talanta*, 14, 733, 1969.
76. **Levinson, W., Faras, A., Woodson, B., Jackson, J., and Bishop, J. M.**, Inhibition of the RNA-dependent DNA polymerase of Rous sarcoma virus by thiosemicarbazone and several cations, *Proc. Natl. Acad. Sci. U.S.A.*, 70, 164, 1973.
77. **Schulman, A. and Dwyer, F. P.**, in *Chelating Agents and Metal Chelates*, Dwyer, F. P. and Mellor, D. P., Eds., Academic Press, New York, 1964.
78. **White, D. O., Harris, A. W., Dwyer, F. P., and Schulman, A.**, *Aust. J. Exp. Biol.*, 41, 517, 1963.
79. **Ohki, K. and Hori, K.**, Replication of viral nucleic acids. II. Inhibition of QB replicase by sulphydryl-blocking reagents, *Biochim. Biophys. Acta*, 281, 233, 1972.
80. **Billard, W. and Peets, E.**, Sulphydryl reactivity. Mechanism of action of several antiviral compounds – selenocystine, 4′-(2-propinyloxy)beta-nitrostyrene and acetylaranotin, *Antimicrob. Agents Chemother.*, 5, 19, 1976.
81. **Scrutton, M. C., Wu, C. W., and Goldthwait, D. A.**, The presence and possible role of zinc in RNA polymerase obtained from *E. coli*, *Proc. Natl. Acad. Sci. U.S.A.*, 68, 2497, 1971.
82. **Slater, J. P., Mildvan, A. S., and Loeb, L. A.**, Zinc in DNA polymerases, *Biochem. Biophys. Res. Commun.*, 44, 37, 1971.
83. **Auld, D. S., Kawaguchi, H., Livingston, D. M., and Vallee, B. L.**, RNA dependent DNA polymerase (reverse transcriptase) from avian myeloblastosis virus – a zinc metalloenzyme, *Proc. Natl. Acad. Sci. U.S.A.*, 71, 2091, 1974.
84. **Tonew, E., Lober, G., and Tonew, M.**, The influence of antiviral isatin iso-thiosemicarbazones on RNA dependent RNA polymerase in Mengovirus infected FL cells, *Acta Virol.*, 18, 185, 1974.
85. **Virelizier, J. L., Oxford, J. S., and Schild, G. C.**, The role of humoral immunity in host defense against influenza A infection in mice, *Postgrad. Med. J.*, in press. 52, 332, (1976).
86. **Singer, S. H., Noguchi, P., and Kirschstein, R. L.**, Respiratory diseases in cyclophosphamide treated mice. II. Decreased virulence of PR8 influenza virus, *Infect. Immun.*, 5, 957, 1972.
87. **Suzuki, F., Ohya, J., and Ishida, N.**, Effect of antilymphocyte serum on influenza virus infection in mice, *Proc. Soc. Exp. Biol. Med.*, 146, 78, 1974.
88. **Hurd, J. and Heath, R. B.**, Effect of cyclophosphamide on infections in mice caused by virulent and avirulent strains of influenza virus, *Infect. Immun.*, 11, 886, 1975.
89. **Shore, S. L., Potter, C. W., and McLaren, C.**, Immunity to influenza in ferrets. IV. Antibody in nasal secretions, *J. Infect. Dis.*, 126, 394, 1972.
90. **Rouse, B. T. and Ditchfield, W. J. B.**, The response of ponies to myxovirus influenza A equi 2. II. Immunoglobulin classes of antibody to the virus in serum or nasal secretions, *Can. J. Comp. Med.*, 34, 7, 1970.
91. **Fujita, H., Seto, Y., and Toyoshima, S.**, Mechanism of action of anti-influenza benzamidine derivatives, *Antimicrob. Agents Chemother.*, 7, 426, 1975.
92. WHO Expert Committee, Revised system of nomenclature for influenza viruses, *Bull. WHO*, 45, 119, 1971.
93. **Finter, N. B.**, Quantitative haemadsorption, a new assay technique. I. Assay of interferon, *Virology*, 24, 589, 1964.
94. **Fazekas de St. Groth, S. and White, D. O.**, An improved assay for the infectivity of influenza viruses, *J. Hyg.*, 56, 151, 1958.
95. **Stebbings, N.**, personal communication.
96. **Steward, D.**, 3rd Int. Conf. on Antiviral Substances, New York Academy of Sciences, February 1976.
97. **Grunert, R. R.**, Symp. on Antivirals with Clinical Potential, Stanford University, Stanford, California, 1975.

98. Grunert, R. R., personal communication.
99. Agarwal, A., personal communication.
100. Sidwell, R., personal communication.

Chapter 6

INHIBITORS OF VIRAL NEURAMINIDASE AS POTENTIAL ANTIVIRAL DRUGS

Peter Palese and Jerome L. Schulman

TABLE OF CONTENTS

I. INTRODUCTION

Influenza and parainfluenza viruses are unique among animal viruses by virtue of their possession of a virus-coded neuraminidase (enzyme commission E.C. 3.2.1.18). This enzyme, present in a variety of bacteria as well as in mammalian cells, cleaves terminal α-ketosidically bound neuraminic acid at carbon atom 2 (Figure 1) from a variety of compounds to which it is bound. Included among these compounds are synthetic sugars (such as galactose, galactosamine, and other neuraminic acid molecules) and glycoproteins with a molecular weight of up to several hundred thousand daltons.

In the following chapter, we will review recent evidence concerning the structure of viral neuraminidase and theories concerning its function in virus replication and discuss a variety of inhibitors of neuraminidase with respect to their potential as antiviral compounds. Further technical details of neuraminidase enzyme assay techniques relevant to antiviral screening tests are given in the Appendix of Chapter 5.

II. INFLUENZA VIRUS NEURAMINIDASE

The earliest recognition that influenza viruses contain an enzyme which splits receptors from cells came in 1942; Hirst observed that red cells agglutinated by influenza virus eventually disagglutinated and after addition of fresh virus did not agglutinate again.[1] Perceiving that an enzymatic reaction was involved, he proposed that both hemagglutination and elution were reflections of

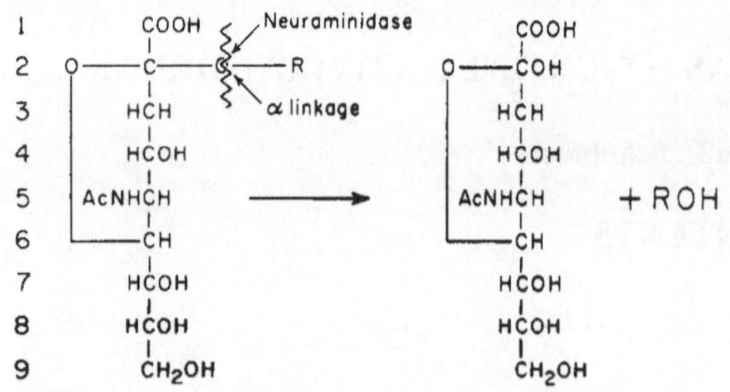

FIGURE 1. N-Acetylneuraminic acid is α-ketosidically bound to the aglycon R. After neuraminidase treatment the carbohydrate, alcohol, or protein portion (ROH) is "freed" from neuraminic acids. The nine carbon atoms containing neuraminic acid molecule is drawn in the pyranose form according to Fischer.

the same enzyme. Since that time, the two properties have been separated, and much has been learned about the structure of neuraminidase.

It is now clear that the neuraminidase of influenza A viruses is composed of a tetramer with a molecular weight of approximately 240,000 daltons (see Figure 4 in Chapter 5). With the addition of reducing agents, monomeric subunits of approximately 60,000 daltons can be isolated, but these lack enzymatic activity. Like hemagglutinin, neuraminidase is glycosylated and incorporated into the viral envelope; together the two proteins comprise the spike-like projections on the surface of influenza virus particles. Estimates of the amount of neuraminidase in each virus particle range from 3 to 14% of the total viral protein and from 50 to 100 molecules per virion.[2] Recently, the gene coding for neuraminidase has been identified as a single RNA segment with a molecular weight of 600,000 to 650,000 daltons.[3]

Antigenically, two neuraminidases (N1 and N2) have been identified among strains of virus isolated from man. Serologic tests of cross reactivity of neuraminidase antigens within the era of a single subtype demonstrate significant antigenic differences. This "antigenic drift" (see Chapter 3) resembles comparable antigenic changes that take place in the viral hemagglutinin, but it occurs independently.[4] Among strains isolated from animals, neuraminidases resembling the human subtypes N1 and N2, as well as two equine and five avian subtypes, have been identified.[5]

The neuraminidases of influenza B viruses have a molecular weight and structure similar to those of influenza A viruses.[6,7] Unlike influenza A and B viruses, influenza C virus does not appear to contain neuraminidase; in this respect, it represents an exception among the orthomyxoviruses[8] (Table 1).

TABLE 1

Viruses that Contain Neuraminidase

A. Orthomyxoviruses
  1. Influenza A virus
  2. Influenza B virus
  (NOT influenza C virus)
B. Paramyxoviruses
  1. Parainfluenza virus 1 to 5
  2. Mumps virus
  3. Newcastle disease virus
  (NOT measles, canine distemper, or rinderpest viruses)

III. PARAINFLUENZA VIRUS NEURAMINIDASE

The second major group of viruses with virion-associated neuraminidase activity is the paramyxoviruses, including parainfluenza virus types 1 to 5, Newcastle disease virus, and mumps virus[9] (see Chapter 4). These viruses are morphologically similar to influenza viruses, consisting of a ribonucleocapsid and a lipid-containing envelope covered with spikes (see Chapter 1). Two major glycoproteins that form these spikes have been identified. The smaller virion glycoprotein F, (derived from a precursor F_0) is most likely involved in virus-induced cell fusion, as has been shown with Sendai virus.[10-13] The larger isolated glycoprotein, (HN of SV5, Newcastle disease virus, and Sendai virus) possesses both neuraminidase and hemagglutinating activity.[10,11,14-18] This indicates that, unlike influenza viruses, hemagglutinating and neuraminidase activities of parainfluenza viruses are located in a single protein subunit. However, it should be noted that some investigators have claimed separation of these two activities.[19-22] If the active centers for neuraminidase and hemagglutinating activity of parainfluenza viruses are on the same polypeptide chain and are identical, compounds that inhibit neuraminidase activity may have an effect on hemagglutinating activity as well. This difference in structure of the neuraminidase also implies that compounds affecting virus replication by their effects on neuraminidase must be studied separately with influenza and parainfluenza viruses.

It should be noted that there is a subgroup of

paramyxoviruses composed of measles, canine distemper, and rinderpest viruses which lacks a virion-associated neuraminidase.

IV. FUNCTION OF VIRAL NEURAMINIDASE

For more than 30 years scientists, intrigued by the unique presence of a virus-coded enzyme on the surface of some viruses, have tried to define the function of neuraminidase in virus replication. Originally it was believed that neuraminidase was essential for adsorption or penetration of the virus.[1] It was suggested that cleavage of neuraminic acid receptors on the cell surface might be required for virus-cell interaction prior to penetration. On the basis of results from a variety of experiments, such a role is unlikely. Among these results are the observations of Schulze[23] demonstrating that virus particles from which neuraminidase has been removed by trypsin are fully infectious. In other experiments it was found that the presence of high concentrations of FANA (a potent neuraminidase inhibitor which will be described later) during the first 3 hr after infection had no effect on infectivity.[84] These observations suggest that neuraminidase activity is not required during the early stages of virus replication (Table 2).

Similarly, no experimental evidence exists to link viral neuraminidase activity with transcription and replication of virus-specific RNA or translation of viral proteins. In fact, experiments with temperature-sensitive mutants with a defect in neuraminidase demonstrate that all virus-specific proteins and RNAs are produced in cells infected at nonpermissive temperatures.[24-27]

As early as 1950, Hoyle proposed that neuraminidase facilitated release of virus from infected cells by removing cellular receptors.[28] According to this hypothesis, completion of the budding process requires cleavage of neuraminic acid receptors on the cell surface to which newly formed virus particles would otherwise remain bound. Support for this theory was subsequently provided by experiments employing antibody to neuraminidase, which prevents the release of newly synthesized virus particles but does not affect the adsorption or penetration of virus.[29-33] However, interpretation of these results is complicated by experiments indicating that the effect of antibody to neuraminidase on virus release might be mediated by bridging budding virus particles to neuraminidase antigens incorporated into the host cell membrane and by flocculation of virus particles.[34,35] In particular, a lack of association between the observed effects of antibody to neuraminidase and the function of neuraminidase activity in virus replication was shown by Becht et al.; they demonstrated that univalent fragments of molecules of antibody to neuraminidase do not reduce virus yields in single-cycle replication, although such molecules retain their enzyme-inhibiting activity in vitro.[36] On the other hand, experiments with univalent fragments of antibody to neuraminidase in plaque-size reduction assay during multicycle replication demonstrated that at least some of the virus-inhibitory activity was retained.[37]

TABLE 2

Proposed Function of Neuraminidase in Virus Replication

Adsorption	Unlikely
Penetration	Unlikely
Viral RNA and protein syntheses	No evidence
Assembly and maturation	No evidence
Release and dispersion of newly formed virus	Has been demonstrated
Destruction of inhibitors in respiratory secretions	Unproven

Recently, on the basis of experiments with influenza virus mutants with specific temperature-sensitive defects of neuraminidase, a new role has been proposed for viral neuraminidase. At nonpermissive temperature (39.5°C) such mutants produce little or no hemagglutination or neuraminidase activity in supernatant fluids; virus yields (as measured by plaque-forming units) are 10,000 times lower than those obtained at the permissive (33°C) temperature.[24,27] Nevertheless, protein analysis of cells infected at the nonpermissive temperature demonstrates that all proteins, including hemagglutinin and neuraminidase, are synthesized. Treatment of the virus yield obtained at 39.5°C with neuraminidase from *Vibrio cholerae* restores the hemagglutinating activity to titers roughly equivalent to those obtained at permissive temperatures and substantially increases infectivity.[24]

Subsequently, electron microscopic examination of cells infected with these mutants at nonpermissive temperatures revealed large aggregates of morphologically intact virus particles at the cell surface (Figures 2 and 3). In contrast, only individual virus particles are observed in cells infected at the permissive temperature with the same mutant viruses (Figure 4, arrows). These aggregated virus particles on cells infected at nonpermissive temperature contain neuraminic acid on the surface, as demonstrated by a colloidal iron hydroxide stain.

Previously, Klenk et al.[38] observed that infectious influenza and parainfluenza viruses do not contain any neuraminic acid on their surfaces; viral neuraminidase removes all the neuraminic acid from the viral membrane and the areas of the host cell membrane into which viral glycoproteins are incorporated. Figures 5 and 6 demonstrate influenza virus particles budding from the cell surface, devoid of neuraminic acid as shown by the absence of granular colloidal iron hydroxide. It is evident that mutant virus grown at nonpermissive temperatures (Figures 7 and 8) autoaggregates because of the continued presence of neuraminic acid in the viral envelope, which serves as a receptor for hemagglutinin molecules on adjacent particles. In mutant virus yields after treatment with bacterial neuraminidase, increased hemagglutinating activity reflects the reversal of the process, which can be achieved by removal of neuraminic acid from the viral envelope.[24] These experiments suggest a heretofore unrecognized function of viral neuraminidase, namely, the prevention of extensive autoaggregation by removal of neuraminic acid from the viral envelope.

In addition it is possible that neuraminidase may have an important function during other stages of virus replication, not readily demonstrable by experiments conducted in cell culture. For example, neuraminidase may facilitate the release of virus from neuraminic acid containing glycoprotein inhibitors in respiratory tract secretions.

V. PATHOGENICITY AND NEURAMINIDASE

Although temperature-sensitive defects in neuraminidase and inhibition of neuraminidase activity by chemicals (see below) are both associated with reduced virus yields, there is no clear linear relationship between the neuraminidase activity, which characterizes a strain of virus, and its replication rate. The replication of strains of recombinant (HON2) influenza A viruses of identical serotype, which have eight-to tenfold differences in the number of neuraminidase molecules per virion, varies according to the host cell system employed (not according to neuraminidase content).[39] The strain with a normal amount of neuraminidase replicates more rapidly in the allantoic cavity of embryonated eggs; the variant with eight-to tenfold lower quantities of neuraminidase has a much higher plaquing efficiency and produces larger plaques in clone 1-5C-4 conjunctival cells. On the other hand, Padgett and Walker found a relationship between neuraminidase activity and growth rate among enzymatic variants of influenza B Lee virus,[40] and McNulty observed that velogenic strains of Newcastle disease virus have threefold greater neuraminidase activity than lentogenic or avirulent strains.[41]

Qualitative differences in viral neuraminidases have been invoked as possibly providing an explanation for neurovirulent properties of some strains of influenza A virus. However, Mayer et al.[42] found no linkage of neurovirulence to either hemagglutinin or neuraminidase. Interesting results using neuraminidase antibodies suggest a role of neuraminidase with respect to the formation of noninfectious influenza virus particles, but further experiments are required for a better understanding of this phenomenon.[43]

To summarize these observations, although neuraminidase activity clearly is required for influenza virus replication, it is not certain whether genetically controlled qualitative or quantitative differences in neuraminidase activity influence virus virulence in any readily apparent way.

VI. INHIBITION OF VIRAL NEURAMINIDASES

For many years, the possibility has been entertained that influenza virus replication might be blocked by agents capable of inhibiting virus neuraminidase activity. On the basis of demonstrations that glycoprotein inhibitors of influenza virus hemagglutination activity could be inactivated by pretreatment with neuraminidase, it was reasoned that decreased neuraminidase activity might increase binding of virus to inhibitors in respiratory tract secretions. Later, with the recognition that neuraminidase is required for replication of influenza and parainfluenza viruses, the search for inhibitors was intensified in an attempt to find compounds that inhibit virus replication by

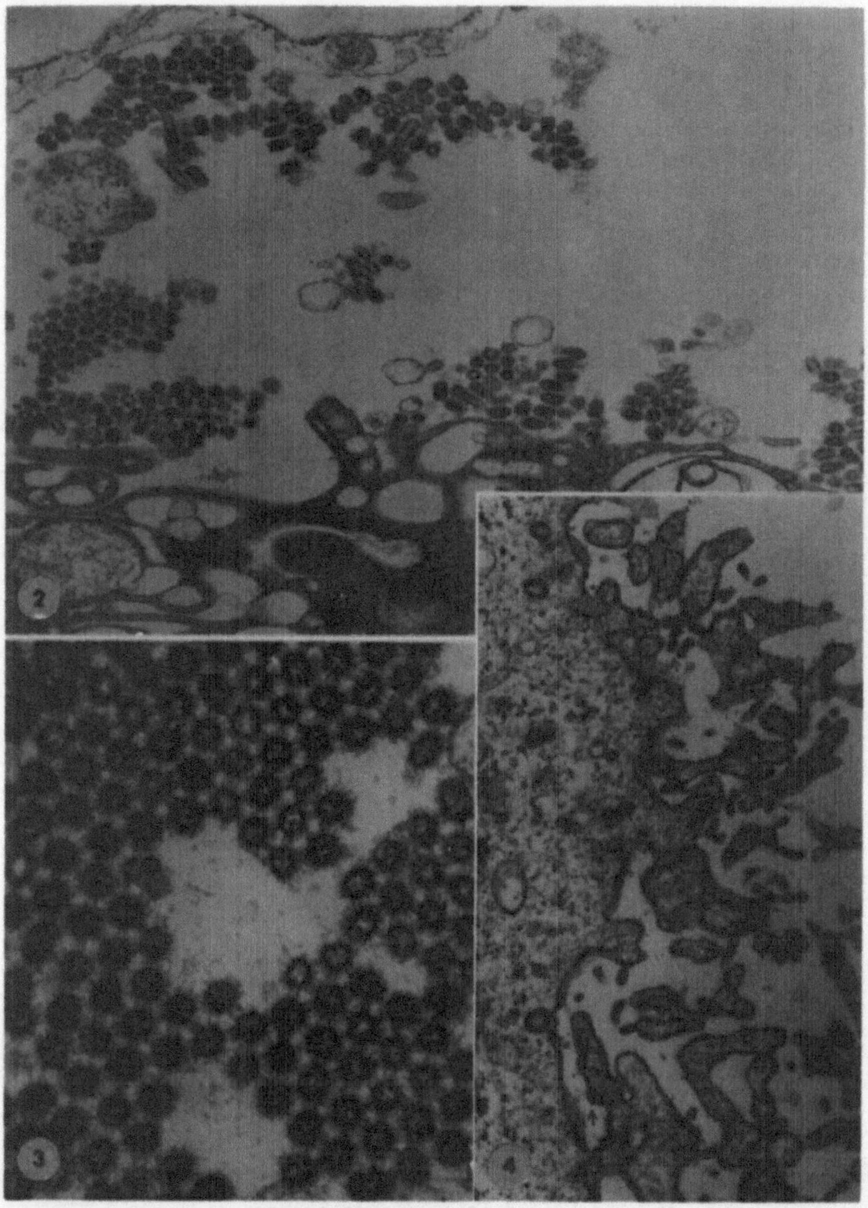

FIGURES 2 thru 4. 2. Clusters of virus particles in MDBK cell cultures infected with influenza A/WSN/ts3 virus, a mutant defective in neuraminidase, and incubated at the nonpermissive temperature (39.5°) for 12.5 hr. Large aggregates of virus particles are formed (compare Figure 4). Magnification × 35,000. (From Palese, P., Tobita, K., Ueda, M., and Compans, R. W., *Virology*, 61, 397, 1974. With permission.) 3. Higher magnification of virus aggregates produced in MDBK cells infected with influenza A/WSN/ts3 virus, a mutant defective in neuraminidase, and incubated at the nonpermissive temperature (39.5°) for 12.5 hr. Magnification × 62,000. (From Palese, P., Tobita, K., Ueda, M., and Compans, R. W., *Virology*, 61, 397, 1974. With permission.) 4. Surface of an MDBK cell infected with influenza A/WSN/ts3 virus and incubated at the permissive temperature (33°) for 12.5 hr. Arrows show single virus particles, some are budding from the cell surface. Note the low magnification as compared to Figures 2 and 3. Magnification × 25,000. (From Palese, P., Tobita, K., Ueda, M., and Compans, R. W., *Virology*, 61, 397, 1974. With permission.)

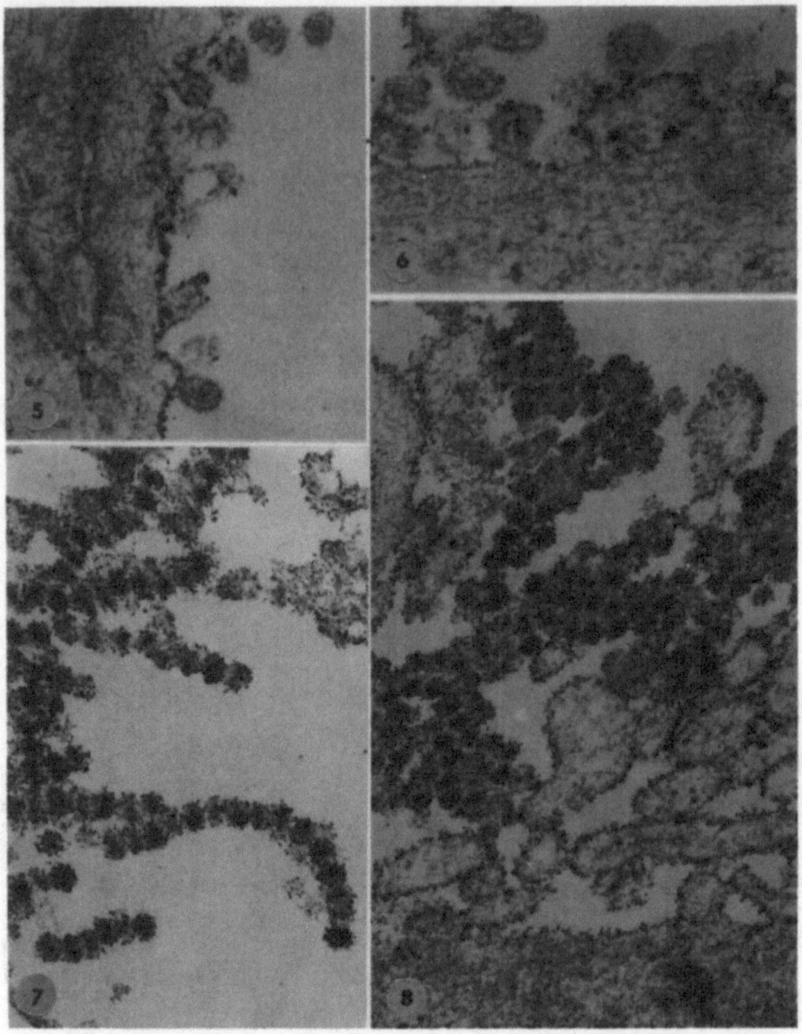

FIGURES 5 thru 8. 5. Colloidal iron hydroxide staining of neuraminic acid residues on the surface of an MDBK cell infected with influenza A/WSN virus at 33° for 15 hr. The budding virus particles are essentially free of colloidal iron hydroxide, indicating that no neuraminic acid is attached to the virus. Some areas of the cell surface clearly contain neuraminic acid, as shown by the dark granular material of colloidal iron hydroxide. Magnification × 60,000. (From Palese, P., Tobita, K., Ueda, M., and Compans, R. W., *Virology*, 61, 397, 1974. With permission.) 6. Colloidal iron hydroxide staining of neuraminic acid residues on the surface of an MDBK cell infected at permissive temperature (33°C) with an influenza A/WSN virus ts mutant defective in neuraminidase (ts11). Virus particles show little staining with colloidal iron hydroxide. Magnification × 74,000. (From Palese, P., Tobita, K., Ueda, M., and Compans, R. W., *Virology*, 61, 397, 1974. With permission.) 7. Colloidal iron hydroxide staining of aggregated virus particles on cells infected with the temperature-sensitive ts11 mutant of influenza A/WSN virus at nonpermissive temperature (15 hr postinfection). The virus particles aggregate in the form of a string of beads and contain neuraminic acid on their surface, as seen by the colloidal iron hydroxide staining. Magnification × 52,000. (From Palese, P., Tobita, K., Ueda, M., and Compans, R. W., *Virology*, 61, 397, 1974. With permission.) 8. Identical to Figure 7. Field demonstrating large clusters of virus particles. Magnification × 52,000. (From Palese, P., Tobita, K., Ueda, M., and Compans, R. W., *Virology*, 61, 397, 1974. With permission.)

their effects on neuraminidase specific virus-coded functions required for replication.

In the following sections a variety of substances, that result in a decrease of neuraminidase activity, will be described. Some of these compounds have been studied solely to increase our understanding of neuraminidases, while others have been specifically designed as candidates for anti-influenza virus agents. No attempt will be made to discuss the effects of antineuraminidase antibodies or their potential antiviral activity.

A. Nonspecific Inhibitors

Synthetic and inorganic polyanions (such as dextran sulfate, trypan red, congo red, and phosphotungstic and silicotungstic acid) inhibit the neuraminidase activity of purified influenza and parainfluenza viruses at concentrations ranging from 10^{-4} to $10^{-7}M$ (Table 3).[44-46] Congo red was shown to inhibit the replication of fowl plague virus and Newcastle disease virus in chick embryo fibroblast tissue culture cells.[46] The mechanism by which this inhibition occurs has not been determined, but it appears to be nonspecific because the inhibitory activity of these compounds can be reversed by the addition of protein or polycations. Furthermore, it is unlikely that compounds such as congo red will be useful as antiviral agents because of their potential carcinogenic properties. Other compounds, such as sulfhydryl reagents or ethylenediaminetetraacetic acid (EDTA), inhibit neuraminidases of influenza and parainfluenza viruses, but their effects are also mediated nonspecifically.[47,48]

An interesting effect of glucose analogs on the replication of influenza and parainfluenza viruses has been observed. D-Glucosamine, 2-deoxy-D-glucose, and 6-amino-6-deoxy-D-glucose have been shown to inhibit the replication of influenza virus in vitro and/or in vivo.[49-52] D-Glucosamine exerts its inhibitory effect by two mechanisms.[53,54] It is assumed that it acts, in part, by depleting the UTP pool. In addition, it interferes with the synthesis of the viral glycoproteins hemagglutinin and neuraminidase by causing the faulty synthesis of the glycoportion of the glycoproteins. Although this effect is nonspecific and affects all glycoproteins, particularly the hemagglutinin,[54] such compounds will prove to be valuable tools in studying the replication of glycoprotein-containing viruses.

B. High Molecular Weight Natural Neuraminidase Inhibitors

Many high molecular weight compounds isolated from natural sources have been shown to inhibit the neuraminidase activity of viruses (see Table 3). Drzeniek demonstrated that RNA, DNA, pig submaxillary mucin, and heparin inhibit neuraminidase activity in vitro.[44,45] Recently, Lin et al. isolated a potent high molecular weight glycoprotein, NI #289, which inhibits influenza virus neuraminidase in vitro at a 10^{-6} *M* concentration.[55,56] Nothing is known about the effects of this glycoprotein on virus replication, but many mucoproteins are potent inhibitors of hemagglutinating activity and may influence virus replication by that mechanism. Campbell demonstrated that ovine α_1-glycoprotein inhibits Newcastle disease virus in a plaque reduction test.[57] Some of these glycoproteins have been shown to have protective effects in vivo. For example, Schmidt and Janssen reported lower mortality in mice infected with influenza virus after administration of sialomucins.[58]

Other compounds that inhibit neuraminidase activity (such as concanavalin A)[59,60] are lectins that bind nonspecifically to both cellular and viral glycoproteins. Although this compound has been shown to inhibit fowl plague virus replication in chick embryo fibroblasts,[61] the mechanism of inhibition is nonspecific.

C. Small Natural Neuraminidase Inhibitors

The neuraminidase inhibitor panosialin (Table 3), isolated from streptomyces strains, is a highly negatively charged compound with a detergent-like chemical structure.[62] Its biological activity is thought to be mediated by a nonspecific mechanism similar to that described for the polyanionic compounds (discussed earlier). Other neuraminidase inhibitors isolated from streptomyces strains are siastatin A and B.[63,64] Siastatin B has some structural relationship to neuraminic acid, but strong inhibitory effects are limited to neuraminidases from nonviral sources. Even if a lasting antiviral activity cannot be demonstrated in vivo, these compounds provide a useful tool in studying the specificity of neuraminidases from different sources.[64]

D. Small Synthetic Neuraminidase Inhibitors (Unrelated to Neuraminic Acid)

Thousands of synthetic compounds belonging to different groups of chemicals have been exa-

TABLE 3

Inhibitors of Viral Neuraminidase

Inhibitor	Structure	Remarks	References
	Nonspecific Inhibitors		
Polyanions			
Dextran sulfate	Polysaccharide: glucose with sulfate groups	Nonspecific inhibitors of influenza and parainfluenza virus neuraminidases	44–46
Trypan red	azo dye, poly cyclic aromatic compound (mol wt 1002)		
Congo red	Azo dye, polycyclic aromatic compound (mol wt 697)		
Phosphotungstic acid, silicotungstic acid			
Sulfhydryl reagents			
Thioglycolate, merthiolate Hg^{++}, etc.		50% neuraminidase inhibition varies between 10^{-2} to 10^{-3} *M*; nonspecific	47, 48
Ethylenediaminetetraacetic acid (EDTA)			
Glucose analogs			
D-Glucosamine, 2-deoxy-D-glucose, 6-amino-6-deoxy-D-glucose		Interference with UTP pool; nonspecific inhibition of synthesis of glycoproteins	49–54
	High Molecular Weight Inhibitors (Natural)		
RNA, DNA, pig submaxillary mucin, heparin	Polyanionic compounds	Nonspecific inhibition of neuraminidase	44
Neuraminidase inhibitor #289	Neuraminic acid-free glycoprotein (88% sugar, 12% protein), mol wt 100,000, isolated from streptomyces #289 fluids	Potent neuraminidase inhibitor in vitro, antiviral activity unknown	55, 56

TABLE 3 (continued)

Inhibitors of Viral Neuraminidase

Inhibitor	Structure	Remarks	References
Low Molecular Weight Inhibitors (Natural)			
Panosialin (isolated from streptomyces strains)	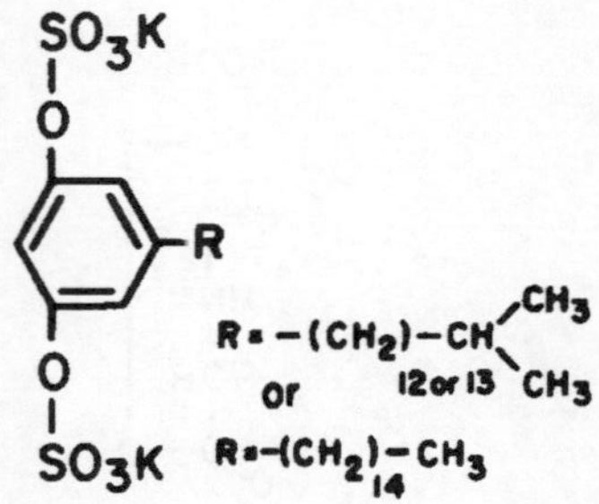	Active against intracerebral infection of mice with NWS virus	62
Siastatin A	Unknown	Potent neuraminidase inhibitors of nonviral neuraminidases	63, 64
Siastatin B	2-Acetamido-3,4-dihydroxy-piperidine-5-carboxylic acid	Antiviral activity not yet examined	
Low Molecular Weight Synthetic Neuraminidase Inhibitors (Unrelated to Neuraminic Acid)			
Oxamic acid derivatives, e.g., N-(*m*-trifluoromethylphenyl)-oxamic acid	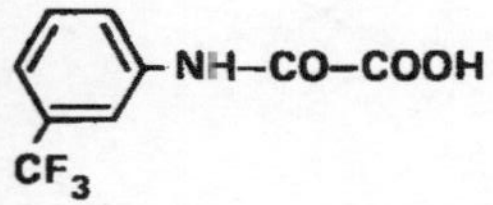	Neuraminidase inhibitor inhibits replication of influenza A/PR/8 virus in the chick embryo membrane system; no effect in embryonated eggs	65
β-Aryl-α-mercapto-acrylic acids, e.g., β-(*p*-nitro-phenyl) α-mercapto acrylic acid	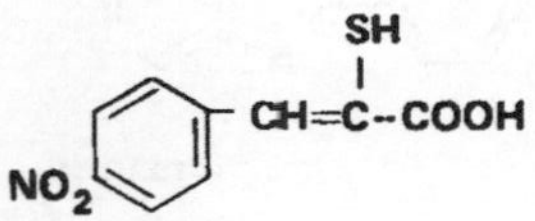	Reported to be active in plaque reduction assay using the HA-1 parainfluenza 3 virus	66

TABLE 3 (continued)

Inhibitors of Viral Neuraminidase

Inhibitor	Structure	Remarks	References
2-(*o*-Aminophenyl) benzimidazole	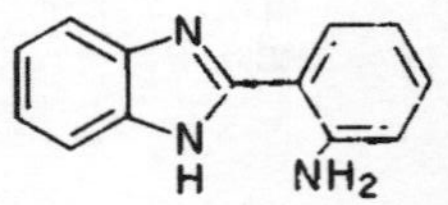	Some benzimidazole and imidazole derivatives inhibit neuraminidase activity, no antiviral activity demonstrated	66
1-(α-Aminophenyl)-3,4-dihydroiso-quinoline	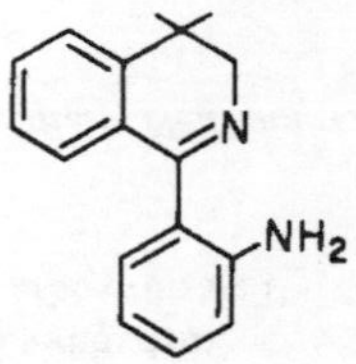	Reported to inhibit neuraminidase in vitro and to be active in vivo, in vitro activity may be artifact	66–68
			69

Low Molecular Weight Synthetic Neuraminidase Inhibitors (Related to Neuraminic Acid)

Inhibitor	Structure	Remarks	References
3-Aza-2,3,4-tri-deoxy-4-oxo-D-arabino-octonic acid δ lactone	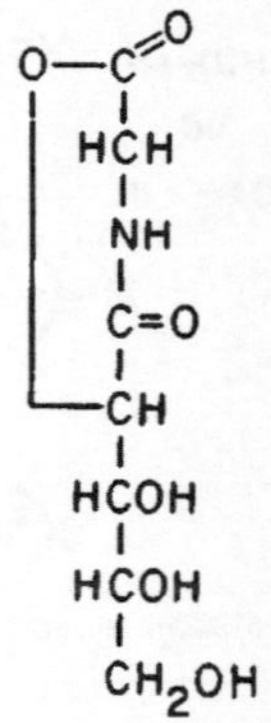		71

TABLE 3 (continued)

Inhibitors of Viral Neuraminidase

Inhibitor	Structure	Remarks	References
2-Deoxy-2-*p*-nitrophenylthio-N-acetyl-α-D-neuraminic acid	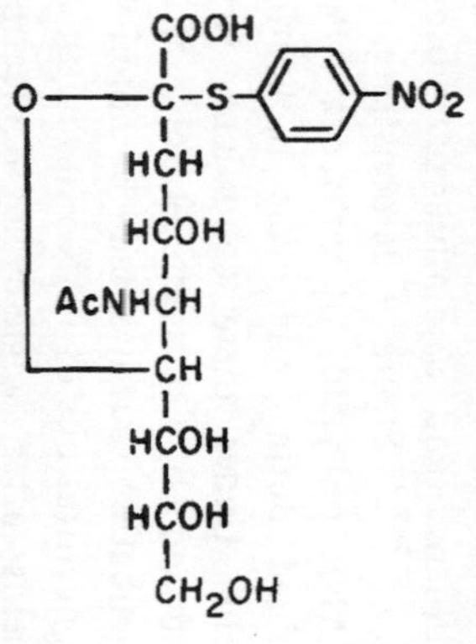	Potent inhibitors of influenza virus neuraminidases; no effect on parainfluenza virus neuraminidases	71
2-Deoxy-2,3-dehydro-N-acetylneuraminic acid and derivatives, e.g., 2-deoxy-2,3-dehydro-N-trifluoroacetyl-neuraminic acid (FANA)	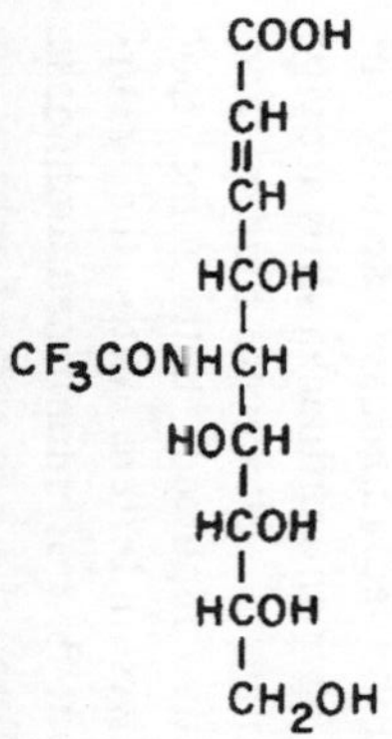	Potent inhibitors of influenza and parainfluenza virus neuraminidases, antiviral effect on influenza, and parainfluenza viruses in tissue culture	37, 73–81

mined for their ability to block neuraminidase activity and to inhibit the replication of influenza viruses. Among the most interesting of these are the oxamic acid derivatives which inhibit neuraminidases at relatively high concentrations.[65] N-(*m*-trifluoromethylphenyl)-oxamic acid inhibits 80% of the activity of influenza A/Eng/1/61 (H2N2) virus neuraminidase at $5 \times 10^{-4} M$. Replication of influenza A/PR/8/34 virus in the chick embryo membrane system is markedly inhibited by this compound. However, all antiviral activity disappears when the virus is grown in the amniotic cavity of embryonated eggs,[65] suggesting that it may be rapidly metabolized or excreted in vivo. Neuraminidase inhibitory activity of oxamic acids seems to be associated with an α-carbonyl group next to a carboxyl group.

Another group of compounds, substituted β-phenyl-α-mercaptoacrylic acids, was shown by Haskell et al. to inhibit neuraminidase activity in vitro and to inhibit parainfluenza virus replication in tissue culture.[66] Haskell et al. also examined basic inhibitors such as benzimidazole and imidazole derivatives, none of which was active in plaque reduction assay.[66]

Dihydroisoquinoline derivatives have been claimed to inhibit neuraminidase activity in vitro[66-68] and to have anti-influenza virus activity in mice and humans.[67] However, Shinkai and Nishimura recently reported that some isoquinoline derivatives interfere with the neuraminidase assay but do not inhibit neuraminidase activity.[69] Therefore, there is some doubt as to the validity of the interpretation of these earlier reports.

E. Small Synthetic Neuraminidase Inhibitors (Related to Neuraminic Acid)

Many attempts have been made to obtain specific neuraminidase and/or influenza virus inhibitors which are structural analogs of neuraminic acids. N-Acetylneuraminic acid itself is a very weak neuraminidase inhibitor[47,70] and does not interfere with the replication of influenza or parainfluenza viruses.

3-Aza-2,3,4-trideoxy-4-oxo-D-arabino-octonic acid ∂-lactone and an analog of a neuraminic acid ketoside, 2-deoxy-2-*p*-nitrophenylthio-N-acetyl-α-D-neuraminic acid, are potent inhibitors of influenza A and B virus neuraminidases (Table 3). Surprisingly, however, these two compounds are inactive against the neuraminidases of parainfluenza viruses.[71] Experiments demonstrating an inhibitory activity of these compounds, and of related ketosides of neuraminic acid bound to glucose or galactose,[72] on the replication of influenza viruses have not yet been reported.

The most potent inhibitors of neuraminidases which thus far have been synthesized are derivatives of 2-deoxy-2,3-dehydro-N-acetylneuraminic acid, first synthesized by Meindl and Tuppy.[73-75] The parent compound has been shown to be active against a variety of neuraminidases derived from bacterial, viral, and mammalian cell sources (Table 3). A comparison of its activity with that of a variety of its derivatives against influenza, parainfluenza, and *Vibrio cholerae* neuraminidases, as summarized in Table 4, demonstrates that substitution of the N-acetyl group by N-fluoro-, N-difluoro-, N-trifluoro-, or N-chloroacetyl groups is associated with greater inhibitory activity.[76]

The parent compound and its trifluoroacetyl derivative, 2-deoxy-2,3-dehydro-N-trifluoroacetylneuraminic acid (FANA), inhibit the replication of influenza and parainfluenza viruses in cell culture.[77-79] Figure 9 demonstrates the effects of FANA on yields of influenza A/WSN virus obtained during multicycle replication in MDBK cells. Under these conditions, virus yields are reduced by at least 60-fold after 32 hr of incubation, but it should be noted that equivalent concentrations of FANA have little effect on virus yields after single-cycle replication. Evidently, even in the presence of inhibitor concentrations which inhibit virtually all neuraminidase activity in vitro, only slight reductions in virus yields are produced during a single cycle of replication. However, after repeated cycles of replication, these effects are magnified.[37,78]

Similar inhibitory effects of FANA can be demonstrated in cells infected under agar, as reflected by reduction in plaque size.[78] With the latter assay system, strain-related differences in susceptibility were noted. Influenza A/NWS (HON1) virus and recombinant strains which derive their neuraminidase from NWS virus are 50- to 100-fold more sensitive to FANA than strains of virus containing N2 neuraminidase. These differences in susceptibility are related only to the neuraminidase contained by the virus and are not influenced by the nature of the virus hemagglutinin.[79] Further proof of the specificity of the inhibitory effects of FANA is provided by its lack

TABLE 4

Inhibition of Neuraminidases from Different Sources by 2-Deoxy-2,3-dehydro-N-acylneuraminic Acids[a]

2-Deoxy-2,3-dehydro-N-acylneuraminic acid, amino group substituted	Concentration in mol/l for 50% neuraminidase inhibition[b]				
	V. cholerae	Newcastle disease virus	Influenza A Melbourne	Influenza A Hong Kong	Influenza B Lee
N-Acetyl	3.0×10^{-5}	1.3×10^{-5}	3.0×10^{-5}	1.9×10^{-5}	9.0×10^{-5}
N-Carbamoyl	4.8×10^{-4}	5.2×10^{-4}	d	1.7×10^{-3}	8.5×10^{-3}
N-Formyl	8.0×10^{-4}	1.1×10^{-3}	4×10^{-3}	1.3×10^{-3}	c
N-Fluoroacetyl	9.5×10^{-6}	1.4×10^{-5}	4.2×10^{-5}	1.2×10^{-5}	4.4×10^{-5}
N-Difluoroacetyl	3.2×10^{-6}	4.4×10^{-6}	1.1×10^{-5}	6.6×10^{-6}	2.5×10^{-5}
N-Trifluoroacetyl- (FANA)	2.5×10^{-6}	1.9×10^{-6}	5.0×10^{-6}	5.0×10^{-6}	2.1×10^{-5}
N-Chloroacetyl	1.0×10^{-5}	1.3×10^{-5}	3.4×10^{-4}	4.4×10^{-4}	9.4×10^{-4}
N-Iodoacetyl	8.0×10^{-5}	1.8×10^{-5}	c	1.2×10^{-3}	c
N-Cyanoacetyl	4.4×10^{-6}	1.7×10^{-5}	c	c	c
N-Aminoacetyl	d	2.0×10^{-4}	d	c	c
N-Mercaptoacetyl	4.0×10^{-5}	c	c	2.0×10^{-3}	c
N-Thiodiacetyl	4.0×10^{-5}	c	c	3.0×10^{-3}	c

[a]Data taken from Meindl et al.[76]
[b]Substrate is 2-phenylneuraminic acid.
[c]Not done.
[d]No inhibition at 5×10^{-3} *M*.

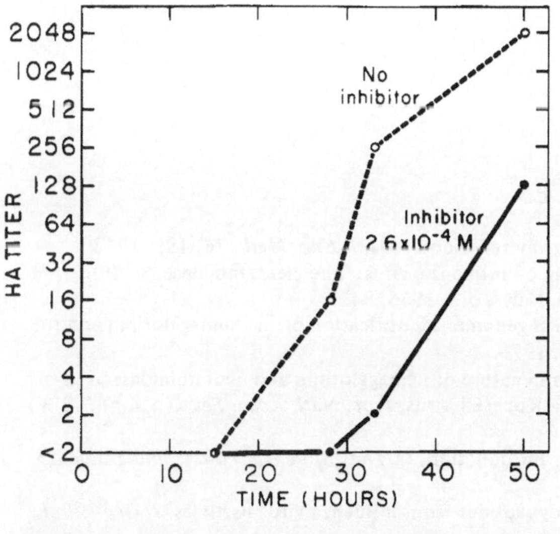

FIGURE 9. Effect of FANA on replication of influenza A/WSN (HON1) virus in MDBK cells during multicycle growth. (From Kilbourne, E. D., Palese, P., and Schulman, J. L., *Perspect. Virol.*, 9, 99, 1975. With permission.)

of effect on the replication of measles virus or vesicular stomatitis virus, both of which are enveloped glycosylated viruses that do not contain neuraminidase.[37,78,80]

Despite its potent inhibitory effects against neuraminidase containing viruses replicating in cell culture, no antiviral effect is detectable in mice infected with influenza A/NWS virus. In separate experiments FANA was administered intranasally or subcutaneously in doses calculated to achieve concentrations in extracellular fluid equivalent to those that produce plaque size reduction in tissue culture cells infected under agar.[85] However, in these experiments FANA was administered at 12-hr intervals, and it is possible that it may have been rapidly degraded or excreted so that effective concentrations were present only for transitory periods. Similar results have been obtained by Meindl and Tuppy.[86]

As has been described previously, influenza virus mutants that are defective in neuraminidase aggregate when grown at nonpermissive temperature, and particles formed at this temperature carry neuraminic acid on their envelopes.[24] A similar result is obtained when cells are infected with wild-type virus in the presence of FANA. Electron microscopic examination reveals a large aggregate of virus particles containing neuraminic acid on their envelopes.[81]

Parenthetically, Kamerling et al. recently reported the detection of 2-deoxy-2,3-dehydro-N-acetylneuraminic acid in the urine of a mentally retarded child with sialuria.[82] This surprising finding disproves the previously held assumption that this compound does not occur naturally and raises the possibility that its presence in high concentrations may be related to the child's illness.

VII. SUMMARY AND OUTLOOK

Influenza and parainfluenza viruses are unique in their possession of a virus-coded neuraminidase which is required during the last stages of virus replication. If this enzymatic activity is inhibited by neuraminidase inhibitors or is absent (as in temperature-sensitive mutants with specific defects in neuraminidase activity), virus yields are markedly reduced, and aggregates of virus particles are found at the cell membrane, bound to each other via neuraminic acid on their surfaces.

A variety of natural and synthetic inhibitors of neuraminidase inhibit enzymatic activity in vitro. Some, most notably synthetic analogs of 2-deoxy-2,3-dehydro-N-acetylneuraminic acid, inhibit virus replication in cell culture. Thus far, experiments in animals have not provided reproducible evidence for any protective effect of neuraminidase inhibitors against influenza or parainfluenza virus infection in the intact host. Newer analogs, particularly those which may be less rapidly excreted or metabolized or are actively taken up by cells, may prove to be more effective in vivo.

However, even if such compounds can be isolated, it is possible that they may be toxic. Neuraminidase is widely distributed in mammalian tissues, and inhibition of normal sialylation processes possibly could have profound side effects. Moreover, it is also possible that mutants resistant to the inhibitor would rapidly be selected. Recently, Germanov et al. isolated fowl plague virus mutants whose neuraminidase was much more resistant to the neuraminidase inhibitor 3-aza-2,3,4-trideoxy-4-oxo-D-arabino-octonic acid δ lactone than wild-type virus.[83]

In summary, although the search for potent neuraminidase inhibitors continues, it is uncertain at present that such compounds will prove to be potent chemotherapeutic or chemoprophylactic agents in man.

REFERENCES

1. **Hirst, G. K.,** Adsorption of influenza hemagglutinins and virus by red blood cells, *J. Exp. Med.*, 76, 195, 1942.
2. **Bucher, D. and Palese, P.,** The biologically active proteins of influenza virus: The neuraminidase, in *Influenza Viruses and Influenza,* Kilbourne, E. D., Ed., Academic Press, New York, 1975, 84.
3. **Palese, P. and Schulman, J. L.,** Mapping of the influenza virus genome: Identification of the hemagglutinin and the neuraminidase genes, *Proc. Natl. Acad. Sci. U.S.A.*, 73, 2142, 1976.
4. **Schulman, J. L. and Kilbourne, E. D.,** Independent variation in nature of hemagglutinin and neuraminidase antigens of influenza virus: Distinctiveness of hemagglutinin of Hong Kong/68 virus, *Proc. Natl. Acad. Sci. U.S.A.*, 63, 326, 1969.
5. **Kilbourne, E. D.,** The influenza viruses and influenza – An introduction, in *The Influenza Viruses and Influenza,* Kilbourne, E. D., Ed., Academic Press, New York, 1975, 1.
6. **Laver, W. G. and Baker, N.,** Amino acid composition of polypeptides from influenza virus particles, *J. Gen. Virol.*, 17, 61, 1972.
7. **Lazdins, I., Haslam, E. A., and White, D. O.,** The polypeptides of influenza virus. VI. Composition of the neuraminidase, *Virology,* 49, 758, 1972.
8. **Kendal, A.,** A comparison of "influenza C" with prototype myxoviruses. Receptor destroying activity (neuraminidase) and structural polypeptides, *Virology,* 65, 87, 1975.
9. **Choppin, P. W. and Compans, R. W.,** Reproduction of paramyxoviruses, in *Comprehensive Virology,* Vol. 4, Fraenkel-Conrat, H. and Wagner, R. R., Eds., Plenum Press, New York, 1975, 95.
10. **Scheid, A. and Choppin, P. W.,** Identification of biological activities of paramyxovirus glycoproteins. Activation of cell fusion, hemolysis and infectivity by proteolytic cleavage of an inactive precursor protein of Sendai virus, *Virology,* 56, 475, 1974.
11. **Scheid, A. and Choppin, P. W.,** Isolation of paramyxovirus glycoproteins and identification of their biological properties, in *Negative Strand Viruses,* Barry, R. D. and Mahy, B. W. J., Eds., Academic Press, London, 1975, 177.

12. **Homma, M.,** Host induced modification of Sendai virus, in *Negative Strand Viruses,* Barry, R. D. and Mahy, B. W. J., Eds., Academic Press, London, 1975, 685.
13. **Homma, M. and Ohuchi, M.,** Trypsin action on the growth of Sendai virus in tissue culture cells. III. Structural differences of Sendai viruses grown in eggs and tissue culture cells, *J. Virol.,* 12, 1457, 1973.
14. **Scheid, A., Caliguiri, L. A., Compans, R. W., and Choppin, P. W.,** Isolation of paramyxovirus glycoproteins. Association of both hemagglutinating and neuraminidase activities with the larger SV5 glycoprotein, *Virology,* 50, 640, 1972.
15. **Tozawa, H., Watanabe, M., and Ishida, N.,** Structural components of Sendai virus. Serological and physicochemical characterization of hemagglutinin subunit associated with neuraminidase activity, *Virology,* 55, 242, 1973.
16. **Scheid, A. and Choppin, P. W.,** Isolation and purification of the envelope proteins of Newcastle disease virus, *J. Virol.,* 11, 263, 1973.
17. **Seto, J. T., Becht, H., and Rott, R.,** Effect of specific antibodies on biological functions of the envelope components of Newcastle disease virus, *Virology,* 61, 354, 1975.
18. **Scheid, A. and Choppin, P. W.,** The hemagglutinating and neuraminidase protein of a paramyxovirus. Interaction with neuraminic acid in affinity chromatography, *Virology,* 62, 125, 1975.
19. **Inuma, M., Nagai, Y., Maeno, K., Yoshida, T., and Matsumoto, T.,** Studies on the assembly of Newcastle disease virus: Incorporation of structural proteins into virus particles, *J. Gen. Virol.,* 12, 239, 1971.
20. **Brostrom, M. A., Bruening, G., and Bankowski, R. A.,** Comparison of neuraminidases of paramyxoviruses with immunologically dissimilar hemagglutinins, *Virology,* 46, 856, 1971.
21. **Popa, L. M., Repanovici, R., Samuel, I., Smelt, D., and Portocola, R.,** Identification of Sendai virus neuraminidase and hemagglutinin subunits, *FEBS Lett.,* 51, 270, 1975.
22. **Bucher, D. J. and Schloer, G.,** Annu. Meet. of the American Society for Microbiology, Abstr. S 141, 1975.
23. **Schulze, I. T.,** The structure of virus. I. The polypeptides of the virion, *Virology,* 42, 890, 1970.
24. **Palese, P., Tobita, K., Ueda, M., and Compans, R. W.,** Characterization of temperature sensitive influenza virus mutants defective in neuraminidase, *Virology,* 61, 397, 1974.
25. **Krug, R. M., Ueda, M., and Palese, P.,** Temperature sensitive mutants of influenza WSN virus defective in virus – specific RNA synthesis, *J. Virol.,* 16, 790, 1975.
26. **Sugiura, A., Ueda, M., Tobita, K., and Enomoto, C.,** Further isolation and characterization of temperature sensitive mutants of influenza virus, *Virology,* 65, 363, 1975.
27. **Sugiura, A., Tobita, K., and Kilbourne, E. D.,** Isolation and preliminary characterization of temperature-sensitive mutants of influenza virus, *J. Virol.,* 10, 639, 1972.
28. **Hoyle, L.,** The multiplication of influenza viruses in the fertile egg, *J. Hyg.,* 48, 277, 1950.
29. **Seto, J. T. and Rott, R.,** Functional significance of sialidase during influenza virus multiplication, *Virology,* 30, 731, 1966.
30. **Jahiel, R. T. and Kilbourne, E. D.,** Reduction in size and reduction in plaque number as plaque number and plaque size and differing indices of influenza virus-antibody reactions, *J. Bacteriol.,* 92, 1521, 1966.
31. **Webster, R. G. and Laver, W. G.,** Preparation and properties of antibody directed specifically against the neuraminidase of influenza virus, *J. Immunol.,* 99, 49, 1967.
32. **Kilbourne, E. D., Laver, W. G., Schulman, J. L., and Webster, R. G.,** Antiviral activity of antiserum specific for an influenza virus neuraminidase, *J. Virol.,* 2, 281, 1968.
33. **Compans, R. W., Dimmock, N. J., and Meier-Ewert, H.,** Effect of antibody to neuraminidase on the maturation and hemagglutinating activity of an influenza A2 virus, *J. Virol.,* 4, 528, 1969.
34. **Dowdle, W. R., Downie, J. C., and Laver, W. G.,** Inhibition of virus release by antibodies to surface antigens of influenza viruses, *J. Virol.,* 13, 269, 1974.
35. **Kendal, A. P. and Madeley, C. R.,** Flocculation of influenza virus by specific antineuraminidase antibody, *Arch. Gesamte Virusforsch.,* 31, 219, 1970.
36. **Becht, H., Hammerling, U., and Rott, R.,** Undisturbed release of influenza virus in the presence of univalent antineuraminidase antibodies, *Virology,* 46, 337, 1971.
37. **Kilbourne, E. D., Palese, P., and Schulman, J. L.,** Inhibition of viral neuraminidase as a new approach to the prevention of influenza, *Perspect. Virol.,* 9, 99, 1975.
38. **Klenk, H. D., Compans, R. W., and Choppin, P. W.,** An electron microscopic study of the presence or absence of neuraminic acid in enveloped viruses, *Virology,* 42, 1158, 1970.
39. **Palese, P. and Schulman, J.,** Isolation and characterization of influenza virus recombinants with high and low neuraminidase activity. Use of 2-(3'-methoxyphenyl)-N-acetylneuraminic acid to identify cloned populations, *Virology,* 58, 457, 1974.
40. **Padgett, B. L. and Walker, D. L.,** Enzymatic variants of influenza virus. III. Function of neuraminidase in the viral growth cycle, *J. Bacteriol.,* 87, 363, 1964.
41. **McNulty, M. S., Gowans, E. T., Houston, M. T., and Fraser, G.,** Neuraminidase content of strains of Newcastle disease virus which differ in virulence, *J. Gen. Virol.,* 27, 399, 1975.
42. **Mayer, V., Schulman, J. L., and Kilbourne, E. D.,** Nonlinkage of neurovirulence exclusively to viral hemagglutinin or neuraminidase in genetic recombinants of A/NWS (HON1) influenza virus, *J. Virol.,* 11, 272, 1973.
43. **Meier-Ewert, H. and Dimmock, N.,** The role of the neuraminidase of the infecting virus in the production of noninfectious (von Magnus) influenza virus, *Virology,* 42, 794, 1970.

44. Drzeniek, R., Inhibition of neuraminidase by polyanions, *Nature*, 211, 1205, 1966.
45. Drzeniek, R., Interactions between myxoviruses and cell surfaces. III. Neuraminidase of myxoviruses, *Z. Med. Mikrobiol. Immunol.*, 156, 1, 1970.
46. Becht, H. and Drzeniek, R., The effect of azo dyes on myxovirus neuraminidase and on virus multiplication, *J. Gen. Virol.*, 2, 261, 1968.
47. Rafelson, M. E., Jr., Schneir, M., and Wilson, V. W., Jr., Studies on the neuraminidase of influenza virus. II. Additional properties of the enzymes from the Asian and PR_8 strains, *Arch. Biochem.*, 103, 424, 1963.
48. Boschman, T. A. C. and Jacobs, J., The influence of ethylene-diaminetetra-acetate on various neuraminidases, *Biochemistry*, 342, 532, 1965.
49. Kilbourne, E. D., Inhibition of influenza virus multiplication with a glucose antimetabolite (2-deoxy-D-glucose), *Nature*, 183, 271, 1959.
50. Kaluza, G., Scholtissek, C., and Rott, R., Inhibition of the multiplication of enveloped RNA-viruses by glucosamine and 2-deoxy-D-glucose, *J. Gen. Virol.*, 14, 251, 1972.
51. Hruskova, J., Jary, T., Danes, L., Jelinková, A., and Novak, M., Effect of 6-amino-6-deoxy-D-glucose hydrochloride on influenza infection in mice, *Acta Virol. (Engl. Ed.)*, 19, 47, 1975.
52. Ghandi, S. S., Stanley, P., Taylor, S. M., and White, D. O., Inhibition of influenza virus glycoprotein synthesis by sugars, *Microbios*, 5, 41, 1972.
53. Scholtissek, C., Rott, R., and Klenk, H. D., Two different mechanisms of the inhibition of the multiplication of enveloped viruses by glucosamine, *Virology*, 63, 191, 1975.
54. Klenk, H. D., Scholtissek, C., and Rott, R., Inhibition of glycoprotein biosynthesis of influenza virus by D-glucosamine and 2-deoxy-D-glucose, *Virology*, 49, 723, 1972.
55. Lin, W., Oishi, K., and Aida, K., A new screening method of viral neuraminidase inhibitors, *Agric. Biol. Chem.*, 39, 759, 1975.
56. Lin, W., Oishi, K., and Aida, K., Isolation purification and physical and chemical properties of neuraminidase inhibitor No. 289, *Agric. Biol. Chem.*, 39, 923, 1975.
57. Campbell, B. J., Schneider, A., Howe, D., and Durand, D., Structural aspects of ovine α_1-glycoprotein, an inhibitor of Newcastle disease virus, *Biochim. Biophys. Acta*, 148, 137, 1967.
58. Schmidt, E. and Janssen, M., The protective effect of mucoids from human and cow's milk in experimental infection with grippe virus in mice, *Z. Kinderheilkde.*, 98, 307, 1967.
59. Palese, P., Bucher, D., and Kilbourne, E. D., Applications of a synthetic neuraminidase substrate, *Appl. Microbiol.*, 25, 195, 1973.
60. Zalan, E., Wilson, C., and Freitag, R., The role of hemagglutinin in the inhibition of neuraminidase activity of influenza virus by concanavalin A, *Arch. Virol.*, 47, 177, 1975.
61. Rott, R., Becht, H., Klenk, H. D., and Scholtissek, C., Interactions of concanavalin A with the membrane of influenza virus infected cells and with envelope components of the virus particle, *Z. Naturforsch. Teil B*, 27, 227, 1972.
62. Aoyagi, T., Yagisawa, M., Kumagal, M., Hamada, M., Okami, Y., Takeuchi, T., and Umezawa, H., An enzyme inhibitor, panosialin, produced by streptomyces. Biological activity isolation and characterization of panosialin, *J. Antibiot.*, 24, 860, 1971.
63. Umezawa, H., Aoyagi, T., Komiyama, T., Morishima, H., Hamada, M., and Takeuchi, T., Purification and characterization of a sialidase inhibitor, siastatin, produced by streptomyces, *J. Antibiot.*, 27, 963, 1974.
64. Aoyagi, T., Komiyama, T., Nerome, K., Takeuchi, T., and Umezawa, H., Characterization of myxovirus sialidase, *Experientia*, 31(8), 896, 1975.
65. Edmond, J. D., Johnston, R. G., Kidd, D., Rylance, H. J., and Sommerville, R. G., The inhibition of neuraminidase and antiviral action, *Br. J. Pharmacol.*, 27, 415, 1966.
66. Haskell, T. H., Peterson, F. E., Watson, D., Plessas, N. R., and Culbertson, T., Neuraminidase inhibition and viral chemotherapy, *J. Med. Chem.*, 13, 697, 1970.
67. Brammer, K. W., McDonald, C. R., and Tute, M. S., Antiviral properties of 1-phenoxy-methyl-3,4-dehydro and 1,2,3,4-tetra-hydroisoquinolines, *Nature*, 219, 515, 1968.
68. Tute, M. S., Brammer, K. W., Kaye, B., and Broadbent, R. W., The inhibition of viral neuraminidase by phenoxy-methyl-3,4-dihydroisoquinolines. I. Steric effects, *J. Med. Chem.*, 13, 44, 1970.
69. Shinkai, K. and Nishimura, T., Inability of isoquinoline derivatives to inhibit virus neuraminidase activity, *J. Gen. Virol.*, 16, 227, 1972.
70. Walop, J. N., Boschman, T. A. C., and Jacobs, J., Affinity of N-acetylneuraminic acid for influenza virus neuraminidase, *Biochim. Biophys. Acta*, 44, 185, 1960.
71. Khorlin, A. Ya., Privalova, I. M., Zakstelskaya, L. Ya., Molibog, E. V., and Evstigneeva, N. A., Synthetic inhibitors of *V. cholerae* neuraminidase and neuraminidases of some influenza virus strains, *FEBS Lett.*, 8, 17, 1970.
72. Zakstelskaya, L. Ya., Molibog, E. V., Yakhno, M. A., Evstigneeva, N. A., Privalova, I. M., and Khorlin, A. Ya., Use of synthetic inhibitors of neuraminidase and hemagglutinin for the study of the functional role of membranes of myxo- and paramyxoviruses, *Vopr. Virusol.*, 2, 223, 1972.
73. Meindl, P. and Tuppy, H., 2-Deoxy-2,3-dehydrosialic acids. I. Synthesis and properties of 2-deoxy-2,3-dehydro-N-acylneuraminic acids and their methyl esters, *Monatsh. Chem.*, 100, 1295, 1969a.

74. **Meindl, P. and Tuppy, H.**, 2-Deoxy-2,3-dehydrosialic acids. III. Synthesis and properties of 2-deoxy-2,3-dehydroneuraminic acid and of new N-acyl derivatives, *Monatsh. Chem.*, 104, 402, 1973.
75. **Meindl, P. and Tuppy, H.**, 2-Deoxy-2,3-dehydrosialic acids. II. Competitive inhibition of *V. cholerae* neuraminidase by 2-deoxy-2,3-dehydro-N-acylneuraminic acids, *Hoppe-Seyler's Z. Physiol. Chem.*, 350, 1088, 1969.
76. **Meindl, P., Bodo, G., Palese, P., Schulman, J., and Tuppy, H.**, Inhibition of neuraminidase activity by derivatives of 2-deoxy-2,3-dehydro-N-acetylneuraminic acid, *Virology*, 58, 457, 1974.
77. **Meindl, P., Bodo, G., Lindner, J., and Palese, P.**, Influence of 2-deoxy-2,3-dehydro-N-acetylneuraminic acid on myxovirus neuraminidases and the replication of influenza and Newcastle disease virus, *Z. Naturforsch.*, 263, 792, 1971.
78. **Palese, P., Schulman, J. L., Bodo, G., and Meindl, P.**, Inhibition of influenza and parainfluenza virus replication in tissue culture by 2-deoxy-2,3-dehydro-N-trifluoroacetylneuraminic acid (FANA), *Virology*, 59, 490, 1974.
79. **Schulman, J. L. and Palese, P.**, Susceptibility of different strains of influenza A virus to the inhibitory effects of 2-deoxy-2,3-dehydro-N-trifluoroacetylneuraminic acid (FANA), *Virology*, 63, 98, 1975. The requirement of neuraminidase activity for influenza virus replication.
80. **Palese, P., Tobita, K., and Schulman, J. L.**, *Behring Inst. Res. Commun.*, 55, 11, 1974.
81. **Palese, P. and Compans, R. W.**, Inhibition of influenza virus replication in tissue culture by 2-deoxy-2,3-dehydro-N-trifluoroacetylneuraminic acid (FANA): Mechanism of action, *J. Gen. Virol.*, in press.
82. **Kamerling, J. P., Vliegenthart, F. G., Schauer, R., Strecker, G., and Montreuil, J.**, Isolation and identification of 2-deoxy-2,3-dehydro-N-acetylneuraminic acid from the urine of a patient with sialuria, *Eur. J. Biochem.*, 56, 253, 1975.
83. **Germanov, A. B., Zaktsielskaya, L. Ya., Molibog, E. V., Evstigneeva, N. A., Parasivk, N. A., and Kornilaeva, G. V.**, Neuraminidase mutants of fowl plague virus, *Intervirology*, 2, 273, 1973–74.
84. **Schulman, J. L. and Palese, P.**, unpublished results.
85. **Schulman, J. L. and Palese, P.**, unpublished.
86. **Meindl, P. and Tuppy, H.**, personal communication.

Guide to Index

Page numbers followed by **I** indicate Volume I.
Page numbers followed by **II** indicate Volume II.

INDEX

A

B

C

D

E

S

T

[illegible] OF [illegible]

[illegible]

[illegible] HANDBOOK [illegible] CHROMATOGRAPHY [illegible]

[illegible]

[illegible] OF [illegible] NATURAL SCIENCE

[illegible]

[illegible] HANDBOOK OF MICROBIOLOGY

[illegible]

[illegible] PROGRESS [illegible]

[illegible]

[illegible] BIOCHEMISTRY OF [illegible]

[illegible]

METHODS OF DETECTION AND IDENTIFICATION OF BACTERIA

[illegible]

[illegible] INHIBITORS [illegible]

[illegible]

[illegible]

[illegible]

[illegible]

[illegible]